Applied Physics for Engineering

Second Edition

BHAVANA BUTEY
Professor
Department of Applied Physics
G. H. Raisoni College of Engineering
Nagpur

OXFORD
UNIVERSITY PRESS

OXFORD
UNIVERSITY PRESS

Oxford University Press is a department of the University of Oxford. It furthers the University's objective of excellence in research, scholarship, and education by publishing worldwide. Oxford is a registered trade mark of Oxford University Press in the UK and in certain other countries.

Published in India by
Oxford University Press
22 Workspace, 2nd Floor, 1/22 Asaf Ali Road, New Delhi 110002, India

First published in 2017
Second published in 2023

ISBN-13: 978-93-5497-861-6
ISBN-10: 93-5497-861-4

eISBN-13: 978-93-5497-886-9
eISBN-10: 93-5497-886-X

Typeset in Times New Roman
by Ideal Publishing Solutions, Delhi
Printed in India by M P Printer (a unit of D B Corp. Ltd.), Noida 201 305

Cover image: abyrvalg00 / Shutterstock

Dedicated to
My guruji
Param Sant Sri Sri Sri Ravi Shankarji
and my late parents
Dr. Gajendra Fukey and Mrs. Kirti Fukey.

Special Thanks

Dr Bhavana Butey would wish to register special gratitude to Dr Sachin Untawale, Director, G H Raisoni College of Engineering, Nagpur for his encouragement in the pursuit of higher learning.

Preface

The second edition of Applied Physics for Engineering is written in a comprehensive manner, which will encourage students to learn the basics of engineering science. The topics are thoroughly discussed for conceptual understanding and appended with checkpoints and exercises for developing logical and analytical skills. The summary and self-study questions will help students in studying the subject systematically.

Science and technology has revolutionized our life. Day in and day out we get information about some or the other invention from different parts of the globe. We fantasize about some unreal things in science but are sure that they will become real soon; looking at the pace at which research and innovations are taking place. We encourage young minds to think outside of the box and motivate them to take challenges. To make these young minds tread the less travelled path we will have to equip them with fundamentals and nuances of science besides the current trends in the development of applications.

Key Features

- Presents a detailed discussion on applications like electron microscopy, nanomaterials, and fabrication of ICs
- Includes features such as checkpoints and solved problems
- Includes chapter-end exercises and practice problems in each chapter

Content and Structure

The book contains eight chapters, which optimally fulfills the requirement of first year undergraduate students of all branches.

Chapter 1 deals with the basic concepts of uniform electric and magnetic field and the behaviour of the charged particle in it. Chapter 2 covers the non-uniform electric field and related applications. Chapter 3 is related to atomic or modern physics that talks about relevance of quantization in atomic world. Chapters 4, 5 and 6 cover significant portion in optics, namely Laser, Interference and Polarization illustrating fundamental concepts of these phenomena exhibited by light along with their relevant applications. Semiconductor Physics and its contribution to the development of electronics is described in Chapter 7. In Chapter 8 on advanced trend, the already learnt concepts are taken to a higher level and are fine-tuned with the current techniques for some fields. In Chapters 9 and 10 fundamentals of quantum computing and challenges for the researchers and engineers across the globe to make quantum technology commercially viable are discussed.

Acknowledgement

I am indebted to my family for their patience during the entire process of writing of this book. I feel grateful to my Director Dr Sachin Untawale for his support and constant encouragement. I am also thankful to my colleagues of the first-year department for their cooperation during the development of this book. I would also like to acknowledge with gratitude the unflinching support and guidance provided by the editorial team at Oxford University Press, India who made it possible to publish this book in time. Suggestions and feedback for the book are welcomed at butey.bhavana@raisoni.net or bpbuety@gmail.com.

Bhavana P. Butey

Features of

Learning Objectives Each chapter of the book has a section 'Learning Objectives', which briefs about all the topics discussed in the chapter.

Learning Objectives

After going through this chapter, you will be able to:

- Explain the basics of non-uniform electric field
- Explain the working of electron gun which is based on the concept of electron lens
- Explain the functioning of the fundamental parts of CRT
- Explain the working of CRO based on the concept of uniform and non-uniform electric fields
- Highlight the applications of CRO

Figures Numerous well-illustrated figures are added to the book for better understanding of concepts.

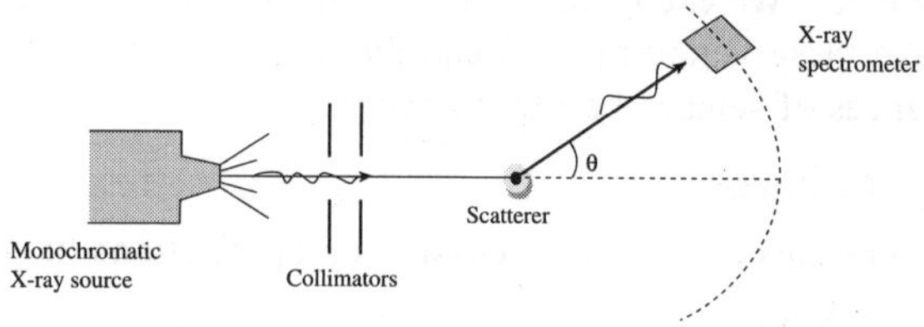

Fig. 3.2 Experimental arrangement of Compton effect

Checkpoint 3

1. In the static magnetic field, work done by the field is always
   (a) maximum (c) zero
   (b) negative (d) positive
2. When an electron moving with velocity $\vec{v}$ enters a uniform transverse magnetic field the Lorentz force alters the
   (a) magnitude of $\vec{v}$ (c) magnitude and direction of $\vec{v}$
   (b) direction of $\vec{v}$ (d) magnitude and direction of $\vec{v}$ remains unaffected
3. When a charged particle enters a uniform magnetic field at right angle to the direction of field, it will move along
   (a) helical path (c) circular path
   (b) straight line (d) its motion is unaffected
4. A charged particle will travel along helical path in a magnetic field only when its velocity vector is
   (a) along the field direction (c) perpendicular to the field direction
   (b) opposite to the field direction (d) at an angle with the field direction

Checkpoint Each chapter is interspersed with sets of multiple choice questions to check the understanding of concepts.

Solved Numericals Each chapter supports numerous well-illustrated numerical examples.

Solved Numericals

1. Find the ratio of population of the two states in He–Ne laser that produces light of wavelength 6328 Å at 27°C.

Solution: Eq. 4.14 gives the ratio of population as

$$\frac{N_2}{N_1} = \exp\ \exp\left[-\frac{E_2 - E_1}{kT}\right]\exp\left[\frac{\Delta E}{kT}\right]$$

Value of kT at room temperature (that is 27°C) is 0.026 eV. The difference in energy levels emitting light of wavelength λ is given by

$$\Delta E = h\nu = \frac{hc}{\lambda}$$

Substituting the values of $h = 6.63 \times 10^{-34}$ J-s, $c = 3 \times 10^8$ m/s and 1 eV = 1.6×10^{-19} J, we get

3. Calculate l_{coh} for Hg vapour lamp emitting in the green portion of spectrum at wavelength of 546.1 nm with emission bandwidth $\Delta\nu = 6 \times 10^{+8}$ Hz.

Solution: $\tau = \frac{1}{\Delta\nu} = 16.6 \times 10^{-10}$ s

The coherence length, l_{coh} is given by

$$l_{coh} = c\tau = 3 \times 10^8 \times 16.6 \times 10^{-10}\text{m}$$
$$= 0.49 \text{ m}$$

4. If the ray width of the 10.6 μm transition of a CO_2 laser is 60 MHz; calculate the coherence length of the laser. If the cavity length is 2m, show that not more than one mode will oscillate.

the Book

Summary Summary at the end of each chapter enables quick revision of important concepts discussed in the chapter.

Summary

- Laser light is different from ordinary source of light in view of its exceptional directionality, high monochromaticity, intensity, and remarkable coherence.
- Laser light can be obtained by exceeding the rate of stimulated emission over stimulated absorption and spontaneous emission.
- The rate of stimulated emission can be maximized by creating population inversion in the medium and enclosing the medium between two mirrors.
- Out of three pumping schemes, the four-level scheme is more efficient than the three-level pumping scheme.
- Choice of pumping mechanism depends on the state of the material (solid, liquid, or gas) in which laser is to be obtained.
- Ruby laser is a solid-state laser that uses the optical pumping mechanism and belongs to the three-level pumping scheme. It is a high wattage laser source that gives out highly directional and intense light.
- Helium–Neon laser is a gas laser, which uses electrical pumping mechanism and belongs to the four-level pumping scheme. It is a low-wattage laser source that gives out highly directional, monochromatic, and coherent light.
- Semiconductor laser is a low-wattage, highly compact laser source, which uses p-n junction diode, made of semiconducting material.
- Laser holograms contain three-dimensional information of the object. This is a reliable method of storing large

Self-study Questions

1. (a) What is laser? How does it differ from an ordinary source of light?
   (b) Discuss the properties of laser.
2. Explain spatial coherence and temporal coherence.
3. Explain the terms coherence length and coherence time. Derive the expression for the coherence length of a wave in terms of line width $\Delta\lambda$ corresponding to frequency band $\Delta\nu$.
4. (a) What is stimulated absorption?
   (b) Explain the three quantum processes. Which processes are maximized in laser operation and how?
5. Explain the terms: stimulated emission, population inversion, pumping and metastable states. What is their

   (b) What is the role of resonant cavity length in supporting different frequency modes?
9. (a) What is the role of metastable state in laser media? Is it a necessary for occurrence of lasing?
   (b) In laser, active media should have preferably broad absorption band. Explain why?
10. (a) Describe the construction of a solid state ruby laser.
    (b) Describe the working of ruby laser using energy-level diagram.
    (c) Why are the end faces of the ruby rod silvered?
11. (a) Explain the construction and working of He–Ne laser. Explain the role of end mirrors in laser. What frequencies are amplified by this mirror system?

Review Questions Each chapter supports a wealth of questions to help students during exam preparation.

Numerical Problems Numerical problems are also given in the book in relevant chapters.

Numerical Exercises

1. Suppose that an X-ray with a wavelength of 3.75 pm is fired at a stationary electron such that the scattering angle is 85°. What is the resulting energy of the electron?
   (Ans. 1.96×10^{-14} J or 1.74×10^{5} eV)
2. During Compton collision with an electron, a photon of violet light (λ = 400 nm) is backward scattered through an angle 180°. How much energy is transferred to the electron in this collision?
   (Ans. 1.9×10^{-21} J or 3.76×10^{-5}ev)
3. Incident X-rays have a wavelength of 0.31 nm and are scattered by the free electrons in graphite. The scattering angle in is θ = 135.0°. What is the magnitude of the momentum of (a) the incident photon and (b) the scattered photon?
   (Ans. h = 6.626×10^{-34}J.s and c = 2.998×10^{8} m/s.) [(a) 21.23×10^{-25} kg.m/s (b) 21.11×10^{-25} kg. m/s]

   Calculate (a) wavelength of scattered radiation at 60°. (b) energy of recoiled electron.
   [Ans. (a) 0.0372 A (b) 166.67 KeV])
5. A photon collides on the graphite target at right angles. Find the percentage change in the energy of the photon when (a) Gamma ray photon with wavelength 0.013 Å is used (b) visible photon of wavelength 6000 Å is used.
   [Ans. (a) 65.14 % (b) 4.05×10^{-4} %])
6. A bacterium (mass = 2×10^{-15} kg) in the blood is moving at 0.33 m/s. What is the de Broglie wavelength of this bacterium? (Ans. 10.04×10^{-19} m)
7. How fast does a proton have to move in order to have the same de Broglie wavelength as an electron moving at the speed of 4.5×10^{6} m/s?
   (Ans. 2.45×10^{3} m/s)

Contents

Kinematics of Electron in Electric and Magnetic Field

Learning Objectives

After going through this chapter, you will be able to:

- Gain knowledge about the various concepts of electric field, electric potential, and magnetic field
- Understand the motion of electron in uniform electric field
- Learn the motion of electron in uniform magnetic field
- Solve problems related to uniform electric and magnetic fields
- Solve application-based problems related to uniform electric and magnetic fields

1.0 Introduction

Today's technology is strongly driven by electronics. A majority of instruments, appliances, and gadgets is based on the customized electron movement in the circuits due to application of electric or magnetic field. Charged particles travel in a straight-line path in vacuum and behave just as any point mass particle, when it is in a free state. Under the influence of the electric and magnetic fields, the charged particles can be subjected to the desired forces and hence, predictable trajectories/behaviour can be obtained.

In this chapter, the particle properties of electron and the basic concepts of electric and magnetic fields are discussed. How various combinations and configurations of electric and magnetic field can be used to control the particle trajectories is discussed as cases using simple mathematics. The applications of various cases are also stated.

1.1 Electric Field - Basic Concepts

We are surrounded by many devices that work on the phenomenon of electromagnetism. Electromagnetism involves the interaction of electric and magnetic fields. Let us first learn the basics of electric field.

1.1.1 Electric Charge

There exist two kinds of electrical charges in nature, namely positive charges and negative charges. Like charges repel and unlike charges attract each other. Matter in its neutral state contains equal amounts of positive charges and negative charges. The charges, either positive or negative, are actually a collection of elementary charges carried by fundamental particles, namely the proton and electron. Therefore any positive or negative charge Q can be detected and can be written as

$$Q = \pm ne$$

where n is an integer and takes values 1, 2, 3…. (1.1)

The charge is a scalar quantity and its SI unit is coulomb. In Eq. (1.1), 'e' is the elementary charge that has value $e = 1.6 \times 10^{-19}$ C. Both electron and proton have the charge of magnitude 'e'. When a physical quantity such as charge can have only discrete values rather than any value, we say that the quantity is *quantized.* Electrical charge is therefore quantized. Thus for example, a particle can have a charge of $+11e$ or $-5\,e$ or no charge at all; but it cannot have a charge of $3.57e$.

1.1.2 Coulomb Force

Charles Augustin Coulomb in 1785 first measured the electrical attraction and repulsion quantitatively and deduced the law that governs them. Thus, according to Coulomb, the force of attraction or repulsion between unlike and like charges is inversely proportional to the square of the distance between them.

$$\vec{F} = \frac{Q_1\, Q_2}{4\pi\varepsilon_0\,\varepsilon_r\, r^2} \tag{1.2}$$

where Q_1 and Q_2 are point charges separated by a distance r, ε_0 is called the permittivity of free space and its value is $\varepsilon_0 = 8.85 \times 10^{-12}$ F/m, and ε_r is called the relative permittivity of the medium in which the charges are kept. The force between two static charged particles is often referred to as *Coulomb force* and it is a vector quantity. Coulomb force will be maximum when the charges are in vacuum and it gets reduced due to the presence of medium as $\varepsilon_r > 1$ for all medium except vacuum and air. The Coulomb force is specified in Newton and acts along the line joining the two point charges, Q_1 and Q_2.

1.1.3 Electric Field

Many common forces might be referred to as contact forces. We exert force on an object by coming into contact with it; for example, you push or pull a box. Similarly, a tennis racket exerts force on a ball when they make contact with each other. On the other hand, both gravitational force and electrical force act over the distance. A force exists between two objects even when they are not in contact with each other. Any charge sets up its electric field in the space surrounding it. Thus, the electrical interaction takes place through the electric field. Hence, electrostatic force can come into play even when charges are not in actual contact.

We can investigate the electric field surrounding a charge or group of charges by measuring the force on a small positive test charge. By a test charge we mean a charge so small that the force it exerts does not significantly alter the distribution of the other charges that cause the field which is being measured.

The force on a tiny positive charge q_0 placed at point P in the vicinity of a charged object is shown in Fig. 1.1.

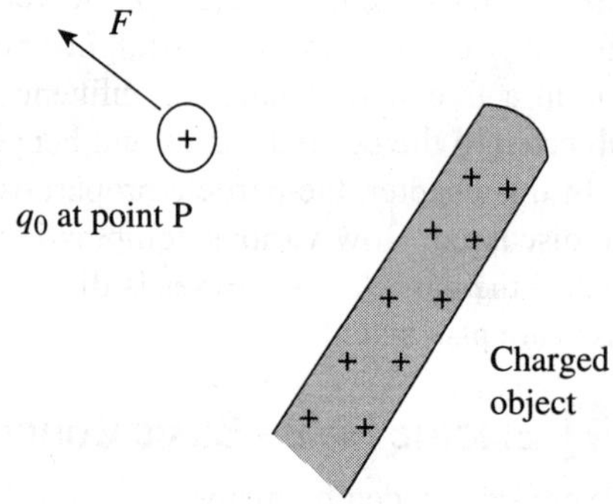

Fig. 1.1 A positive test charge q_0 placed at point P near a charged object

The electric field is a vector quantity and is defined in terms of the force on such a positive test charge. In particular, the electric field $\vec{E}$ at any point in space is defined as the force $\vec{F}$ exerted on a tiny positive test charge at that point divided by the magnitude of the test charge q_0.

$$\vec{E} = \frac{\vec{F}}{q_0} \tag{1.3}$$

Electric field has magnitude as given by Eq. (1.3), and its direction is that of the force $\vec{F}$ that acts on the positive charge (see Fig. 1.2). To define the electric field within some region we must similarly measure it at all points in the region. The SI unit of electric field is Newton per coulomb (N/C). An electric field has *unit intensity* if it exerts a force of one Newton upon one Coulomb. Although we use a positive test charge to define the electric field of a charged object, that field exists independently of the test charge.

Ideally, $\vec{E}$ is defined as the limit of F/q_0 as q_0 is taken smaller and smaller, approaching zero. The reason for defining $\vec{E} = \dfrac{\vec{F}}{q_0}$ as (with $q_0 \to 0$) is that $\vec{E}$ does not depend on the magnitude of the test charge q_0.

This means that $\vec{E}$ describes only the effect of the charges creating the electric field at that point.

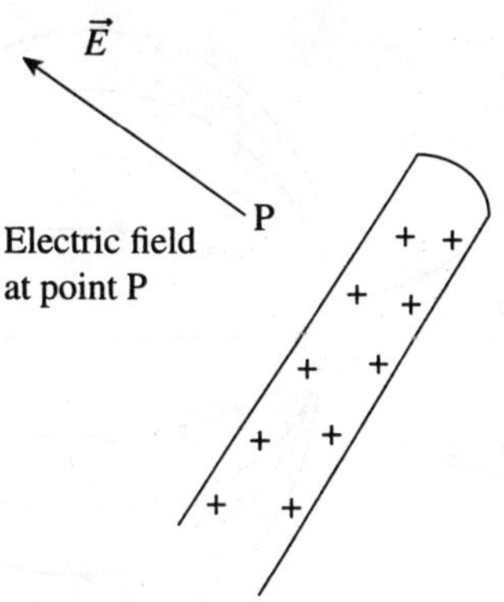

Fig. 1.2 Electric field E at point P produced by the charged object

To find the electric field at distance r from a single point charge Q, we assume a test charge q at that point. The coulomb force on test charge q is

$$\vec{F} = \frac{qQ}{4\pi\varepsilon_0\,\varepsilon_r\,r^2} \tag{1.4}$$

Therefore, electric field at that point would have magnitude

$$\vec{E} = \frac{qQ}{4\pi\varepsilon_0\,\varepsilon_r\,r^2} \times \frac{1}{q} \tag{1.5}$$

$$\vec{E} = \frac{Q}{4\pi\varepsilon_0\,\varepsilon_r\,r^2} \tag{1.6}$$

Thus, $\vec{E}$ is independent of test charge q; it depends only on the charge Q which produces the field.

1.1.4 Field Lines

Michael Faraday, who introduced the idea of electric fields in the 19th century thought of the space around the charged body as filled with lines of force. It is found to be convenient to represent and visualize electric field in terms of lines of force. The lines of force start from positive charge and end upon negative charge.

The relation between field lines and electric field is (1) the tangent to the line of force at any point gives the direction of the electric field $\vec{E}$ at that point. (2) The number of lines of force at any point passing through the cross-sectional area is proportional to the magnitude of electric field in that region as shown in Fig 1.6. The second relation means, field $\vec{E}$ is said to be strong if field lines are close and weak when the lines are far apart in that region. In a uniform electric field, the lines are straight, parallel, and uniformly spaced. In non-uniform fields, the field lines are curved and spaced non-uniformly. Figures 1.3, 1.4, and 1.5 show the electric field lines in different cases.

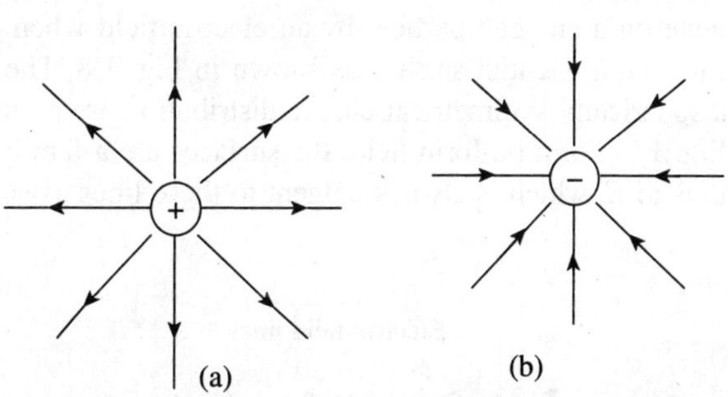

Fig. 1.3 Electric field lines (a) near a single positive point charge (b) near a single negative point charge

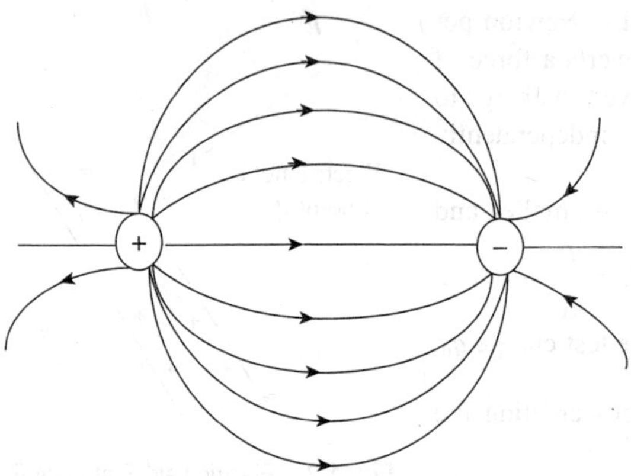

Fig. 1.4 Electric field lines near two equal charges of opposite sign

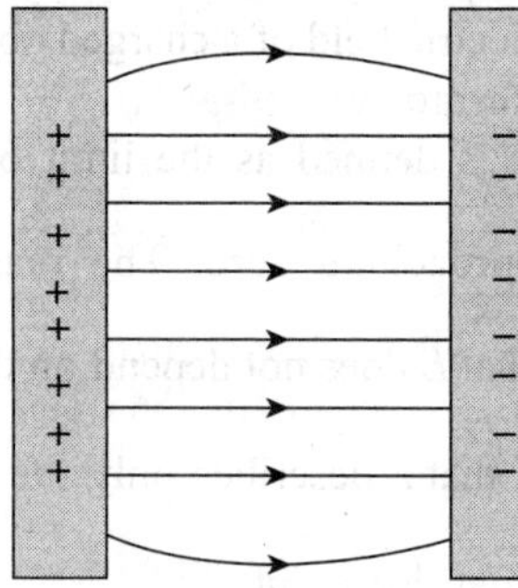

Fig. 1.5 Uniform electric field produced by a parallel plate capacitor. Field is non-uniform at the edges

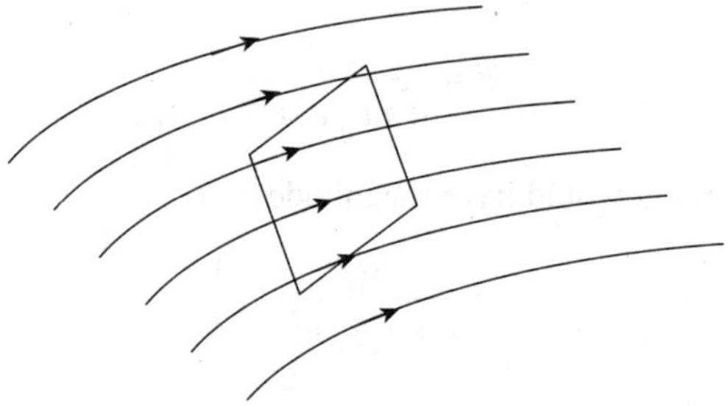

Fig. 1.6 Conventional representation of electric field intensity by the number of lines of force passing perpendicularly through a unit area

1.1.5 Equipotential Surfaces

An equipotential surface is a surface on which the potential has the same value at all points. No work is done on a charged particle by an electric field when the particle moves between two points i and f on the same equipotential surface as shown in Fig. 1.8. The equipotential surfaces produced by a point charge or a spherically symmetrical charge distribution are a family of concentric spheres as shown by dotted lines in Fig. 1.7. For a uniform field, the surfaces are a family of planes perpendicular to the electric field lines and thus to E, which is always tangent to these lines as shown in Fig. 1.8.

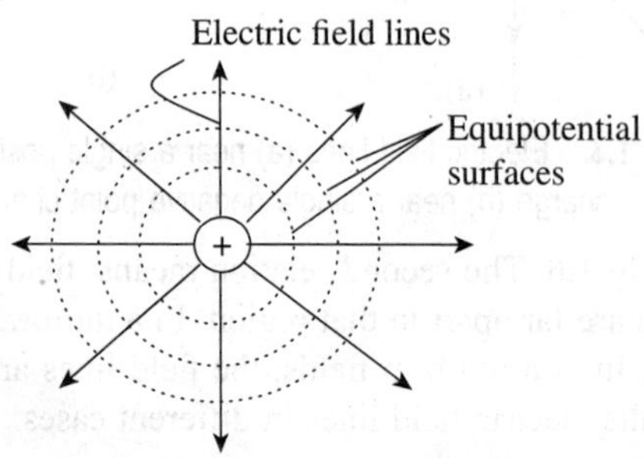

Fig. 1.7 Field of a point charge

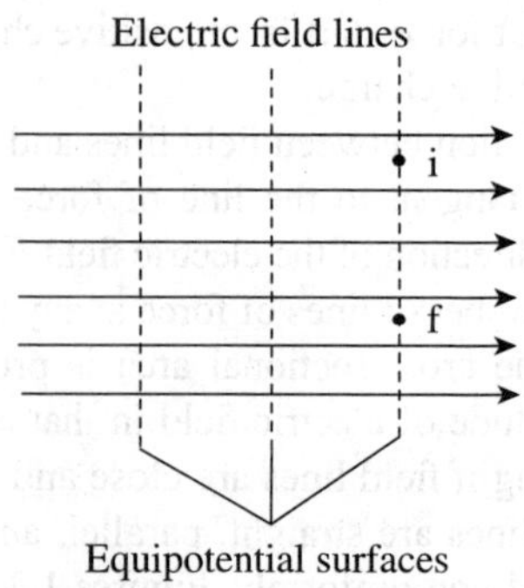

Fig. 1.8 Electric field lines and equipotential surfaces

1.2 Concepts of Electric Potential

In the previous sections we have identified one of the basic forces in the world, that is, the electric force. In this section we will associate energy viz. potential energy with the electrical force.

1.2.1 Electric Potential

Imagine a test charge q_0, for example, a proton in 'free fall' in an electric field. The motion of the test charge should be analysed in terms of energy transfer. So in particular we must be able to assign to the test charge an electric potential energy U, whose value for a given test charge depends only on the position of that charge in the electric field. In case of gravitation, if a test body falls or moves from an initial point i to a final point f, we define the difference in its gravitational potential energy as

$$\Delta U = U_f - U_i = -w_{if} \tag{1.7}$$

where w_{if} is the work done by the gravitational force on the body.

The same equation can be used as a definition of the difference in the electric potential energy of a test charge q_o as it moves from an initial point i to a final point f, in an electric field. The difference in the electric potential energy of a test charge between these points is the negative of the work done by the electrostatic force via the electric field on the test charge during its motion. Thus, the field itself does the work on the charge. Now from definition of difference in electric potential energy of a test charge between two points, we move to the definition of electric potential energy of a test charge at a single point. We will first assign the value zero to potential energy at point i which is at infinity.

Now

$$U = -w_{\infty f} \tag{1.8}$$

Thus, the potential energy U of a test charge q at any point is equal to the negative of the work done $-w_{\infty f}$ on the test charge by the electric field. Electric potential energy is the energy of a charged object in an external electric field and is measured in joules. Now the potential energy of a point charge in an electric field depends on the magnitude of the charge as well as the field. However, potential energy per unit charge has a unique value at any point in an electric field.

Thus the potential energy per unit charge $\frac{U}{q}$ is independent of the magnitude of test charge q_0 and is characteristic of any electric field. Electric potential is a property of the field itself, whether or not a charged object has been placed in it. It is measured in J/C or volt. The quantity $\frac{U}{q}$ is called *electric potential* which is a scalar quantity.

$$V = \frac{U}{q} \tag{1.9}$$

$$\Delta V = V_f - V_i \tag{1.10}$$

$$= \frac{U_f}{q} - \frac{U_i}{q}$$

$$= \frac{\Delta U}{q} \tag{1.11}$$

$$\Delta V = V_f - V_i = \frac{-w_{if}}{q} \tag{1.12}$$

Thus, if $-w_{if}$ is positive, negative, or zero, then potential at f will be less, greater than, or same as potential at i.

1.2.2 Calculating the Potential Difference from Field

We can calculate the potential difference between any two points i and f in an electric field if we know the electric field vector all along any path connecting these points. To find the potential from $\vec{E}$ field, we find the work done on a positive test charge by the field as the charge moves from i to f.

Consider an arbitrary electric field represented by field lines as shown in Fig. 1.9 and a positive test charge q_0 that moves along the path from point i to point f. At any point on the path, an electrostatic force $q_0\vec{E}$ acts on the charge as it moves through the differential displacement ds.

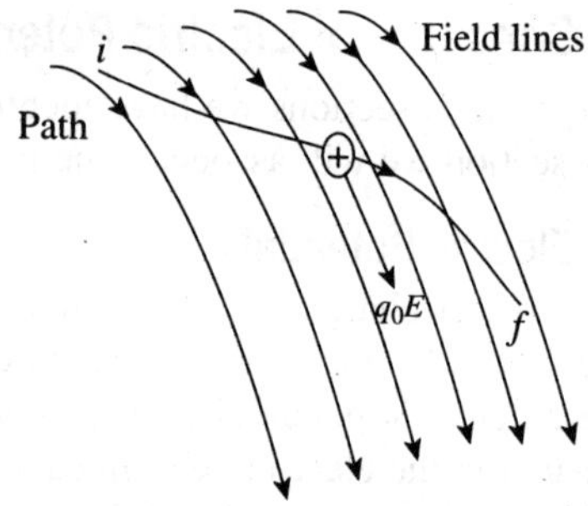

Fig. 1.9 A test charge q_0 moves from point i to point f along the path shown in a non-uniform electric field. During a displacement ds, an electrostatic force q_0E acts on the test charge. This force points in the direction of the field line at the location of the test charge

The differential work dw done on a particle by a force $\vec{F}$ during displacement ds is

$$dw = \vec{F}.ds \tag{1.13}$$

Using Eq. (1.3)

$$dw = q_0\,\vec{E}.ds \tag{1.14}$$

To find the total work w_{if} done by the field on the particle as it moves from point i to point f, we add via integration the differential work done on the charge for all the differential displacements ds along the path.

$$w_{if} = q_0 \int_i^f E.ds \tag{1.15}$$

If we substitute the total work w_{if} from Eq. (1.15) in Eq. (1.12), we get

$$V_f - V_i = -\int_i^f E.ds \tag{1.16}$$

Thus the potential difference $V_f - V_i$ between any points i and f in an electrical field is equal to the negative of the line integral (meaning the integral along the path) of $E.ds$ from point i to point f. If the electric field is known throughout a certain region, Eq. (1.16) allows us to calculate the difference in potential between any two points in the field. As the electrostatic force is conservative, all paths give the same result. If we choose the potential at point i to be zero and at point f as V, then Eq. (1.16) becomes

$$V = -\int_i^f E.ds \tag{1.17}$$

If the field is uniform, that is, $\vec{E}$ is constant over the path and can be moved outside the integral then we get

$$V = -Ed$$

$$\text{or } E = -\frac{V}{d} \tag{1.18}$$

Here d is the distance between the initial point 'i' and final point 'f'.

Potential due to point charge

Consider a point P at a distance r from a fixed point charge of magnitude q as shown in Fig. 1.10. In order to use Eq. (1.17) we imagine that a positive test charge moves from infinity to point P. As the path followed by the test charge does not matter, we consider a line that extends from infinity to point P along the radius from the point charge q.

We now evaluate the dot product $(\vec{E}.ds)$ in Eq. (1.17) along the path taken by the test charge. The test particle is at some intermediate point, at distance r' from the point charge. The angle between $\vec{E}$ and ds is 180 degree. Using this angle and writing the magnitude of the displacement ds as dr' we can write the dot product of Eq. (1.17) as

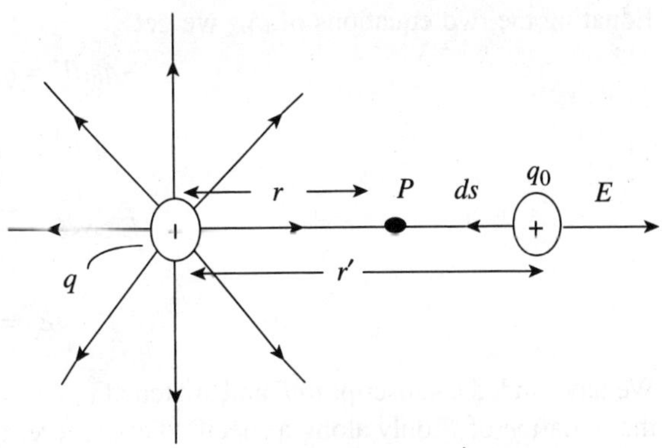

Fig. 1.10 The positive point charge q produces an electric field E and an electric potential V at point P

$$V_f - V_i = -\int_i^f \vec{E}.ds \tag{1.19}$$

$$V = -\int_i^f \vec{E}.ds \tag{1.20}$$

$$= -\int_\infty^r \frac{q}{4\pi\varepsilon_0\varepsilon_r r'^2} dr'$$

$$= \frac{-q}{4\pi\varepsilon_0\varepsilon_r}\left[\frac{-1}{r'}\right]_\infty^r = \frac{-q}{4\pi\varepsilon_0\varepsilon_r}\left[-\frac{1}{r}\right]$$

$$\therefore V = \frac{q}{4\pi\varepsilon_0\varepsilon_r r} \tag{1.21}$$

Thus, the potential V at any point around a positive point charge is positive, relative to zero potential at infinity.

1.2.3 Calculating Field from the Potential

Let us see how to find the electric field when we know the potential. Figure 1.11 shows the cross sections of a family of closely spaced equipotential surfaces. The potential difference in between each pair of adjacent surfaces is 'dV'.

The field $\vec{E}$ at any point P is perpendicular to the equipotential surface through P. Suppose that a positive test charge q_0 moves through a displacement ds from one equipotential surface to the adjacent surface, then by using Eq. (1.12), we get

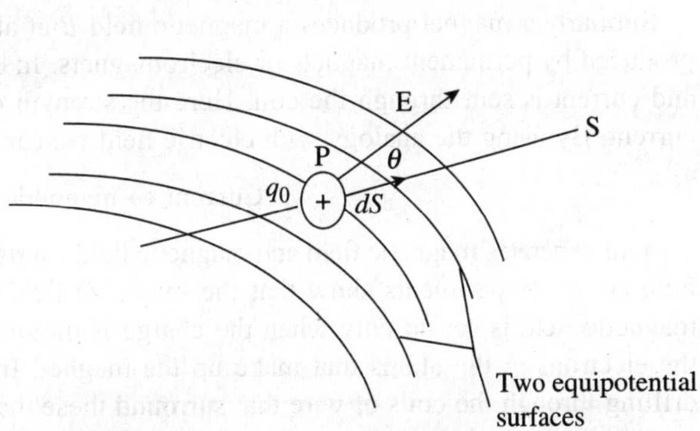

Fig. 1.11 A test charge q_0 moves distance ds from one equipotential surface to another. The displacement ds makes an angle θ with the direction of the electric field $\vec{E}$

$$dw = -q_0\, dV$$

Work done by the field may also be written using Eq. (1.14) as

$$dw = q_0 \vec{E}.ds$$

Equating the two equations of dw, we get

$$-q_0 dV = q_0 \vec{E}.\, ds$$

$$-q_0 dV = q_0 E(\cos\theta) ds$$

$$E\cos\theta = \frac{-dV}{ds} \tag{1.22}$$

$$E_s = -\frac{dV}{ds} \tag{1.23}$$

We have added a subscript to $\vec{E}$ and switched to partial derivative to emphasize that Eq. (1.23) involves only the variation of V only along a specified axis (here it is S axis). The component of E in any direction is the negative of the rate of change of electric potential with distance in that direction. Thus,

$$E_x = -\frac{dV}{dx},\ E_y = -\frac{dV}{dy},\ \text{and}\ E_z = -\frac{dV}{dz} \tag{1.24}$$

Thus, if we know the value of V for all points in the region around a charge direction, that is, $V(x,y,z)$ we can find the component of $\vec{E}$ at any point by taking its partial derivatives.

1.3 Basic Concepts of Magnetic Field

So far we have examined the fundamental concepts related to electric field. Now in this section we will learn how a magnetic field can be set up in a region and the magnitude and direction of the force produced by it.

1.3.1 Current and Magnetic Field

Charge generates electric field and electric field exerts forces on charges. This can be written as

$$\text{Charge} \leftrightarrow \text{electric field} \leftrightarrow \text{Charge}$$

Similarly a magnet produces a magnetic field B at all points in the space around it. Magnetic fields are produced by permanent magnets or electromagnets. In electromagnets a coil is wound around an iron core and current is sent through the coil. Here the strength of the magnetic field depends on the magnitude of current. By using the analogy with electric field we can write,

$$\text{Current} \leftrightarrow \text{magnetic field} \leftrightarrow \text{Current}$$

Current generates magnetic field and magnetic field exerts force on current. But from where does the magnetic field come? Experiments show that the magnetic field is developed from the moving electric charges. A magnetic field is set up only when the charge is moving. In a permanent magnet, the moving charges are the electrons in the atoms that make up the magnet. In electromagnets the moving charges are electrons drifting through the coils of wire that surround these magnets. Thus, moving charges in an atom or current set up in a wire produces the magnetic field. If we place a moving charge or a wire carrying current in the magnetic field, the magnetic field will act on it. We can define magnetic field by finding the force acting on a test charge. When a test charge q moves with velocity $\vec{v}$ in the magnetic field of strength $\vec{B}$, then the force exerted by the magnetic field is given by

$$\vec{F}_L = q(\vec{v} \times \vec{B}) \tag{1.25}$$

This equation represents

(1) Magnetic force F_L always acts perpendicular to the velocity vector, that is, it can neither speed up nor slow down a moving charge particle, but can only deflect its direction of motion. Thus, magnetic force can change only the direction of the particle's velocity v but cannot change the kinetic energy of the particle.

(2) A magnetic field exerts no force on the charge that is at rest or moves parallel or anti-parallel to the field.

$$F_L = qvB\sin\theta = 0 \text{ when } \theta = 0° \text{ or } 180°.$$

(3) The maximum value of deflecting force occurs when the test charge is moving perpendicular to the magnetic field, that is, when $\theta = 90°$.

$$\vec{F}_L = qvB \tag{1.26}$$

(4) The magnitude of deflecting force is directly proportional to q and $\vec{v}$. The greater the charge of the particle and the faster it is moving, the greater the magnetic deflecting force. If the particle is stationary or electrically neutral then no magnetic force acts on it.

(5) The direction of magnetic deflecting force depends on the polarity of the charge, q. The direction of force is determined for the positive charge using the right-hand thumb rule for cross product (see Fig. 1.12). The curling of fingers indicates the velocity vector turned into the magnetic field vector and the thumb denotes the direction of the magnetic force. For the negative charge the direction of magnetic force will be in opposite direction as given by the thumb rule.

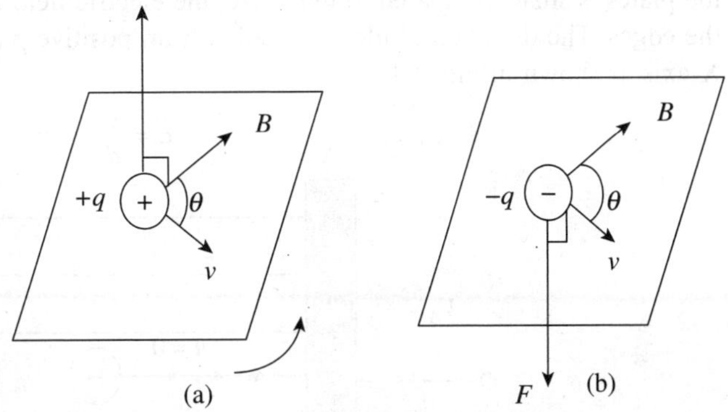

Fig. 1.12 Right hand rule to determine the direction of Lorentz force – (a) On a positive charge (b) On a negative charge

The SI unit of $\vec{B}$ is tesla or wb/m^2 (1 Tesla = 10^4 gauss).

If a charged particle moves through a region in which both electric and magnetic fields are present, then the resultant force is found by combining both the forces.

$$\vec{F} = q\vec{E} + q\,(\vec{v} \times \vec{B}) \tag{1.27}$$

This is sometimes called the *Lorentz equation.*

In next sections we will learn the behaviour of a charged particle in uniform electric and magnetic field which forms the basis of the countless electromagnetic devices.

Checkpoint 1

1. When an electron moves on an equipotential surface, its potential energy
   (a) increases
   (b) decreases
   (c) remains constant
   (d) becomes zero
2. In an uniform electric field, the field lines are
   (a) parallel
   (b) perpendicular
   (c) curved
   (d) none of these
3. If an electron moves through a potential difference of 100V, its kinetic energy is
   (a) 100 J
   (b) 16×10^{-19} J
   (c) 100 eV
   (d) none of these
4. An electric field has unit intensity if it
   (a) exerts a force of one Newton upon one Coulomb.
   (b) takes one joule to carry one Coulomb through it.
   (c) exerts a force of one Dyne upon one electron.
   (d) exerts a force of 2×10^{-7} Newton on 1 one metre long wire carrying one ampere.

1.4 Motion of Electron In Uniform Electric Field

1.4.1 Electron Motion in Longitudinal Uniform Electric Field

A uniform electric field can be generated with the help a pair of parallel plates A and B of area '*a*' separated by a small distance '*d*'. The length of the plates should be large as compared to '*d*'. If a dc voltage source is connected between the plates, then one plate acquires a negative charge and the other a positive charge of equal magnitude and an electric field is produced in the region between the plates. If the separation *d* of the plates is small compared to their size, the electric field in the region will be fairly uniform except near the edges. The direction of electric field is from positive plate A towards the negative plate B or along the X-axis as shown in Fig. 1.13.

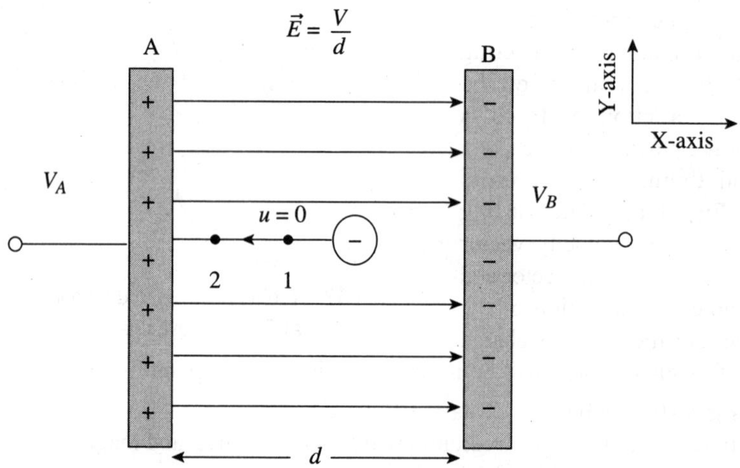

Fig. 1.13 Motion of electron in uniform longitudinal electric field

If the potential difference between the plates A and B is *V* volts, the electric field strength *E* is given by Eq. (1.18), that is, $E = \frac{V}{d}$. These two metal plates which form a parallel plate capacitor are kept in an evacuated space so that the particle paths are not distorted by the collisions with atmospheric gas molecules. Let us consider an electron of mass 'm' and charge '*e*', which is initially at rest in the uniform electric field. As soon as the field is established, the electron experiences a force due to the electric field, which is given by

$$\vec{F} = -e\vec{E} \tag{1.28}$$

The negative sign indicates that the force $\vec{F}$ accelerates the electron in a direction opposite to that of $\vec{E}$. According to the second law of Newton, the acceleration is given by

$$\vec{a} = \frac{\vec{F}}{m}$$

$$\vec{a} = \frac{-e\vec{E}}{m} \tag{1.29}$$

The negative sign in Eq. (1.29) is dropped in subsequent expressions as it indicates only that the direction of acceleration is opposite to that of the electric field. In the above equation $e, \vec{E}$, and *m* are constants and therefore the electron acceleration is constant, that is, the electron is moving with uniform acceleration. As the electron has constant acceleration, we can apply the equations of kinematics to know the motion of uniformly accelerated electron. The motion resembles that of a freely falling body in a gravitational field. By using the kinematic equation for finding the final velocity and initial condition that at $x = 0$, $u = 0$ (electron is initially at rest), we can write the final velocity as

$$v = u + at = 0 + at$$

Using Eq. (1.29) for acceleration, the final velocity becomes

$$v = \frac{eEt}{m} \tag{1.30}$$

Using kinematic equation for displacement, we can find the displacement along X-direction at any time 't' as given by

$$x = ut + \frac{1}{2}at^2 = 0 + \frac{1}{2}at^2$$

Substituting for acceleration we get

$$x = \frac{eEt^2}{2m} \tag{1.31}$$

To find the kinetic energy of an electron after it has travelled some distance x, using the kinematic equation, we get

$$v^2 = u^2 + 2as = 0 + 2ax$$

Substituting for acceleration we get

$$v^2 = \frac{2Eex}{m}$$

Therefore kinetic energy of an electron is given by

$$\text{Kinetic energy} = \frac{1}{2}mv^2 = \frac{1}{2}m \times \frac{2Eex}{m} = Eex \tag{1.32}$$

Thus, kinetic energy of an electron in longitudinal electric field is directly proportional to the distance travelled, because electric field E and charge of electron e are constants.

1.4.2 Energy Acquired by Electron in Electric Field

Referring to Fig.1.13 the electric field $\vec{E}$ causes electron motion towards the positively charged plate A. The force on the electron can be written by using Eq. (1.24) as

$$F = -eE = e\frac{dV}{dx} \tag{1.33}$$

Work is done on the electron by the electric field in moving it towards the plate A. The work done dw in moving the electron through an infinitesimal distance dx is given by

$$dw = F.dx$$

Total work done in moving the electron from point 1 to point 2 by using Eq. (1.33) is given by

$$w_{12} = \int_1^2 F.dx = \int_1^2 e\frac{dV}{dx}.dx$$

$$w_{12} = \int_1^2 edV = e(V)_1^2$$

Therefore, $$w_{12} = e(V_2 - V_2) \quad (1.34)$$

In Eq. (1.34) V_1 and V_2 are the potentials at positions 1 and 2 respectively. The electron moves in the direction of increasing potential. According to work–energy theorem, the work done by the field in moving the electron is equal to the kinetic energy acquired by the electron.

According to Newton's Second Law $$\vec{F} = ma$$

$$= \frac{d}{dt}(mv) = m.\frac{dv}{dt}$$

As the work done is given by $dw = F.dx$

Substituting force in terms of velocity by using the above equation, we get

$$W_{12} = \int_1^2 F.dx = \int_1^2 m\frac{dv}{dt}dx = \int_1^2 m\frac{dv}{dx}\times\frac{dx}{dt}dx$$

$$= \int_1^2 mvdv = m\left(\frac{v^2}{2}\right)_1^2$$

$$\therefore\ W_{12} = \frac{1}{2}m(v_2^2 - v_1^2) \quad (1.35)$$

where v_1 and v_2 are electron velocities at positions 1 and 2, respectively.

Equating Eqs (1.34) and (1.35), we get

$$e(V_2 - V_1) = \frac{1}{2}m(v_2^2 - v_1^2) \quad (1.36)$$

Equation (1.36) indicates that the electron gains energy from the electric field. Assuming that the electron starts from rest and accelerates through a potential difference of V volts, we can rewrite Eq. (1.36) as

$$eV = \frac{1}{2}mv^2 \quad (1.37)$$

Equation (1.37) gives us an important relation between the kinetic energy and the potential energy of any charged particle accelerated through potential difference of V volts. As the amount of kinetic energy acquired is usually very small compared to a joule, the energy is expressed as an atomic unit, viz. electron volt. One *electron volt* is defined as the energy acquired by an electron when it is accelerated through a potential of one volt and is given by

$$1\ \text{eV} = 1.602\times10^{-19}\ \text{J} \quad (1.38)$$

For expressing the energy of atomic particles, electron volt unit is widely used. The higher values of energies are expressed as

$$1\ \text{MeV}(1\ \text{million eV}) = 10^6\text{eV and 1B eV}(1\ \text{billion eV}) = 10^9\ \text{eV}$$

The final or the terminal velocity v is given as

$$v = \sqrt{\frac{2eV}{m}} \tag{1.39}$$

For electron, substituting the values of its charge and mass, viz. 1.6×10^{-19} C and 9.1×10^{-31} kg we get

$$v = 5.93\sqrt{V} \times 10^5 \text{ m/s} \tag{1.40}$$

The application of longitudinal uniform electric field is to accelerate charged particles, that is, to impart kinetic energy to the charged particles.

1.4.3 Electron Motion in Transverse Uniform Electric Field

Let 'A' and 'B' be two plane parallel metal plates of length 'l' oriented horizontally and separated by distance 'd' as shown in Fig. 1.14. A potential difference of V volts is applied between the plates which produces a vertically acting uniform electric field (along negative Y-direction) directed from plate A to B. The electron enters the electric field in the positive X-direction with an initial velocity v_0. The initial velocity of the electron is acquired when it is accelerated through a potential difference V_A as given by

$$v_0 = \sqrt{\frac{2eV_A}{m}} \tag{1.41}$$

As the electron enters the plates, it will experience a force due to electric field along Y-direction which is given by

$$\vec{F} = e\vec{E}$$

As the electric field acts only in the Y-direction, there will be no force along either X- or Z-direction. Therefore acceleration of the electron along the X- and Z-directions is zero and thus velocity is constant. Since the electrons move initially along the X-direction with an initial velocity, v_0 it will continue to travel

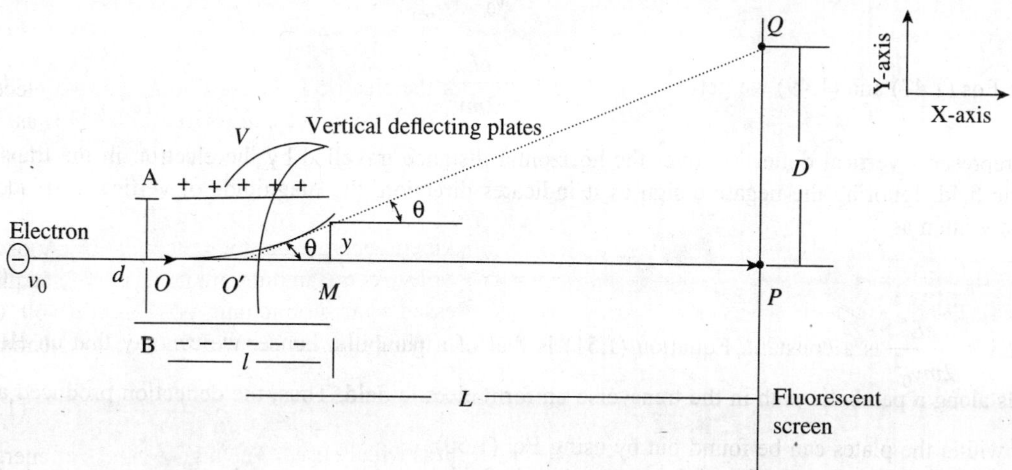

Fig. 1.14 Electron motion in transverse uniform electric field

along the X-axis with the same velocity. As electric field is acting along Y-direction, the force experienced by electron along Y-direction is

$$\vec{F}_y = -m\vec{a}_y \tag{1.42}$$

$$e\vec{E} = -m\vec{a}_y$$

$$\vec{a}_y = \frac{-e\vec{E}}{m} = \text{constant} \tag{1.43}$$

Using initial conditions of zero velocity along Y-direction, velocity of the electrons at any time 't' by using kinematic equation for velocity is

$$v_y = 0 + a_y t \tag{1.44}$$

Substituting Eq. (1.43) in Eq. (1.44) we get

$$v_y = \frac{-eEt}{m} \tag{1.45}$$

Using the initial condition of zero displacement $y_0 = 0$ and $u_y = 0$ the displacement of the electron in the Y-direction is

$$y = 0 + \frac{1}{2}a_y t^2 \tag{1.46}$$

$$y = \frac{-eEt^2}{2m} \tag{1.47}$$

Along X-direction $a_x = 0$ so the distance x travelled by the electron in the time interval t is given by

$$x = v_0 t + 0$$

$$x = v_0 t \tag{1.48}$$

From Eq. (1.48), we can write

$$t = \frac{x}{v_0} \tag{1.49}$$

Using Eqs (1.47) and (1.49) we get

$$y = -\frac{eEx^2}{2mv_0^2} \tag{1.50}$$

This represents vertical deflection over the horizontal distance travelled by the electron in the transverse electric field. Ignoring the negative sign as it indicates direction, the magnitude of vertical displacement can be written as

$$y = kx^2 \tag{1.51}$$

where $k = \dfrac{eE}{2mv_0^2}$ is a constant. Equation (1.51) is that of a parabola; hence, we can say that an electron travels along a parabolic path in the transverse uniform electric field. Thus, the deflection produced at any point within the plates can be found out by using Eq. (1.50).

The vertical deflection produced due to the electric field cannot be measured as we cannot see electrons. If we keep a fluorescent screen in the path of the electron when it ejects out of the electric field, then we can measure the deflection produced by the applied electric field. The whole assembly is kept in an evacuated glass tube called cathode ray tube abbreviated as CRT of which the front face is the screen.

After leaving the deflection plates, that is, electric field, the electron travels along a straight line as shown in Fig. 1.14. The electrons hit the fluorescent screen enabling the measurement of the deflection produced due to transverse electric field. The straight line along which electrons move after exiting the deflecting plates is a tangent to the curve at the end of the plates. If this tangent is extended backwards, then it intersects the X-axis at point O' as shown in Fig. 1.14. This point O' is called the *virtual source of electrons* as they appear to come from this point. This apparent origin of electrons is at the centre of the deflection plates and the screen is kept at a distance 'L' from it. Let us obtain an expression for the deflection produced on the screen.

The slope of the tangent to the parabolic path is obtained by differentiating Eq. (1.50) and dropping the negative sign,

$$\frac{dy}{dx} = \frac{eEx}{mv_0^2} \tag{1.52}$$

The slope of the parabola at the end of the plates is obtained by using $x = l$ (where l is the length over which electric field acts) in the above equation,

$$\therefore \frac{dy}{dl} = \frac{eE}{mv_0^2} l \tag{1.53}$$

$$\frac{dy}{dl} = \tan\theta = \frac{eE}{mv_0^2} l \tag{1.54}$$

From the geometry of the Fig. 1.14, the deflection 'D' produced on the screen is given by

$$D = L \tan\theta \tag{1.55}$$

Substituting the value of tan θ by using Eq. (1.54), we get

$$D = \frac{LeEl}{mv_0^2} \tag{1.56}$$

Here v_0 is the initial velocity acquired by an electron when accelerated through a potential V_A. So substituting Eq. (1.41), we get

$$D = \frac{LeEl}{m}\left(\frac{m}{2ev_A}\right) = \frac{LEl}{2V_A} \tag{1.57}$$

As electric field $\vec{E}$ is given by $$\vec{E} = \frac{V}{d}$$

Equation (1.57) can also be written as $$D = \frac{LlV}{2dV_A} \tag{1.58}$$

Angular displacement of the electron path in the electric field is given by

$$\theta = \tan^{-1}\left(\frac{D}{L}\right) \tag{1.59}$$

Substituting value of D from Eq. (1.58), we get

$$\theta = \tan^{-1}\left(\frac{lV}{2\,dV_A}\right) \tag{1.60}$$

Time spent or transit time of the electron in the electric field is given by

$$t = \frac{l}{v_0} \tag{1.61}$$

The deflection sensitivity, 'S' is defined as the deflection produced per unit deflecting voltage applied to deflecting plates. It is given by

$$S = \frac{D}{V} = \frac{lL}{2\,dV_A} \tag{1.62}$$

From the expression of sensitivity it is clear that we can increase it by increasing the values of L and l. But we cannot increase L (distance of screen from the centre of the deflecting plates) beyond a certain limit as it will make the cathode ray tube longer which will make it difficult to handle. Similarly we cannot increase the length of the plates l, because a large deflection of electrons is not possible as electrons may hit the plates instead of coming out. The distance between the plates d cannot be reduced for the same reason. Hence, optimum values of L, l and d (these are called tube constants) are selected by the manufacturer for the desired sensitivity of the tube. From Eq. (1.62) sensitivity is inversely proportional to the accelerating potential.

$$S \propto \frac{1}{V_A} \tag{1.63}$$

According to Eq. (1.63) sensitivity will decrease if the accelerating potential is more. This is because the electrons will be moving very fast and will not stay in the field for sufficient time for field to show its effect. Thus, the sensitivity is affected by the accelerating potential.

The reciprocal of the deflection sensitivity is called *deflection factor* 'G', which is given by

$$G = \frac{1}{s} = \frac{2dV_A}{lL} \tag{1.64}$$

The application of transverse uniform electric field is to deflect the charged particle along the desired trajectory. Thus, we can manipulate the path of the charged particle by using the transverse electric field.

1.4.4 Electron Projected at an Angle into Uniform Electric Field

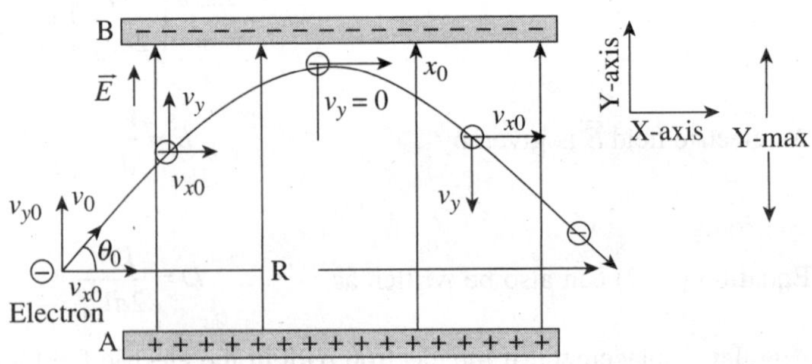

Fig. 1.15 Projectile motion of an electron in a uniform electric field.

Let 'A' and 'B' be two plane parallel metal plates oriented horizontally and separated by distance 'd' as shown in

Fig. 1.15. A potential difference of V volts is applied between the plates which produces a vertically acting uniform electric field (along Y-direction) directed from plate A to B. Suppose an electron is projected into a uniform electric field at an angle θ_0 with the X-axis and with an initial velocity v_0. The electric field acts in the positive Y-direction and the electron gets accelerated in the negative Y-direction. It means that the electron is projected such that it travels against the force due to electric field. The acceleration of electron is constant and the motion of electron will be very much similar to that of a projectile in a gravitational field. The equations of kinematics in two dimensions can be extended to the present case.

The velocity component in X-direction v_0 remains constant as no field is acting along X-direction, while v_y decreases initially and again increases when the electron reverses its path. The initial velocity components along X- and Y-directions are,

$$v_{x0} = v_0 \cos \theta_0$$

$$v_{y0} = v_0 \sin \theta_0 \tag{1.65}$$

The components of final velocity at time t along X- and Y-directions are

$$v_x = v_{x0} = v_0 \cos \theta_0 = \text{constant} \tag{1.66}$$

$$v_y = v_{y0} + at = v_0 \sin \theta_0 + at \tag{1.67}$$

From Eqs (1.66) and (1.67) we can obtain X- and Y-coordinates for the electron at any time by using kinematic equation as follows.

X-coordinate:

$$S = ut + \frac{1}{2}at^2$$

$$\therefore x = v_{x0}t = v_0 \cos \theta_0 t \tag{1.68}$$

Y-coordinate:

$$y = v_{y0}t + \frac{1}{2}at^2 = (v_0 \sin \theta_0)t + \frac{1}{2}at^2 \tag{1.69}$$

From Eq. (1.68)

$$t = \frac{x}{v_0 \cos \theta_0}$$

Substituting t in Eq. (1.69), we get

$$y = v_0 \sin \theta_0 \left(\frac{x}{v_0 \cos \theta_0} \right) + \frac{1}{2} a \left(\frac{x}{v_0 \cos \theta_0} \right)^2$$

$$y = (\tan \theta_0)x + \frac{1}{2} a \left(\frac{x^2}{v_0^2 \cos^2 \theta_0} \right)$$

Rearranging the terms we can rewrite the equation as

$$y = (\tan \theta_0)x + \left(\frac{a}{2v_0^2 \cos^2 \theta_0} \right) x^2$$

$$y = ax + bx^2 \tag{1.70}$$

The quantities in the bracket are constants, $a = \tan\theta_0$ and $b = \dfrac{a}{2v_0^2\ \cos^2\theta_0}$.

Equation (1.70) is the equation of a parabola. Therefore the trajectory of an electron projected into a uniform electric field is a parabola. Various parameters of motion of a charged particle can be obtained similar to the projectile in a gravitational field.

The maximum distance that the charged particle reaches in Y-direction, Y_{max}

Using kinematic equation, $v^2 = u^2 + 2as$ along Y-direction, we get

$$v_y^2 = v_{y0}^2 + 2ay \tag{1.71}$$

At maximum height the Y-component of velocity reduces to zero.

$$0 = v_{y0}^2 + 2\,a\,Y_{max}$$

$$0 = (v_0 \sin\theta_0)^2 - 2aY_{max}$$

The negative sign is due to the deceleration

$$2aY_{max} = v_0^2\ \sin^2\theta_0$$

$$Y_{max} = \frac{v_0^2\ \sin^2\theta_0}{2a} \tag{1.72}$$

Time taken by the charged particle to reach the maximum distance in the Y-direction

Using kinematic equation, $v_y = v_{y0} + at$ along Y-direction and as at maximum height the Y-component of velocity reduces to zero, we get

$$0 = v_0\ \sin\ \theta_0 + at$$

$$\therefore t = \frac{v_0 \sin\theta_0}{a} \tag{1.73}$$

Time of flight T It is the time taken by the charged particle to return to its original position along the X-direction which is exactly twice the time required to reach the maximum height.

$$T = 2t$$

Substituting the value of t from Eq. (1.73), we get

$$T = \frac{2v_0 \sin\theta_0}{a} \tag{1.74}$$

Range (R) It is the horizontal distance travelled (along X-direction) by the charged particle when it returns to its initial launch height. As the distance travelled along X-direction is due to X-component of the velocity, the distance covered along X-direction is the product of the X-component of the velocity and the time of flight.

$$\therefore \mathrm{R} = v_x T \tag{1.75}$$

Substituting Eq. (1.74) for T (time of flight), we get

$$R = v_0 \cos\theta_0 \left(\frac{2v_0 \sin\theta_0}{a}\right) \tag{1.76}$$

$$R = \frac{v_0^2}{a}(2 \sin \theta_0 \cos \theta_0)$$

$$R = \frac{v_0^2 \sin 2\theta_0}{a} \tag{1.77}$$

Checkpoint 2

1. Electric field lines are always ...... to equipotential surfaces.
   (a) perpendicular
   (b) parallel
   (c) opposite
   (d) none of the above
2. The application of longitudinal electric field is to
   (a) deflect the particle
   (b) decelerate the particle
   (c) accelerate the particle
   (d) none of the above
3. The deflection sensitivity of CRT in electrostatic case is
   (a) inversely proportional to V_A
   (b) does not depend on V_A
   (c) inversely proportional to $\sqrt{V_A}$
   (d) directly proportional to V_A
4. When an electron moves towards the negatively charged plates, it will
   (a) accelerate
   (b) decelerate
   (c) stops abruptly
   (d) none of the above
5. The terminal velocity of an electron is:
   (a) directly proportional to accelerating potential
   (b) indirectly proportional to accelerating potential
   (c) directly proportional to square root of accelerating potential
   (d) indirectly proportional to square root of accelerating potential
6. In an electric field the range covered will be maximum for the particles projected at same angle and with same velocity for particles with
   (a) higher specific charge
   (b) lower specific charge
   (c) zero specific charge
   (d) does not depend on specific charge
7. The electrostatic deflection sensitivity is independent of
   (a) tube constants
   (b) electric field
   (c) accelerating potential
   (d) none of the above
8. The deflection of an electron beam on a CRT screen is 10 mm. If the pre-accelerating anode voltage is halved and the potential between the deflection plates is doubled, the deflection of the electron beam will be
   (a) 80 mm
   (b) 20 mm
   (c) 40 mm
   (d) 10 mm
9. A charged particle projected at an angle with electric field will follow
   (a) circular path
   (b) helical path
   (c) parabolic path
   (d) straight line path
10. For an electron travelling with initial velocity 2×10^3 m/s, the time taken by the electron to travel the electric field region of 200 mm width is
    (a) 10^{-4} s
    (b) 10^2 s
    (c) 10^4 s
    (d) 200 s

1.5 Motion of Electron in Uniform Magnetic Field

A static magnetic field does not exert any force on a charged particle at rest. When the charged particle moving with a velocity v enters a magnetic field, it experiences a magnetic force as given by Eq. (1.25)

$$\vec{F}_L = q(\vec{v} \times \vec{B})$$

The infinitesimal work done by the magnetic field in moving the charged particle through a small distance dx is dw which is calculated as follows.

$$dw = \vec{F}_L \cdot dx = q(\vec{v} \times \vec{B}) \cdot dx$$

As velocity is distance per unit time, distance dx can be written as $dx = vdt$. Substituting this value of dx in the above equation we get

$$dw = q(\vec{v} \times \vec{B}) \cdot v\,dt = 0 \tag{1.78}$$

Thus, from Eq. (1.78) it is proved that a steady magnetic field does no work on the moving charge particles.

From the above result and work–energy theorem, the kinetic energy of the electron does not change due to the action of magnetic field. As force F_L and velocity are perpendicular to each other, we can write

$$F_L \cdot v = ma \cdot v = m\left(\frac{dv}{dt}\right) \cdot v = 0 \tag{1.79}$$

Therefore $$\frac{d}{dt}\left(\frac{mv^2}{2}\right) = 0$$

$$\text{or } \frac{1}{2}mv^2 = \text{constant} \tag{1.80}$$

or $$v = \text{constant}$$

Hence, we can say that applied magnetic field cannot change the speed of the electron. It can only change the direction of the velocity vector. Therefore, the force exerted on the moving charged particles by the magnetic field is called *deflecting force*.

1.5.1 Motion in Longitudinal Uniform Magnetic Field

If a charged particle (positive or negative) moves along the magnetic lines of induction, the magnetic force, also called the *Lorentz force*, acting on it will be zero as the velocity vector $\vec{v}$ is parallel to the magnetic field vector $\vec{B}$ as shown in Fig. 1.17.

$$\vec{F}_L = q(\vec{v} \times \vec{B}) = qvB \sin\theta = qvB \sin 0° = 0 \tag{1.81}$$

Similarly, if the charged particle moves (positive or negative) opposite to the field lines, then the magnetic force on it will be again zero as

$$\vec{F}_L = q(\vec{v} \times \vec{B}) = qvB \sin\theta = qvB \sin 180° = 0 \tag{1.82}$$

So, if the charged particle (positive or negative) enters along or opposite to the direction magnetic field, it continues to move along the initial direction of motion without any change in its speed and direction.

1.5.2 Motion in Transverse Uniform Magnetic Field

Let us consider an electron of charge '*e*' entering in a uniform magnetic field $\vec{B}$ with its initial velocity $\vec{v}$ normal to the field. Magnetic field is acting into the plane of the page which is shown by crosses in Fig. 1.18. Due to the transverse magnetic field, maximum force acts on electrons.

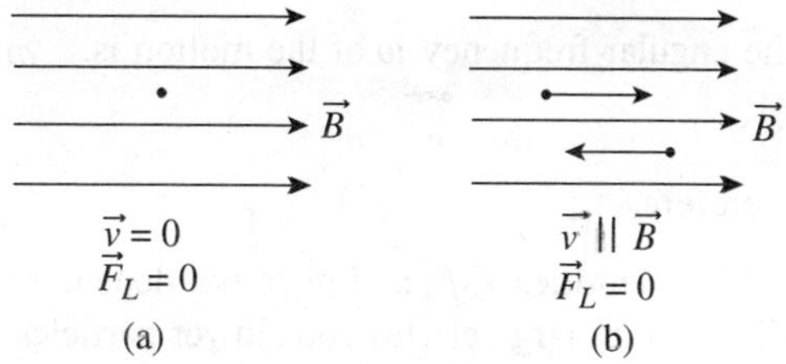

Fig. 1.17 Electron in a uniform magnetic field. (a) When a charged particle (positive or negative) is at rest (b) When a charged particle (positive or negative) is moving parallel or antiparallel

$$\vec{F}_L = q(\vec{v} \times \vec{B}) = qvB \sin\theta = evB \sin 90° = evB \quad (1.83)$$

A magnetic force F_L continually deflects the electrons and because $\vec{v}$ and $\vec{B}$ are always perpendicular to each other, this deflection causes the electrons to follow a circular path of radius 'R'. Let us determine the parameters that characterize the circular motion of an electron.

The magnetic force exerted on a charged particle in a transverse magnetic field is always directed towards the centre. Hence, it is similar to the centripetal force which is given by

$$F_c = \frac{mv^2}{R} \quad (1.84)$$

Equating the two forces we get

$$F_L = F_c$$

$$evB = \frac{mv^2}{R}$$

Solving for R

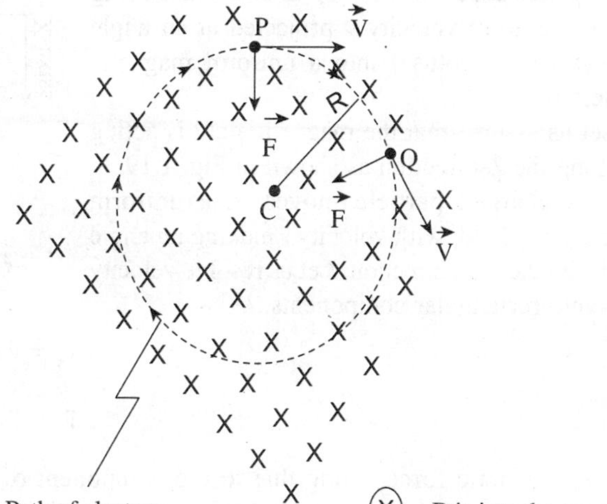

Fig. 1.18 Force exerted by a uniform magnetic field on a moving charged particle (in this case, an electron) produces a circular path

$$R = \frac{mv}{eB} \quad (1.85)$$

The above equation gives the radius of the circular path. The radius of the circular path depends on the particle's momentum. The period '*T*', that is, the time required to complete one revolution is equal to the circumference divided by the speed is obtained as follows.

$$T = \frac{2\pi R}{v} \quad (1.86)$$

Substituting for R from Eq. (1.85), we get

$$T = \frac{2\pi m}{eB} \quad (1.87)$$

The frequency f is given by

$$f = \frac{1}{T} = \frac{eB}{2\pi m} \quad (1.88)$$

The angular frequency ω of the motion is $\omega = 2\pi f = 2\pi \times \frac{eB}{2\pi m}$

Therefore, $$\omega = \frac{eB}{m} \tag{1.89}$$

The quantities T, f, and ω do not depend on the speed of the particle (for $v < c$). Hence, faster particles will move in large circles and slower particles in small circles, but all the particles with the same q/m take the identical time T to complete one revolution. The sense of rotation will be clockwise for electrons and it is counterclockwise for a positive charge. The application of transverse uniform magnetic field is to change the path of the charged particle along the curved path, which is employed in devices such as cyclotron and Bainbridge mass spectrograph.

Projected case Motion of an electron moving with uniform velocity v projected at an angle θ (acute or obtuse) into a uniform magnetic field $\vec{B}$.

Let us assume that the magnetic field is acting along the Z-direction as shown in Fig. 1.19.

A charged particle moves in a uniform magnetic field, with velocity $\vec{v}$ making an angle θ with the field direction. Let us resolve velocity $\vec{v}$ into rectangular components.

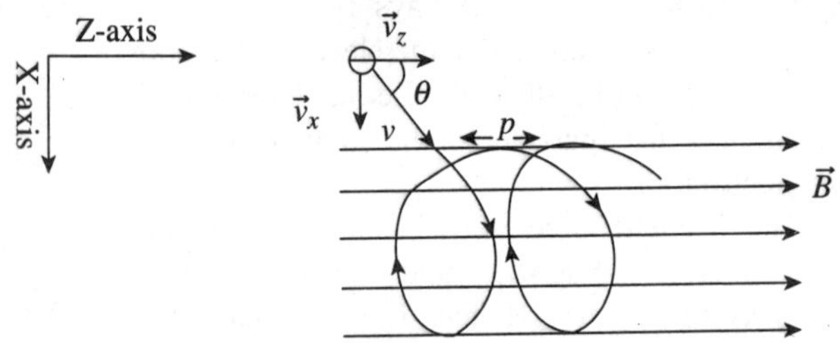

Fig. 1.19 An electron entering in a magnetic field at an angle θ spirals about the field lines

$$\vec{v}_x = \vec{v}\sin\theta \tag{1.90}$$

$$\vec{v}_z = \vec{v}\cos\theta \tag{1.91}$$

The magnetic force acting due to the component of velocity of the charged particle that is parallel to the magnetic field $\vec{v}_z$, is zero. Hence, the parallel component of velocity $\vec{v}_z$, is unaffected. Therefore, the charged particle moves at a constant speed along the direction of the magnetic field. The magnetic force due to the component of the velocity which is perpendicular to the magnetic field $\vec{v}_x$, is maximum. This force will cause the particle to execute uniform circular motion in a plane perpendicular to the magnetic field.

$$\vec{F}_L = q(\vec{v}_x \times \vec{B}) = qv_x B\sin\theta = ev_x B\sin 90° = ev_x B$$

Thus, the parallel component of velocity $\vec{v}_z$ gives rise to the linear motion, whereas the perpendicular component of velocity $\vec{v}_x$ is responsible for the circular motion of the electron. The resultant motion of the electron will be superposition of the linear and circular motion which is a helical motion. Hence, due to the component of velocity parallel to the magnetic field, circular motion will drift at a constant speed along the magnetic field producing an overall helical motion.

The radius of the circular path can be found by equating the magnetic force on the charge with the centripetal force.

$$ev_x B = \frac{mv_x^2}{R} \tag{1.92}$$

$$R = \frac{mv_x}{eB}$$

$$R = \frac{mv\sin\theta}{eB} \tag{1.93}$$

The periodic time T for one revolution is $\quad T = \dfrac{2\pi R}{v_x}$

Substituting for R from Eq. (1.93) we get $\quad T = \dfrac{2\pi mv\sin\theta}{eBv\sin\theta}$

$$T = \frac{2\pi m}{eB} \tag{1.94}$$

The axis of the helix is along the field direction and the pitch of the helix is the linear distance covered by the electron in one revolution along the field direction.

$$p = v_z \times T = v\cos\,\theta \times T \tag{1.95}$$

$$p = \frac{2\pi mv\cos\theta}{eB} \tag{1.96}$$

The magnetic field acting at an angle with respect to the path of the charged particle is best used for focusing the charged particle. This principle finds application in electron microscopy.

1.6 Velocity Filter/Selector

A combination of electric and magnetic fields applied at right angles is called *crossed field configuration.* The crossed field configuration forms what is known as velocity selector and is a common occurrence in mass spectrometers and particle accelerators. The underlying physical principle is explained below.

In the crossed field configuration, a charged particle moving with velocity $\vec{v}$ enters in a direction perpendicular to both the fields. The fields are applied in such a way that for a given charged particle, the fields will exert the force in the same plane but in the opposite direction. A typical configuration which is shown in Fig 1.20 can be used for both positively and negatively charged particles. This configuration can be used to find the unknown velocity of the charged particle.

When the charged particle of unknown velocity enters into the configuration, both the fields will show their existence by exerting the force on it. Electric field $\vec{E}$ will exert a force of magnitude qE on the charged particle, while the magnetic field exerts a force qvB on the same particle. The electric force is in opposite direction to the magnetic force as shown in the figure. When these two forces are made equal in magnitude, they will cancel out each other and the charged particle will move undeflected in a straight line. Its velocity will be given by

$$F_E = F_L$$

Therefore, $\quad qE = qvB$

Therefore, $$v = \frac{E}{B} \tag{1.97}$$

By adjusting the magnitude of any one of the fields, the forces can be made equal. Thus, this crossed field configuration can be used for determining the unknown velocity of a charged particle by knowing the magnitudes of the electric and magnetic fields.

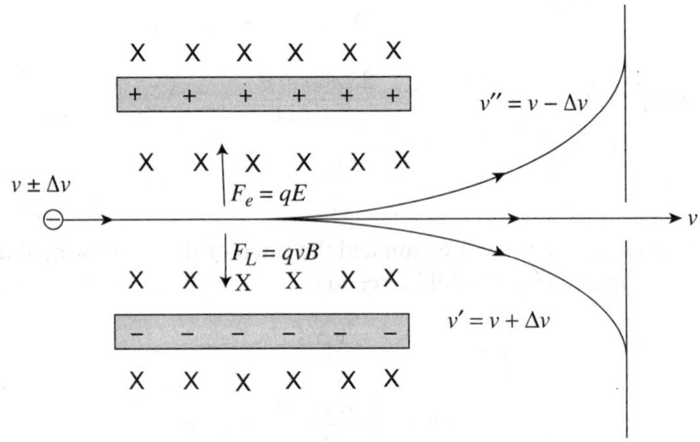

Fig. 1.20 Crossed field configuration as a velocity filter/selector

The crossed field configuration can also be used to filter out the charged particles with desired velocity from the stream of particles having spread in their velocity. The working of the crossed field configuration as a velocity filter or selector is explained below.

Only those crossed configurations can be used as a filter/selector, where the electric and magnetic field directions are such that the forces exerted by them on the charged particle are in the same plane, but are acting in opposite directions.

The desired velocity of the charged particle to be selected is obtained by adjusting the magnitudes of electric and magnetic field, such that $v = \frac{E}{B}$. The stream of negatively charged particles having a small velocity spread is allowed to enter the configuration as shown in Fig. 1.20. The electric force is independent of the velocity of charged particles, but the magnetic force is dependent on the magnitude of velocity; hence, it will be more if velocity is more and vice versa. Hence, charged particles having more velocity than the selected velocity, that is, for $v' > v$, the force due to magnetic field is greater than the force due to electric field. Hence, charged particles with velocity higher than the selected velocity will be deflected along the direction of magnetic force which is in downward direction. For charged particles having velocity less than the selected velocity, that is, $v'' < v$, the magnetic force will reduce. Hence, electric force will be more and the charged particles will be deflected along the direction of the electric force which acts in the upward direction as shown in Fig. 1.20. Thus, the crossed configuration can be used to filter out or select the charged particles with desired velocity. Thus, a mono-velocity beam can be obtained by using a simple combination of static electric and magnetic field which finds application in many devices.

Checkpoint 3

1. In the static magnetic field, work done by the field is always
   (a) maximum (c) zero
   (b) negative (d) positive
2. When an electron moving with velocity $\vec{v}$ enters a uniform transverse magnetic field the Lorentz force alters the
   (a) magnitude of $\vec{v}$ (c) magnitude and direction of $\vec{v}$
   (b) direction of $\vec{v}$ (d) magnitude and direction of $\vec{v}$ remains unaffected
3. When a charged particle enters a uniform magnetic field at right angle to the direction of field, it will move along
   (a) helical path (c) circular path
   (b) straight line (d) its motion is unaffected
4. A charged particle will travel along helical path in a magnetic field only when its velocity vector is
   (a) along the field direction (c) perpendicular to the field direction
   (b) opposite to the field direction (d) at an angle with the field direction
5. The pitch of the helix described by a proton in uniform magnetic field will
   (a) decrease with increase in magnetic field strength (c) does not depend on magnetic field strength
   (b) decrease with decrease in magnetic field strength (d) none of these
6. For measuring the pitch in helical motion we use the velocity of an electron
   (a) along the field direction (c) perpendicular to the field direction
   (b) opposite to the field direction (d) none of the above
7. If an electron and a proton enter the same transverse uniform magnetic field, then
   (a) electron will make more revolutions (c) both electron and proton will make same number of revolutions
   (b) proton will make more revolutions (d) none of the above
8. The time required for the slower particle than the faster particle for completing one revolution in a transverse uniform magnetic field is
   (a) large (c) same
   (b) small (d) zero
9. In a transverse uniform magnetic field the charged particle will always describe the path in a plane
   (a) of the magnetic field (c) which contains magnetic field direction and velocity vector
   (b) perpendicular to the magnetic field (d) which contains magnetic field direction and force vector
10. A helium ion and a proton are travelling in the transverse uniform magnetic field along the circular path with the same tangential speed. The periodic time of the helium ion will be
    (a) half as that of proton (c) double as that of proton
    (b) equal to that of proton (d) one-third as that of proton
11. In velocity selector electric field acts along Z-direction and magnetic field along X-direction. A negatively charged particle moving faster than the selected velocity will deflect
    (a) along positive Z-direction (c) along Y-direction
    (b) along X-direction (d) along negative Z-direction

12. In a velocity filter, if $E = 100$ V/m and $B = 0.01$ T, the speed of the particle that passes undeflected through the field is
    (a) 10^4 m/s
    (b) 10^{-4} m/s
    (c) 1 m/s
    (d) 10^3 m/s
13. To select a charged particle moving with monovelocity $\vec{v}$ in velocity selector, we have to adjust
    (a) magnitude of E = magnitude of B
    (b) $E > B$
    (c) $B > E$
    (d) none of the above

Solved Numericals

1. A proton accelerates from rest in a uniform electric field of 1000 N/C. After some time it attains the speed of 5×10^5 m/s.
   (a) Find the acceleration of the proton.
   (b) How long does it take the proton to reach the above velocity?
   (c) How far has it moved in this time?
   (d) What is its kinetic energy at this time?

Given data: Electric field strength E = 1000 N/C, Initial velocity of the proton $u = 0$

Proton velocity after time t is $v = 5 \times 10^5$ m/s

Proton charge, $e = 1.602 \times 10^{-9}$ C

Proton mass $m = 1.67 \times 0^{-27}$ kg

Solution:

(a) $a = \dfrac{eE}{m} = \dfrac{(1.602 \times 10^{-19} c)\,(1000\text{N/C})}{1.67 \times 10^{-27}\,\text{kg}}$

$= \dfrac{1.602 \times 10^{-16}}{1.67 \times 10^{-27}\,\text{kg}}$

$\therefore a = 9.6 \times 10^{10}\,\text{m/s}^2$

(b) $t = \dfrac{v}{a} = \dfrac{5 \times 10^5\,\text{m/s}}{9.6 \times 10^{10}\,\text{m/s}^2}$

$t = 5.2 \times 10^{-6}\,\text{S}$

(c) $s = \dfrac{1}{2}at^2 = \dfrac{1}{2}(9.6 \times 10^{10})\,(5.2 \times 10^{-6})^2$

$s = 1.29$ m

(d) $\text{K.E.} = \dfrac{1}{2}mv^2 = \dfrac{1}{2}(1.67 \times 10^{-27})\,(5 \times 10^5)^2$

$= 20.87 \times 10^{-17}\,\text{kg.}\dfrac{m^2}{s^2} = 20.87 \times 10^{-17}\,\text{j}$

$\text{K.E.} = 1.3$ keV

2. A proton has an initial velocity of 2×10^5 m/s sin the X-direction. It enters a uniform electric field of 2.5×10^4 N/C in a direction perpendicular to the field lines.
   (a) Find the time it takes for the proton to travel 0.03 m in the X-direction.
   (b) Find the vertical displacement of the proton after it has travelled 0.03 m in the X-direction.
   (c) Determine the components of proton velocity after it has travelled 0.03 m in the X-direction.

Given data: Initial velocity $v_0 = 2 \times 10^5$ m/s, Electric Field $E = 2.5 \times 10^4$ N/C, Proton charge, $e = 1.602 \times 10^{-19}$c, Proton mass $m = 1.67 \times 10^{-27}$ kg

Solution:

(a) Proton is not accelerated along x-direction. Time taken by proton to cover distance l is

$t = \dfrac{1}{v_0} = \dfrac{0.03\,\text{m}}{2 \times 10^5\,\text{m/s}} = 1.5 \times 10^{-7}\,\text{s}$

(b) $y = \dfrac{eEt^2}{2\,\text{m}}$

$= \dfrac{(1.602 \times 10^{-19}\text{C})\,(2.5 \times 10^4 N/c)\,(1.5 \times 10^{-7}\text{s})^2}{2(1.67 \times 10^{-27}\,\text{kg})}$

$= 2.697 \times 10^{-2}\,\text{m} = 2.7\,\text{cm}$

(c) $v_y = \dfrac{eEt}{m}$

$= \dfrac{(1.602 \times 10^{-19}\text{C})\left(2.5 \times \dfrac{10^4 N}{c}\right)(1.5 \times 10^{-7}\text{s})}{(1.67 \times 10^{-27}\,\text{kg})}$

$v_y = 3.59 \times 10^5$ m/s

$v_x = v_0 = 2 \times 10^5$ m/s

3. Hundred-volt electrons are introduced at A into a uniform electric field of 10^5 V/m, as shown in Fig. SN.1 The electrons emerge at the point B in time 5.39 ns.

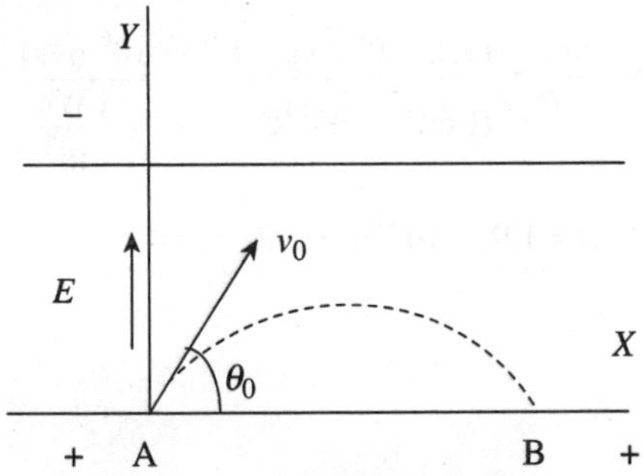

Fig. SN.1 Parabolic path of an electron in uniform electric field

(a) What angle does the electron make with the horizontal?

(b) What is the distance AB?

Given data: Electric field $E = 10^5$ V/m,

Time of flight $T = 5.39$ ns

Solution: The path of the electrons will be a parabola, as shown by the dashed curve in Fig. SN.1. The initial electron velocity is found using the equation

$$v_0 = 5.93 \times 10^5 \sqrt{V_A}\, \frac{\text{m}}{\text{s}} = 5.93 \times 10^5 \sqrt{100}$$

$$= 5.93 \times 10^6 \text{ m/s}$$

The acceleration due to electric field is

$$a = \frac{eE}{m} = \frac{1.6 \times 10^{-19} \times 10^5}{9.1 \times 10^{-31}} = 1.76 \times 10^{16} \text{ m/s}^2$$

(a) Here we must find θ_o first. The time given in the problem is the time of flight. The formula for time of flight is

$$T = \frac{2v_0 \sin \theta_0}{a}$$

$$\theta_0 = \sin^{-1}\left(\frac{T \times a}{2 \times v_0}\right) = \sin^{-1}\left(\frac{5.39 \times 10^{-9} \times 1.76 \times 10^{16}}{2 \times 5.93 \times 10^6}\right)$$

$$= \sin^{-1}(0.799)$$

$$\theta_0 = 53.12^0$$

(b) The distance AB is the range which is given by

$$R = v_0 \cos \theta_0 \times T$$

$$R = (5.93 \times 10^6 \cos 53.12)\,(5.39 \times 10^{-9})$$

$$= 19.18 \times 10^{-3} \text{ m} = 19.18 \text{ mm}$$

4. An electron with a velocity of 4.6×10^7m/s enters a uniform magnetic field of induction 0.085 T perpendicular to the field lines. Determine (a) the Lorentz force acting on the electron and (b) radius of the circle in which it moves.

Given data: Charge of electron $e = 1.6 \times 10^{-19}$ C, Mass of electron $\mathbf{m} = 9.1 \times 10^{-31}$ kg,

$v = 4.6 \times 10^7$ m/s, $B = 0.085$ T

Solution: (a)

$$F_L = evB = (1.602 \times 10^{-19}\text{C})\,(4.6 \times 10^7 \text{m/s})\,(0.085\text{T})$$

$$= 6.26 \times 10^{-13} \text{ N}$$

(b) $$R = \frac{mv}{eB} = \frac{(9.8 \times 10^{-31}\text{kg})\,(4.6 \times 10^7 \text{m/s})}{(1.60 \times 10^{-19}\text{C})\,(0.085\text{T})}$$

$$= 3.1 \times 10^{-3} = 3.1\text{mm}$$

5. A uniform magnetic field of 0.03 T exists along the x-axis. A proton is shot into the field with a speed of 8×10^6 m/s at an angle 30° with the x-axis. Find (a) the radius and (b) pitch of the helical path. (Mass of proton m = 1.67×10^{-27} kg)

Given data: Mass of proton = 1.67×10^{-27} kg, Charge of proton = 1.6×10^{-19} C, $v = 8 \times 10^6$ m/s, $B = 0.03$ T and $\theta = 30°$

Solution:

(a) $$R = \frac{mv \sin\theta}{qB} = \frac{(1.67 \times 10^{-27}\text{kg})\,(8 \times 10^6 m/s)\,(0.5)}{(1.602 \times 10^{-19} C)\,(0.3T)}$$

$$= 13.9 \times 10^{-2} \text{ m}$$

$$= 0.14 \text{ m}$$

(b) Pitch of the helical path

$$P = \frac{2\pi mv \cos\theta}{qB}$$

$$= \frac{2(3.143)(1.67 \times 10^{-27}\text{kg})(8 \times 10^6 \text{m/s})(0.866)}{(1.602 \times 10^{-19} C)\,(0.37)}$$

$$= 1.23 \text{ m}$$

6. An electron beam passes through a magnetic field of 4×10^{-3} Wb.m^2 and electric field of 5.4×10^3 V/m both acting simultaneously at the same point. Assuming the path of electrons remains unchanged, (a) calculate the electron speed. (b) If the electric field is switched off, what will be the radius of the circular path?

Given data:

$$B = 4 \times 10^{-3}\ \text{Wb.m}^2\ ;\ m = 9.1 \times 10^{-31}\ \text{kg}$$

$$E = 5.4 \times 10^3\ \text{V/m}\ ;\ e = 1.602 \times 10^{-19}\ \text{C}$$

Solution:

(a) $v = \dfrac{E}{B} = \dfrac{5.4 \times 10^3\ \text{V/m}}{4 \times 10^{-3}\ \text{wb/m}^2} = 1.35 \times 10^6\ \text{m/s}$

(b) $R = \dfrac{mv}{eB} = \dfrac{(9.1 \times 10^{-31}\ \text{kg})\ (1.35 \times 10^6\ \text{m/s})}{(1.602 \times 10^{-19}\ \text{C})\ (4 \times 10^{-3}\ \frac{\text{Wb}}{\text{m}^2})}$

$\therefore R = 1.92 \times 10^{-3}\ \text{m} = 1.92\ \text{mm}$

Numerical Exercises

1. An electron travelling with initial velocity of 4×10^6 m/s enters a region of uniform electric field of strength 4.5×10^3 N/C along the field lines.

(a) Find the deceleration of the electron.

(b) Determine the time it takes for the electron to come to rest after it enters the field.

(c) How far does the electron move in the electric field before coming to rest?

(Ans: 7.92×10^{14} m/s^2, 5.05×10^{-9} s, 0.0106 m)

2. A proton accelerates from rest in a uniform electric field of 600 N/C. After some time its speed is 4.5×10^6 m/s.

(a) Find the acceleration of the proton.

(b) How long does it take the proton to reach the above velocity?

(c) How far has it moved in this time?

(d) What is its kinetic energy at this time?

(Ans: 5.74×10^{10} m/s^2, 7.83×10^{-5} s, 175.45 m, 16.9×10^{-15} J)

3. An electron placed in an electric field experiences an upward acceleration of 3.6×10^3 m/s^2. Determine the acceleration of a proton in the same electric field.

(Ans: 1.95 m/s^2)

4. A uniform electric field exists in a region between two oppositely charged plates. An electron is released from rest at the surface of the negatively charged plate and strikes the surface of the opposite plate, 2 cm away, in time 1.5×10^{-8} s. What is the speed of the electron as it strikes the second plate? What is the magnitude of the electric field intensity?

(Ans: 2.6×10^6 m/s, 965 V/m)

5. An electron is projected along the axis midway between the plates of a cathode ray tube with an initial velocity 3.8×10^6 m/s. The uniform electric field between the plates has an intensity of 2500 N/C and acts upwards.

(a) How far below the axis has the electron moved when it reaches the end of the plates?

(b) At what angle with the axis is it moving as it leaves the plates?

(c) How far below the axis will it strike the florescent screen?

(Ans: 24 mm, 50.63° 144 mm)

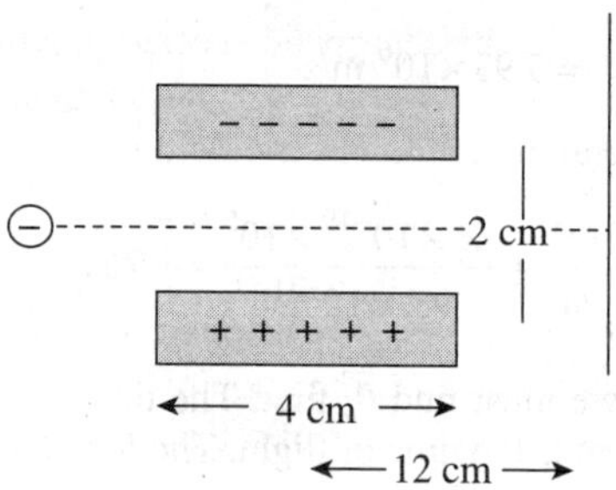

6. An electron beam is accelerated through a potential difference of 3000 V. It is then made to pass through a uniform electric field established perpendicular to its path by a set of rectangular parallel plates of length 5 cm. The distance between the plates is 2.5 cm and the potential difference is 60 V. Determine the angle of deflection of the electron beam and its transit time through the electric field.

(Ans: 1.14°, 1.53×10^{-9} s)

7. Find the electrostatic deflection factor of the cathode ray tube in volts per metre, if the distance between the plates is 0.01 m, length of plates is 0.045 m, and distance from the screen to the centre of plates is 0.33 m, for accelerating voltages of 5000 V and 10000 V.

(Ans: 13468 V/m, 6734 V/m)

8. An electron is projected with an initial velocity of 10^7 m/s and at an angle 30° to the horizontal into a uniform electric field of 5000 N/C.

(a) Find the maximum height to which the electron rises vertically.

(b) After what horizontal distance does the electron return to its original elevation?

(Ans: 1.4 cm, 9.8 cm)

9. A proton is projected at an angle of 47° to the horizontal at an initial speed of 6×10^5 m/s in a region of uniform electric field of 200 N/C.

(a) Find the time taken by the proton to return to its initial height.

(b) Find the horizontal displacement of the proton when it reaches the maximum height.

(Ans: 4.7×10^{-5} s, 9.4 m)

10. A singly charged carbon ion moving with the speed of 3×10^5 m/s enters a magnetic field of 7000 G at right angles to the field. Mass of the ion is 19.9×10^{-27} kg.

(a) What is the force that acts on the carbon ion?

(b) What is the centripetal acceleration of the ion?

(c) What is the radius of the circle in which the ion moves?

(Ans: 3.36×10^{-14} N, 1.68×10^{12} m/s^2, 0.053 m)

11. An electron of KE 5×10^3 eV enters a uniform magnetic field of induction 0.32 Wb/m^2 perpendicular to its direction of motion. Calculate its velocity and radius of the path.

(Ans: 4.2×10^7 m/s, 0.746 mm)

12. An α-particle flies into a homogeneous magnetic field perpendicular to its velocity. The angular momentum of the α-particle in the magnetic field is 1.33×10^{-22} kg.m^2/s. The induction of the magnetic field is 2×10^{-2} Tesla. Find the kinetic energy of the α-particle in eV. (Mass of the α-particle = 6.68×10^{-27} kg., charge = 3.2×10^{-19} C).

(Ans: 398 eV)

13. An electron projected at a velocity of 5×10^7m/s enters a uniform magnetic field at an angle of 30° with it. Calculate

(a) Magnetic flux density required for making the electron to follow a helical path of diameter 0.5 m.

(b) The time taken by the electron for completing one revolution.

(c) Pitch of the helix.

(Ans: 5.68×10^{-4}, 63 ns, 2.72 m)

14. A positive ion beam moving along the x-axis enters a region of uniform electric field of 5 kV/m along *y*-axis and magnetic field of 2 kG along z-axis. Calculate the speed of the ions which pass without deviation. What will happen to ions which are moving (a) faster, (b) slower, than these ions?

(Ans: 2.5×10^4 m/s)

15. An electron is accelerated through a potential difference of 1 kV and directed into a region between two parallel plates separated by 0.02 m with a 100V potential difference between them. If the electron moves perpendicular to the electric field between the plates, what magnetic field is necessary perpendicular to both the electron path and the electric field so that the electron travels in a straight line?

(Ans: 2.67×10^{-4})

Summary

- Electric field is a vector quantity and is defined as the force $\vec{F}$ exerted on a tiny positive test charge q_0 at that point divided by the magnitude of the test charge. The SI unit of electric field is Newton per coulomb (N/C).
- In a uniform electric field, the lines are straight, parallel, and uniformly spaced. In non-uniform fields, the field lines are curved and spaced non-uniformly.
- For a uniform field, the equipotential surfaces are a family of planes perpendicular to the electric field lines and thus to *E*, which is always tangent to these lines.

- The potential energy U of a test charge q at any point is equal to the negative of the work done $-w_{\infty f}$ on the test charge by the electric field. It is measured in joules.
- The potential energy per unit charge U/q is called *electric potential, which* is a scalar quantity.
- The component of E in any direction is expressed as the negative of the rate of change of electric potential with distance in that direction.
- Magnetic field is defined as the magnetic force exerted per unit charge per unit velocity and the SI unit of B is Tesla or wb/m^2.
- In longitudinal uniform electric field, the charged particle moves along straight line. The longitudinal electric field is used to impart kinetic energy to the charged particles.
- In transverse electric field, the charged particle moves along the parabola. The transverse electric field is used to deflect the charged particles horizontally or vertically.
- The deflection sensitivity, 'S' is defined as the deflection produced 'D' per unit deflecting voltage 'V_d' applied to deflecting plates. It is inversely proportional to the accelerating potential V_A.
- The trajectory of an electron projected into a uniform electric field is a parabola. Various parameters of projected motion of a charged particle are similar to the projectile in a gravitational field.
- A static magnetic field does not change kinetic energy of the moving charged particles.
- A static magnetic field does not exert any force on a charged particle at rest or moving parallel or anti-parallel to the field.
- The trajectory of a charged particle in transverse uniform magnetic field is circle. The application of transverse magnetic field is to change the path of the charged particle along the curved path.
- A charged particle projected in a uniform magnetic field at an acute or obtuse angle will move along a helical path. The helical motion in non-uniform magnetic field is used for focusing the charged particle in an electron microscope.
- The crossed field configuration is used as a velocity filter or selector to find unknown velocity or to filter out the charged particles with the desired velocity.

Self-study Questions

1. Show that an electron moves along a parabolic path when it enters a uniform electric field applied perpendicular (transverse) to its motion.
2. What is the force experienced by a charged particle in (a) electric field and (b) magnetic field? Show that the kinetic energy of a charged particle remains constant when it moves in the magnetic field.
3. Describe the motion of an electron when projected at an acute angle with the direction of uniform electric field and determine:
   (a) Maximum distance reached by the electron in the direction of the field.
   (b) Time taken to reach maximum distance.
   (c) Range of charged particle
4. If a moving electron has no deflection in passing through certain region of space, can we be sure that there is no magnetic field in the region? If it is deflected sideways can we be sure that a magnetic field exists in that region?
5. State the shapes of trajectories of a charged particle of velocity $\vec{v}$ moving in an electric field $\vec{E}$ when
   (a) $\vec{v}$ is perpendicular to $\vec{E}$.
   (b) $\vec{v}$ makes acute angle with $\vec{E}$.
   What are these shapes if $\vec{E}$ is replaced by magnetic field $\vec{B}$?
6. State the conditions under which a charged particle moves in a straight line in
   (a) an electric field $\vec{E}$
   (b) a magnetic field $\vec{B}$
   (c) in a region having both $\vec{E}$ and $\vec{B}$
7. In a transverse uniform magnetic field, show that the radius of curvature of the path of a charged particle is proportional to its momentum.
8. How is it possible for a charged particle to pass through a combination of electric and magnetic fields without any deviation? What will be its

velocity? What will happen if the particles are moving slowly or faster than this velocity?

9. When and why does a charged particle entering magnetic field follow a helical path?

10. Magnetic field changes the velocity of a charged particle without changing its speed—explain.

Answers

Checkpoint 1

1(c), 2(a), 3(c), 4(a)

Checkpoint 2

1(a), 2(c), 3(a), 4(b), 5(c), 6(b), 7(b), 8(c), 9(c), 10(a)

Checkpoint 3

1(c), 2(b), 3(c), 4(d), 5(a), 6(a), 7(a), 8(c), 9(b), 10(c), 11(a), 12(a), 13(d)

Electron Optics

2

Learning Objectives

After going through this chapter, you will be able to:

- Explain the basics of non-uniform electric field
- Explain the working of electron gun which is based on the concept of electron lens
- Explain the functioning of the fundamental parts of CRT
- Explain the working of CRO based on the concept of uniform and non-uniform electric fields
- Highlight the applications of CRO

2.0 Introduction

The motion of a charged particle in a uniform electric field is much simpler and can be described through simple mathematical equations. But in most of the practical cases, electric field is non-uniform. The analysis of electron motion in non-uniform fields is tedious. Fortunately, there is an analogy between the motion of electrons in an electrostatic field and the propagation of light in a transparent medium.

In this chapter, motion of electrons in non-uniform electric field is studied by using the concepts of geometrical optics. Hence, the study of electron motion in non-uniform electric field is called *electron optics.* The engineering applications of this concept are also discussed.

2.1 Electron Refraction

If the magnitude of intensity of the electric field in a region changes from one point to another, then such a field is called *non-uniform electric field.* Instead of straight lines, this field is represented by curved electric field lines as shown in Fig. 2.1(a). From the figure, it is clear that the field strength increases from point A to point B.

The motion of an electron beam in a non-uniform electric field can be understood by studying Bethe's law.

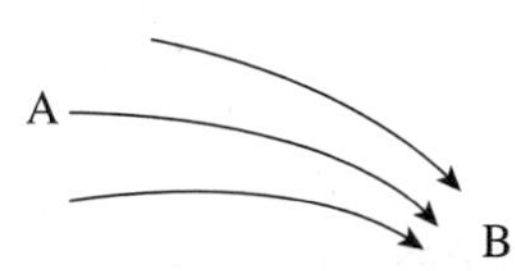

Fig. 2.1(a) Schematic of non-uniform electric field

2.1.1 Bethe's Law

Motion of electrons in a non-uniform electric field is explained with the help of equipotential surfaces. These are surfaces where the electric potential remains constant. These surfaces are perpendicular to the electric field lines and are denoted by V_1, V_2, ... as shown by dotted lines in Fig. 2.1(b). Thus the potential on the surface is constant, but as the strength of the electric field increases, $V_5 > V_4 > V_3 > V_2 > V_1$. In a non-uniform electric field, the direction of electric force acting on an electron is given by the normal to the equipotential surface.

For simplicity we will consider only one equipotential surface across which the potential changes abruptly. In order to visualize the field and the equipotential surface, let us consider that the electric field is set up over an infinitesimal region by using two perforated metal plates A and B which are connected to a dc potential as shown in Fig. 2.2. The electric field is directed from plate A to plate B.

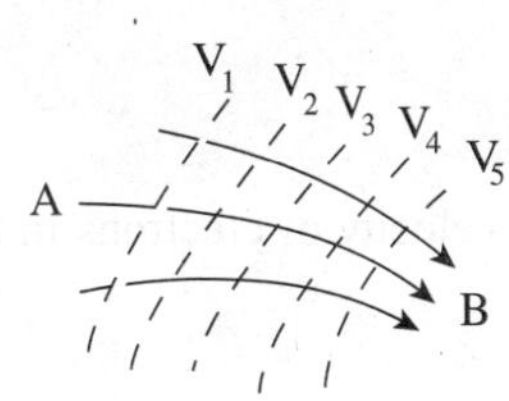

Fig. 2.1(b) Equipotential surfaces

Consider an equipotential surface 'PQ' as shown in Fig. 2.2 above which the potential is V_1 (Region-I) and below is V_2 (Region-II), such that $V_2 > V_1$. Thus, electrons pass from a lower to a higher potential region. 'XY' is the normal drawn to the equipotential surface. An electron moving through region-I has uniform velocity, v. As the electron passes through the equipotential surface PQ, it experiences a force normal to the surface, which will change its velocity from v to v' Thus, in Region II the electron moves with a higher velocity v'. The electric field acts in only the Y direction; hence only the Y component of electron velocity, v_y will change and the tangential component v_x will remain constant as there is no field in the X direction.

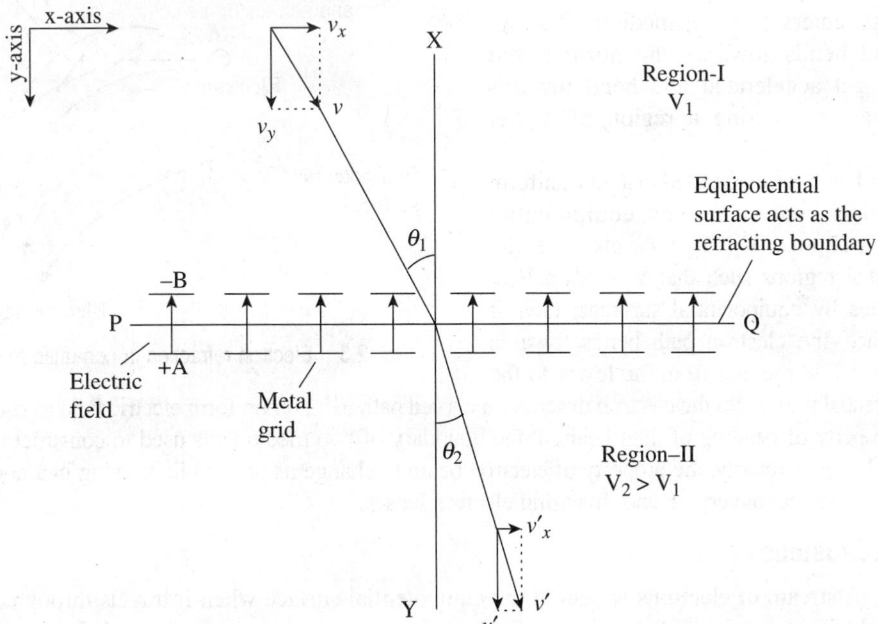

Fig. 2.2 Electron refraction

As $V_2 > V_1$, then v_y increases as the electron beam moves from lower potential to higher potential region, and if $V_2 < V_1$, then v_y decreases as electron beam moves from higher potential to lower potential region. In this case, $V_2 > V_1$; hence, velocity v_y increases and therefore the electron path bends towards the surface normal XY.

As the tangential component of velocity remains constant in regions I and II, we can equate them.

$$v_x = v'_x$$

$$v_1 \sin\theta_1 = v_2 \sin\theta_2$$

$$\frac{\sin\theta_1}{\sin\theta_2} = \frac{v_2}{v_1} \tag{2.1}$$

The velocity of electrons in terms of potential is given by

$$v = \sqrt{\frac{2eV}{m}}$$

Substituting for velocity in Eq. (2.1), we get $$\frac{\sin\theta_1}{\sin\theta_2} = \sqrt{\frac{V_2}{V_1}} \tag{2.2}$$

Equation (2.2) is known as *Bethe's law* for electron refraction. It is identical to the Snell's law of refraction of light beam in geometrical optics. According to Bethe's law the electron bends towards the normal when it encounters an equipotential surface just like light beam, when it goes from lower potential (rarer medium) to the higher potential (denser medium) and vice versa. The difference between Snell's law and Bethe's law is that the order of velocities, v_1 and v_2 are reversed. When light enters a denser medium, it slows down and bends towards the normal; but electrons get accelerated and bend towards the normal on entering a region of higher potential.

Bethe's law can be extended to a non-uniform electric field represented by equipotential surfaces. Suppose V_1, V_2, V_3, etc. are the equipotential regions such that $V_1 < V_2 < V_3 < V_4$ separated by equipotential surfaces; then at each surface, the electron path bends towards the normal while moving from the lower to the higher potential region. So the electron describes a curved path in a non-uniform electric field as shown in Fig. 2.3.

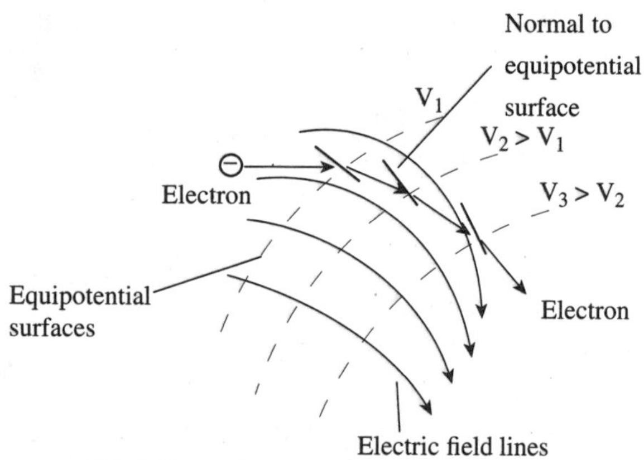

Fig. 2.3 Electron refraction in non-uniform electric field

The property of bending of light beam at the boundary of two mediums is used to construct converging and diverging lenses; similarly, the property of electron beam to change its path while moving in a non-uniform field is used to construct converging and diverging electron lenses.

2.1.2 Electrostatic Lens

Principle A stream of electrons is bent at the equipotential surface when it travels through a non-uniform electric field. Thus, a non-uniform electric field can be used as a lens to focus or defocus electron rays.

Description A simple electrostatic lens or electron lens consists of two short coaxial metal cylinders C_1 and C_2 of same dimensions and is separated by a small distance (a few centimetres). The cylinders are kept at different d.c. potentials and thus, a non-uniform electric field is produced in the gap between the cylinders, as shown in Fig. 2.4. The electric field lines are from cylinder C_2 to C_1 as C_2 is assumed to be at a higher positive potential.

The equipotential surfaces are perpendicular to the electric field lines everywhere, forming a lens in the gap as shown in Fig. 2.4.

Working Let a beam of electrons enter the double cylinder system through the cylinder C_1 which is at lower potential V_1. We will label the electron beam as C which is moving along the axis and for above and below the axis by A and B respectively. Now those electrons which are moving along the axis of the system, (shown by label C) on reaching the equipotential surface will experience electric force acting along the axis in the forward direction. Hence, the electrons are accelerated in the forward direction towards cylinder C_2 along the axis without any deviation.

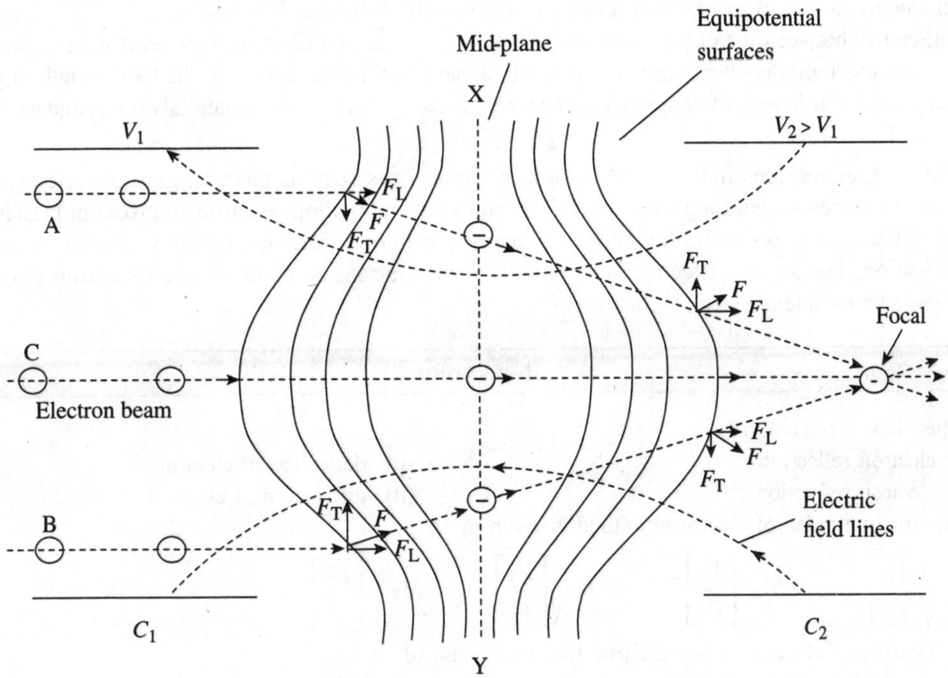

Fig. 2.4 Electrostatic lens

In the first half of the imaginary lens (left side of mid-plane 'XY') formed by the equipotential surface, the curvature of the equipotential surface is convex. All the electrons moving above the axis (shown by label A) on reaching the equipotential surface in the gap experiences an electric force F, acting at an angle to the direction of their motion. The force F, experienced by the electrons at the convex equipotential surface can be resolved into two components, parallel to the axis, that is, the longitudinal component F_L and perpendicular to the axis, that is, the transverse component F_T. Because of the force F_T, electrons are deflected down toward the axis and are simultaneously accelerated in the forward direction towards cylinder C_2 due to the force component F_L.

Similarly, the electron beam moving below the axis (shown by label B) is deflected upwards towards the axis and accelerated in the forward direction. Thus, all the off-axis electrons tend to converge toward the axis.

On crossing the mid-plane XY of the gap, all converging electrons encounter equipotential surfaces of concave shape. In this region the transverse component of the force F_T is directed away from the axis for the entire off-axis electrons and F_L the longitudinal component of force is directed forwards. As a result, electrons are accelerated forward, but tend to diverge. Thus, the first half of the gap acts like a convex lens where the electron rays tend to converge and the second half acts as the concave lens, which defocuses the electron rays. The converging action will be stronger than the diverging action, because electrons spend more time in the lower potential region since they enter with less speed in the first half.

In the second half, electrons move faster because of higher potential; so the time spent by electrons in this region is smaller. Thus, the impulse F_Lt is more for convergence interval than F_Lt in divergence interval. Hence, the converging action is more powerful than the diverging action and the net result is that the entire off-axis electron rays get focused at a point called the *focal point*. Such a double cylinder arrangement is called an *electrostatic lens* or *electron lens* as it focuses all electrons.

The difference between the optical and electrostatic lens is that: (i) Electron rays bend at each equipotential boundary, whereas light rays bend only at the two boundaries of a glass lens. (ii) The focal length of glass lens is fixed and the focal length of electron lens can be varied by varying the potentials on cylinders C_1 and C_2 tubes T_1 and T_2.

Applications Electron lens finds applications in many devices such as electron gun, particle accelerators, electron microscopes—(scanning electron microscope and tunnelling electron microscope), LHC (Large Hadron Collider), electron multiplier tubes for particle counting, microlithography, micromachining, and microfabrication, and so on. Electron lens is the most important component of an electron gun used for producing a narrow intense electron beam.

Checkpoint 1

1. Bethe's law is related to
   (a) electron reflection
   (b) electron refraction
   (c) deflection of electron
   (d) diffraction of electron
2. The refractive index of electrostatic field is given by
   (a) $\sqrt{\left[\frac{V_1}{V_2}\right]}$ (b) $\left[\frac{V_1}{V_2}\right]$ (c) $\sqrt{\left[\frac{V_2}{V_1}\right]}$ (d) $\left[\frac{V_2}{V_1}\right]$
3. For focusing of electron beam, electron lens makes use of
   (a) non-uniform magnetic field
   (b) non-uniform electric field
   (c) uniform electric field
   (d) non-uniform electric and magnetic field
4. The focal length of the electron lens can be varied by varying
   (a) voltage on the metal cylinder
   (b) gap between the two cylinders
   (c) length of the cylinder
   (d) diameter of the cylinder

2.2 Cathode Ray Tube

Cathode ray tube (CRT) is a vacuum tube device which is used in many electrical gadgets for display of images.

Construction Cathode ray tube is a specially constructed vacuum tube; it consists of the following basic parts.

1. An electron gun
2. Horizontal and vertical deflection systems
3. Fluorescent screen
4. Acquadag coating

An *electron gun* consists of a heater, a cathode, a grid, and three anodes—pre-accelerating anode, focusing anode, and an accelerating anode mounted at one end of an evacuated glass tube as shown in Fig. 2.5. Electrical power is supplied to the electrodes (cathode and anodes) and other parts through the base pins.

The *deflection system* consists of two pairs of parallel plates placed in front of the electron gun; the two pairs of plates are mounted perpendicular to each other. The connections to the deflection plates are also given through the base pins.

The *fluorescent screen* consists of a coating of phosphor deposited on the inner surface of the wide end of the vacuum tube. The portion from its neck onwards is coated from inside with a conducting material called *acquadag coating*. It is connected internally to the focusing anode. A power supply provides the required potentials to the various parts of the CRT.

Working The electron gun produces a sharply focused beam of electrons which is accelerated to a high velocity. The beam of electrons is focused on the fluorescent screen which produces a luminous spot on the screen and the electron beam may be moved to any spot on the screen with the help of the deflection system. Each part of the CRT is discussed in detail in the following pages.

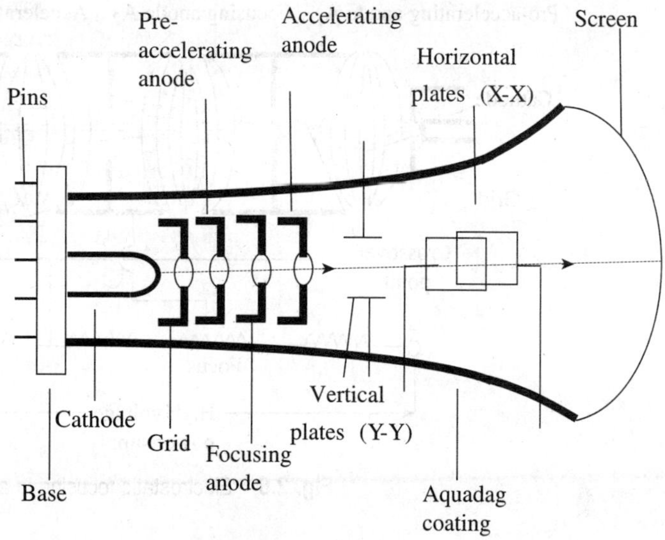

Fig. 2.5 Cathode ray tube

2.2.1 Electron Gun

An electron gun is a device which produces, focuses, and accelerates an electron beam of high intensity. V.K. Zworykin designed the electron gun in 1933.

Principle The electron gun is based on the principle that in a non-uniform electric field, electrons describe a curved path and get focused at a point. This focal point can be adjusted by adjusting the electric field between the deflection plates.

Construction The electron gun consists of a cathode, a grid, a pre-accelerating anode, a focusing anode, and an accelerating anode. The cathode is a short hollow cylinder of nickel, which encloses the heating filament. The front face of the cathode is coated with thoriated tungsten or barium and strontium oxides for copious flow of electrons. A hollow metal cylinder which surrounds the cathode with a small aperture in the front is called *grid*. Beyond the control grid, three short metal cylinders with central apertures are placed co-axially. This entire assembly is kept in an evacuated glass tube.

Working Figure 2.6 shows the functional diagram of electrostatic focusing in an electron gun.

When the power is turned on, the filament gets heated which indirectly heats the cathode. At a moderate temperature of 500° to 600° C, electrons are emitted from the front surface of the cathode and they pass through the control grid. The grid is at a negative potential with respect to the cathode and it controls the number of electrons passing through it. If it is at high negative potential, less number of electrons can pass through it, whereas if it is at a less negative potential then more electrons can pass through it. Thus, the grid with negative potential acts as a gate and controls the number of electrons passing through it.

The anodes of the electron gun provide an electron lens system for focusing the electron beam on the screen. The focusing anode and the accelerating anodes are coaxial with the pre-accelerating anode. The pre-accelerating and accelerating anodes are connected to a high positive potential and are at the same voltage, and the focusing anode is connected to a lower postive potential. Because of the difference of potentials between the cathode and anodes, a non-uniform field exists in the three regions—between the grid and pre-accelerating anode, between the pre-accelerating anode and focusing anode, and between the focusing anode and accelerating anode. The equipotential surfaces act as a convex lens and such electrostatic lenses are formed on each side of the anodes. Electrons coming out of the cathode have a tendency to diverge as they

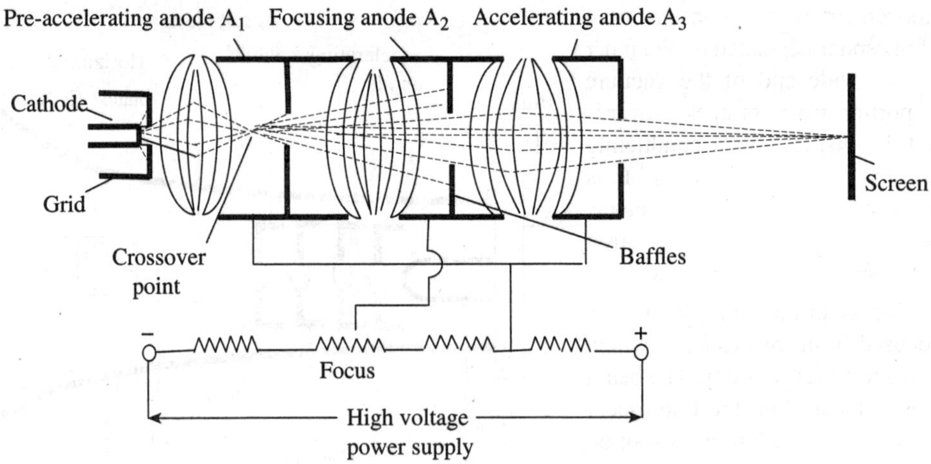

Fig. 2.6 Electrostatic focusing in an Electron Gun

are all like charged particles. The first lens formed between the grid and the pre-accelerating anode acts as a converging lens, as the potential on the grid is negative and that of the pre-accelerating anode is positive. The action of this lens is to converge the diverging electron beam to a point source called cross over point. This lens reduces the cross-sectional area of the source from the broad cathode surface to a point source. It is easy to form a fine beam of electrons that can be focused if the electrons come from a point source instead of the broad one. Now, electrons encounter the second lens formed between the pre-accelerating anode and focusing anode. As the electrons are moving from a higher potential region to lower potential region, diverging action dominates and electrons form an ever diverging beam. Baffles are used to limit the angular divergence of the beam as shown in Fig. 2.6. The electron beam then encounters the third lens formed between the focusing and accelerating anodes. This lens system, where electrons move from the lower potential region to higher potential region, converges the beam on the screen in the form of a spot. By changing the potential of the focusing anode, the focal length of this electron lens can be changed and electron beam can be focused on the screen. An electron gun is one of the important parts of cathode ray tube.

The typical values of current and voltage required by an indirectly heated cathode are 600 mA at 6.3 V. High-efficiency systems use 300 mA at 6.3V. The grid bias is varied between 0 to –50 V. The pre-accelerating anode and the accelerating anode are connected to a common positive high voltage of about 1500V. The focusing anode is connected to a lower adjustable voltage of 500 V.

2.2.2 Deflection System

There can be two types of deflection systems in a CRO, either electrostatic or magnetostatic. The electrostatic system is more popular. Two pairs of metal plates produce the electrostatic deflection. The two plates in each pair are aligned strictly parallel to each other and two pairs of plates are mounted at right angles to each other and also perpendicular to the path of the electrons as shown in Fig 2.7. One pair of plates is arranged horizontally as shown in Fig. 2.7 (a). When potential difference is applied to these plates, a uniform electric field is produced in the vertical direction. The field acts perpendicular to the beam and deflects the beam vertically. Hence, these plates are called vertical deflection plates or Y plates.

The second set of plates is oriented horizontally as shown in Fig. 2.7 (b) and produces the uniform horizontal field when potential difference is applied to the plates. The field acts normal to the beam and deflects the beam horizontally; so this set of plate is called horizontal deflection plates or X plates.

When no voltages are applied to the vertical and horizontal plates, the electron beam travels along the CRT axis and hits the centre of the screen. When a dc voltage is applied to the vertical plates, the spot gets deflected in the direction towards the positive plate and if a dc voltage is applied to horizontal plate, the spot gets deflected in the horizontal direction towards the positive plate. If a dc voltage is applied to both the plates, then the electron beam will be acted upon by the two forces and gets deflected along the direction of their resultant. Thus by changing the voltages on the deflection plates, the electron beam can be moved to any position on the screen. The Y plates always precede the X plates as the distance between the Y plate and screen L will be more which enhances deflection sensitivity and both the pairs of plates are flared up at the end to have bigger displays.

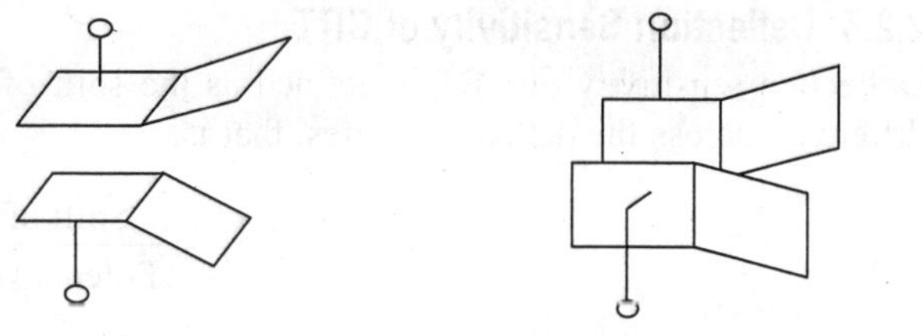

(a) Vertical deflection plates (b) Horizontal deflection plates

Fig. 2.7 Deflection plate

2.2.3 Fluorescent Screen

The inner surface of the circular front face of the CRT is coated with a thin fluorescent layer of phosphors. When the fast moving electrons hit the phosphor coated screen, it produces glow at that point. A phosphor material converts energy of the electrons falling on its surface into visible light. The glow on the screen can be obtained in various colours depending upon the phosphor material as given in Table 2.1. Thus, the fluorescent screen makes the electron beam visible and we can locate the position of the beam on the screen. A graticule (a calibrated scale) is laid on the screen to facilitate measurements along X and Y directions.

Table 2.1 Fluorescent materials and their respective colours

Fluorescent material	Colour
Zinc orthosilicate (Willemite)	Green light
Zinc oxide	Blue light
Zinc beryllium silicate	Yellow light
Zinc sulphide mixed with cadmium zinc sulphide	White light

2.2.4 Acquadag Coating

A thin film of metal such as aqueous graphite (acquadag) is usually deposited on to the non-viewing sidewalls of the tube which has the following three effects.

(1) It acts as a heat sink and reduces the danger of phosphor burn.

(2) The light scatters from phosphor and hence reduces the brightness of the display. The graphite reflects back the light going into the tube towards the viewer, thus increasing the brightness of the displays.

(3) When electron beam strikes the screen it gives energy to the phosphor atoms and thus produces fluorescence. But these electrons tend to accumulate on the screen making it negatively charged and hence repel the incoming electrons reducing the brightness of the display. Moreover, when the electrons strike the screen they not only fluoresce, but also produce secondary electrons which remain glued to the screen. Thus, the electrons deposited on the screen should be removed by providing a conducting path. The electrons accumulated on the screen are attracted by the acquadag coating, which is electrically connected to focusing anode A_2 (maintained at low positive potential). The electrons are returned to the cathode and the cathode is replenished with electrons. Thus, acquadag is a conducting coating used to complete the electrical circuit from screen to cathode and keeps the screen unbiased.

2.2.5 Deflection Sensitivity of CRT

Deflection sensitivity of CRT is defined as the shift of the electron spot on the screen per unit potential difference across the deflection plates, that is,

$$S_E = \frac{\text{Shift of the spot}}{\text{Potential difference}} \tag{2.3}$$

Deflection sensitivity is measured in mm/volt. If voltage is applied across the horizontal plates only then the spot will shift horizontally. If voltage is applied only across the vertical plates then the spot will shift vertically; hence, the total deflection of the spot is the product of the sensitivity and the potential difference.

Checkpoint 2

1. Vertical deflection plates are kept prior to horizontal deflection plates because
   (a) electron beam covers larger area on the CRT screen.
   (b) vertical deflection plates are at a higher potential than that of horizontal deflection plates.
   (c) it enhances deflection sensitivity and area covered on the CRT screen.
   (d) it enhances deflection sensitivity.
2. The function of control grid is to control
   (a) the deflection produced on the screen
   (b) the velocity of the electron beam
   (c) the size of the spot on the screen
   (d) the intensity of the display
3. In a CRT, the path travelled by the electron within the plates to which voltage is given is
   (a) straight
   (b) parabolic
   (c) along the tangential path
   (d) none of these
4. Electric deflection is the deflection of the electron beam caused by the application of
   (a) ac voltage to the deflection plates
   (b) dc voltage to the deflection plates
   (c) both ac and dc voltages to the deflection plates
   (d) dc magnetic field to the deflection plates
5. In an electron gun, the grid potential is
   (a) positive (c) same as cathode
   (b) zero (d) negative
6. The source of emission of electron in a CRT is
   (a) control grid
   (b) a barium and strontium oxide coated cathode
   (c) accelerating anodes
   (d) aquadag coating
7. In the electron gun
   (a) all anodes A1, A2 & A3 are at the same positive potential
   (b) all anodes A1, A2 & A3 are at same negative potential
   (c) two anodes A1 & A2 are at the same positive potential
   (d) two anodes A1 & A3 are at the same positive potential
8. In the electron gun, three anodes are kept
   (a) for proper focusing of beam on screen.
   (b) for accelerating the electron beam.
   (c) to collect the electrons emitted from the cathode.
   (d) for focusing and accelerating the electron beams.

2.3 Cathode Ray Oscilloscope

Cathode ray oscilloscope (CRO) is a very important electronic measuring device. It is used to measure electrical and some non-electrical signals, namely pressure, temperature, and strain. It gives the visual display of the applied electrical signals. CRO mainly consists of the following different components.

1. Cathode ray tube (CRT)
2. Vertical amplifier
3. Horizontal amplifier
4. Delay line
5. Time base circuit
6. Blanking circuit
7. Trigger circuit
8. High-voltage power supply (HVPS)
9. Low-voltage power supply (LVPS)

The block diagram of CRO is shown in Fig. 2.8. The function of each component is discussed below.

2.3.1 Cathode Ray Tube

Cathode ray tube which is the heart of CRO is already discussed in Section 2.2. The basic functions of the CRT are to produce the electron beam and the acceleration and focusing of it on the screen. Different dc potentials are required for the operation of CRT. The grid requires negative potentials, and low and high positive potentials are required by the focusing and accelerating anodes.

Three controls are associated with the operating voltages of CRT—intensity (brightness), focus, and astigmatism. The intensity control varies the potential between the cathode and grid and adjusts the number of electrons flowing through the grid. The application of less negative potential on the grid increases the number of electrons reaching the screen; so the intensity of the spot on the screen increases. Thus, by changing the potential on the grid we can change the intensity of the display formed on the screen. The control knob for varying the grid potential is brought on the front panel of the CRO and is called 'intensity or brightness'. The focus control adjusts the focal length of the electrostatic lens by varying the positive potential on the focusing anode. This control for changing the potential on focusing anode is called 'focus' knob. The astigmatism control adjusts the potential on the accelerating anodes (pre-accelerating and accelerating anodes are at the same positive potential) and thus produces a round spot of electron beam on the screen. The control knob on the front panel of CRO for this is named as 'astigmatism'.

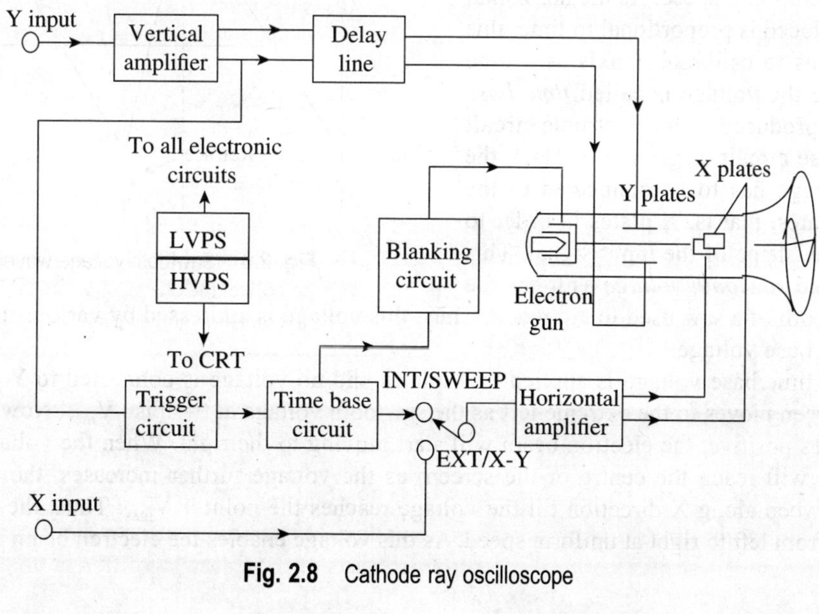

Fig. 2.8 Cathode ray oscilloscope

2.3.2 Vertical Amplifier

The signals, which are under study, are given to the Y plates of CRT through the vertical amplifier. If amplitude of the signal to be studied is very small, then this voltage may not be sufficient to deflect the electron beam. Thus, the weak signals will not be displayed on the screen. Hence the applied voltage to be displayed is first given to a vertical amplifier, which amplifies the signal to the desired level without introducing any distortion in the signal. Thus, the function of the vertical amplifier is to increase the strength of the signal and faithfully reproduce it on the screen. A small part of the vertical amplifier output is also applied to the trigger circuit.

2.3.3 Horizontal Amplifier

Horizontal amplifier amplifies the sweep voltage which is internally produced and gives it to the horizontal deflection plates. When external signals are to be applied to the horizontal plates, they are always given through the horizontal amplifier. Thus, the horizontal amplifier faithfully amplifies the internal or external voltage before connecting it to the horizontal deflection plates.

2.3.4 Time Base Circuit

Any alternating voltage signal is a variation of the amplitude of voltage with respect to time. The signal to be displayed (input signal) is connected to one set of deflection plates, that is, Y plates. Application of only this alternating input voltage to Y plate and no voltage to X plate will cause vertical movement of the electron beam. So, on the screen the spot will be seen moving up and down if the frequency of the input signal is less or a vertical line if the frequency of the signal is more than 10 Hz. Now to reproduce the signal as it is on the screen, the electron beam should also be moved along the horizontal direction with constant speed as the time axis is to be created. The electron beam will move horizontally only when voltage is applied to X plate. A voltage which has characteristics of time, that is, a linearly varying periodic voltage as shown in Fig. 2.9, should be connected to X plate. This voltage when connected to X plate will cause deflection of the electron beam in the horizontal direction with uniform (constant) speed. As the horizontal deflection produced is proportional to time, this voltage helps us to calibrate X axis as a time axis and hence the voltage is called *time base* voltage and is produced by an electronic circuit called time base circuit or generator. Thus, the time base voltage has to be connected to the other set of plates, that is, X plates in order to display the true shape of the input signal. This voltage is called *sawtooth voltage* as it has the shape of the tooth of a saw used to cut wood. Thus, this voltage is addressed by various names as ramp/sawtooth/time base voltage.

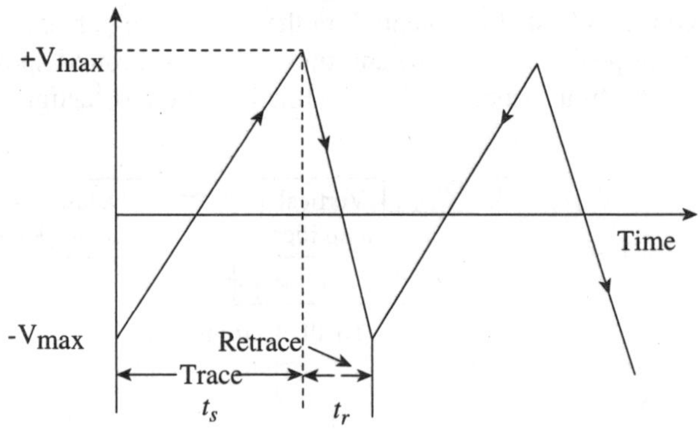

Fig. 2.9 Sawtooth voltage waveform

When only time base voltage is applied to X plates and no voltage is connected to Y plate, then the spot on the screen moves to the extreme left as the sawtooth voltage at t = 0 is $-V_{max}$. Now as the voltage slowly becomes positive, the electron beam will start moving to the right. When the voltage is zero, the electron beam will reach the centre of the screen; as the voltage further increases, the electron beam continues to sweep along X direction till the voltage reaches the point + V_{max}. Thus, the electron beam has traversed from left to right at uniform speed. As this voltage enables the electron beam to sweep along the horizontal, it is also called 'sweep' voltage. Now the voltage suddenly drops from positive maximum to negative minimum which causes the electron beam to suddenly fly back to its original position. The flyback is not instantaneous, but it takes some time. The time taken by the sweep voltage to rise from its maximum negative voltage to its maximum positive voltage is known as *sweep time or trace time* (t_s). The time taken by the sweep voltage to dip from its positive maximum to negative maximum value is called *flyback time* or *retraces time* (t_r). The sum of sweep time and retrace time constitutes the sweep period T_{sweep}.

$$T_{sweep} = t_s + t_r \tag{2.4}$$

Thus, the movement of the electron beam enabled by the sawtooth voltage is similar to the movement of the pen when one writes on the paper. How the alternating signals are displayed on the screen is discussed in detail in Section 2.4.

2.3.5 Blanking Circuit

During the retrace time of sawtooth voltage the electron beam moves from right to left which produces retrace line on the screen. This retrace line is not a part of the signal and hence it gives a bad visual effect. Thus, retrace should be eliminated or blanked out. The retrace is blanked out, or made invisible by applying high negative voltage to the grid during the flyback period. When high negative voltage is applied to the grid of CRT, it turns off the electron beam momentarily as shown in Fig. 2.10; thus the retrace path of the electron beam is made invisible. Hence, only during the rise of the sawtooth voltage as shown by labels Trace-1, Trace-2... in Fig. 2.10, the electron beam and so the signal will be visible on the screen.

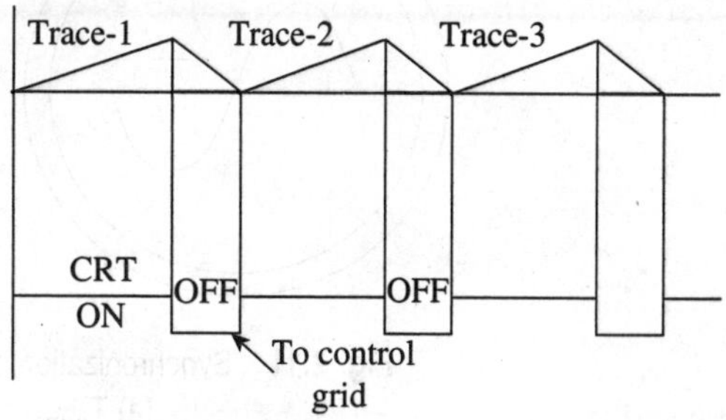

Fig. 2.10 Method of blanking

2.3.6 Trigger or Synchronization Circuit

To display stationary wave pattern on the CRO screen, the sweep voltage should start at the same point of the input signal in each sweep cycle. Thus, when the periodic time of sweep and signal voltages matches, we say that the horizontal sweep voltage is synchronized with the input signal. If synchronization is not provided, the pattern on the screen is not stationary but appears to drift randomly on the screen.

Synchronization is the method of locking (matching) of frequency of sweep voltage to the frequency of the input signal in order to see the stationary wave pattern on the screen of the CRO. The signal is synchronized when its frequency equals the sweep frequency or is an integral multiple of it as shown in Fig. 2.11.

$$F_{signal} - n F_{sweep}$$

$$\text{or } T_{sweep} = n T_{signal} \tag{2.5}$$

The manual synchronization of signals is not possible; hence, trigger circuit is introduced in the CRO circuitry. Trigger circuit takes the sample of the input and shapes a trigger pulse. This trigger pulse is given to the time base circuit which produces ramp voltage of the same frequency as that of the signal or submultiples

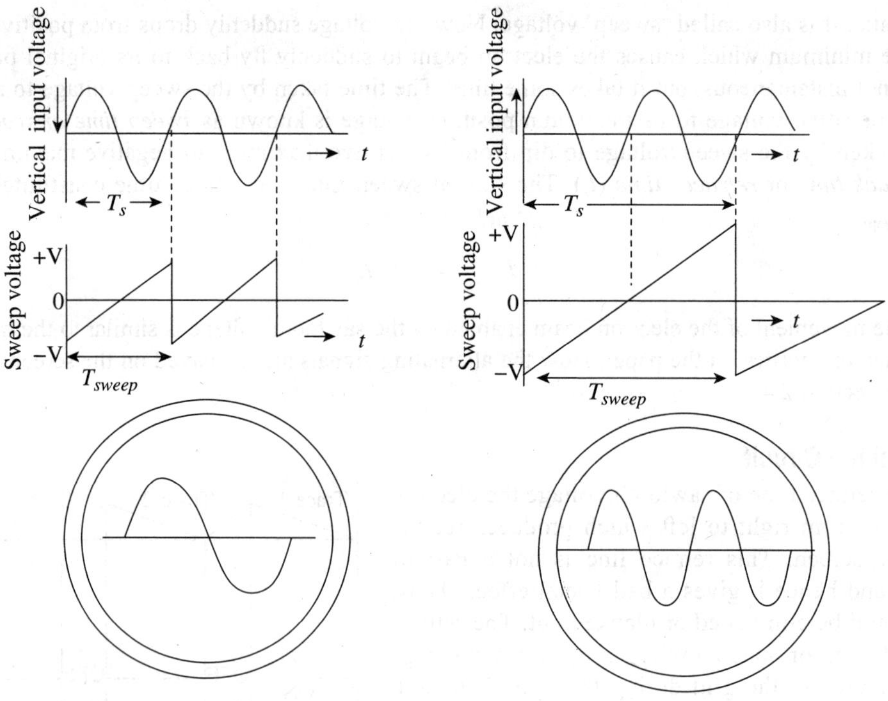

Fig. 2.11 Synchronization of signal and sweep voltages and their display (a) $T_{sweep} = T_{signal}$, (b) $T_{sweep} = 2T_{signal}$

of it. Thus, trigger circuit does the automatic locking (or matching) of the frequency of the signal to that of the time base voltage. This is called *fine synchronization.*

2.3.7 Delay Line

All electronic circuitry in the CRO causes a certain amount of time delay in the transmission of signal voltages to the deflection plates. If we compare the vertical and horizontal deflection circuits in the CRO block diagram, we observe that a portion of the input signal applied to the vertical plates triggers the horizontal signal. Signal processing in the horizontal circuit consists of generating trigger pulse that starts the sweep generator; then the output of sweep generator is given to the horizontal amplifier which is further given to the horizontal plates. Thus, the signal takes some time (appox. 60 to 150sec) to reach the X plates. Hence, the signals of the vertical plates must be delayed by the same duration of time so that both the signals reach the respective plates simultaneously. Thus, delay line ensures that the input signal is displayed as it is on the screen by delaying the Y plate signal.

2.3.8 High-voltage and Low-voltage Power Supply

High-voltage power supply is required to supply voltages of the order of kV to the pre-accelerating and accelerating anodes of the electron gun of CRT.

Low-voltage power supply is given to the amplifiers, trigger circuit, time base generator, and grid of CRT.

2.4 Display of Sinusoidal Input on CRO

The sinusoidal signal to be displayed on the CRO is applied as the vertical input. A sawtooth voltage is applied as the horizontal input. It is important that both the input signal frequency and the linear sweep frequency should be synchronized; otherwise the display will appear moving. For example, if 4 kHz signal is to be displayed and the sweep frequency is also 4 kHz, then a full input cycle will be viewed and if sweep frequency is reduced to 2 kHz, then two input cycles will be seen, and so on.

The process of display can be discussed in three parts.

1. In the first part we will consider that the a.c signal to be displayed is applied to Y plates and no voltage is applied to X plates. The sinusoidal input applied to the vertical plates makes the plates positive and negative alternatively. The spot, as a result, moves upwards and downwards decided by the frequency of input. The position of the spot on the screen at various instants of the signal is shown in Fig. 2.12.

The successive positions of the spot cannot be seen when signal frequencies are large, as the spot moves very fast and due to persistence of vision, a vertical line is seen.

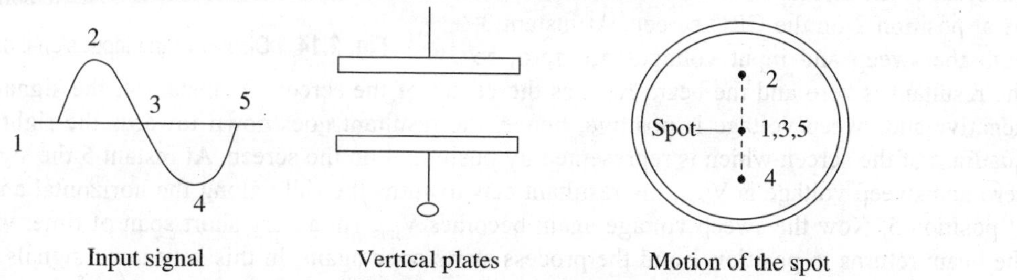

Fig. 2.12 Display on application of sinusoidal voltage to vertical plates

2. In the second part, we assume that only time base voltage is connected to X plates. When a sweep wave (sawtooth voltage) is applied to the horizontal plates, the spot will move from left to right and again will return to left (zero position) to trace the next cycle. The spot will move at uniform speed, as sawtooth wave is a linear curve. For obtaining a full cycle on the screen, the frequency of sweep voltage is kept same as that of input signal. The position of the spot on the screen at various instants of the signal is shown in Fig. 2.13.

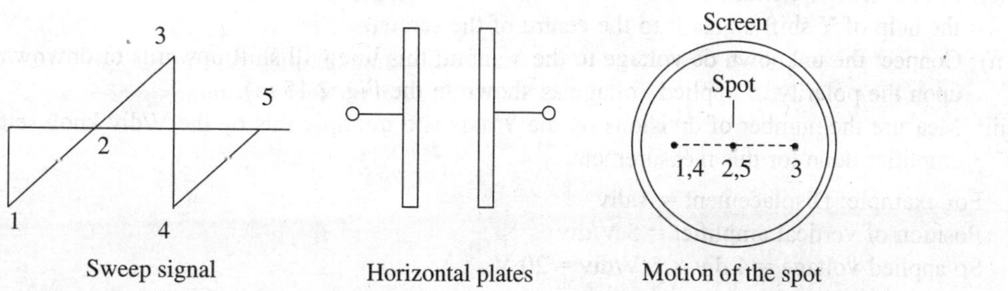

Fig. 2.13 Display on application of sawtooth voltage to horizontal plates

3. In part three, we combine part one and two to get the true shape of the input wave on the screen which is as shown in Fig. 2.14.

As input signal is applied to the vertical plates and sweep signal to the horizontal plates, the electron beam is simultaneously subjected to two forces acting in perpendicular directions. Hence, the deflection of the spot occurs along the direction of the resultant of the two forces.

At instant 1, the magnitude of input signal is zero and that of sweep voltage is $V_{min.}$ Hence, the beam sets at the extreme left. At instant 2, the signal amplitude is positive and sweep voltage is moving towards positive value, so the beam is deflected upwards to the left in the second quadrant of the screen. At this point the beam will be at position 2 on the CRO screen. At instant 3, both the sweep and input voltages are zero, so the resultant is zero and the beam reaches the centre of the screen. At instant 4, the signal amplitude is negative and sweep voltage is positive; hence, the resultant goes down towards the right in the fourth quadrant of the screen which is represented by position 4 on the screen. At instant 5 the signal voltage is zero and sweep voltage is V_{max}, the resultant acts towards the right along the horizontal and the beam is at position 5. Now the sweep voltage again becomes V_{min} (in a very short span of time, viz. μs) and so the beam returns to position 1 and the process starts once again. In this way input signals are displayed on the CRO screen.

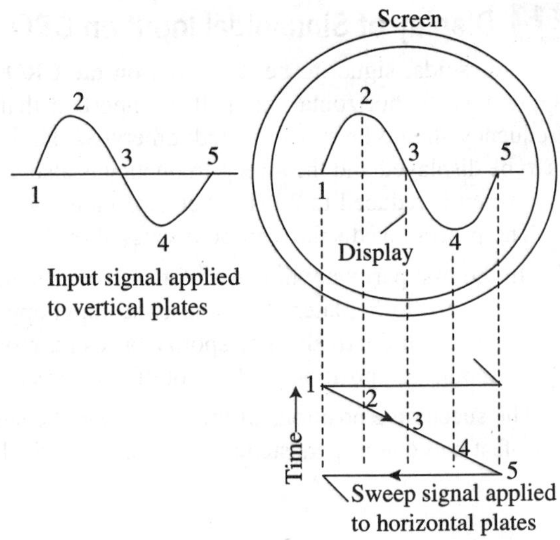

Fig. 2.14 Display of the input signal on CRO screen

2.5 Applications of CRO

Cathode ray oscilloscope is used for a variety of purposes. The measurement of three characteristics of an alternating signal viz. voltage, frequency, and phase are discussed here.

2.5.1 Voltage Measurement

dc Voltage

An unknown dc voltage can be measured using CRO.

(i) First short two terminals of Y input, a horizontal straight line will be observed on the screen. Now with the help of Y shift adjust it to the centre of the screen.

(ii) Connect the unknown dc voltage to the Y input; this line will shift upwards or downwards depending upon the polarity of applied voltage as shown in the Fig. 2.15 (a).

(iii) Measure the number of divisions on the Y axis and multiply this by the V/div knob setting of vertical amplifier done for this measurement.

For example: Displacement = 4 div
Position of vertical amplifier = 5 V/div
So applied voltage = 4 div × 5 V/div = 20 V

ac Voltage

(i) The horizontal trace is adjusted to the centre and then ac voltage is applied to Y input.

(ii) For the measurement of alternating voltages CRO is used in internal mode (or sweep mode). In this mode, the input signal is connected to the Y plate and the ramp voltage generated by time base circuit is connected to the X plate.

(iii) The input waveform is made stationary on the screen by adjusting the position of time/div knob (control of time base circuit) and level knob (control of trigger circuit) on the front panel of CRO. This procedure is called *synchronization*. The waveform may show one or two cycles depending on the settings as shown in Fig. 2.15 (b).

(iv) The sinusoidal signal is amplified to the maximum extent by adjusting the position of V/div knob of the vertical amplifier.

(v) The peak-to-peak vertical distance is measured and then multiplied by V/div knob setting of the vertical amplifier. Thus, V_{p-p} is obtained.

(vi) Calculate peak voltage (V_p) dividing it by 2.

(vii) V_{rms} is calculated using the formula $V_{rms} = V_p \times 0.707$

(viii) The average value of voltage $V_{av} = V_p \times 0.636$

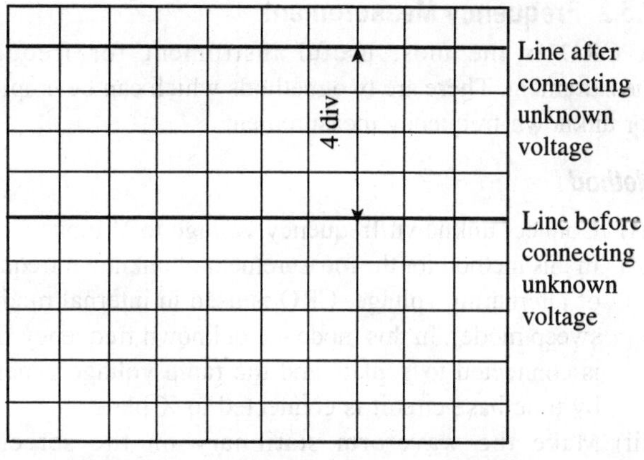

Fig. 2.15a DC voltage measurement

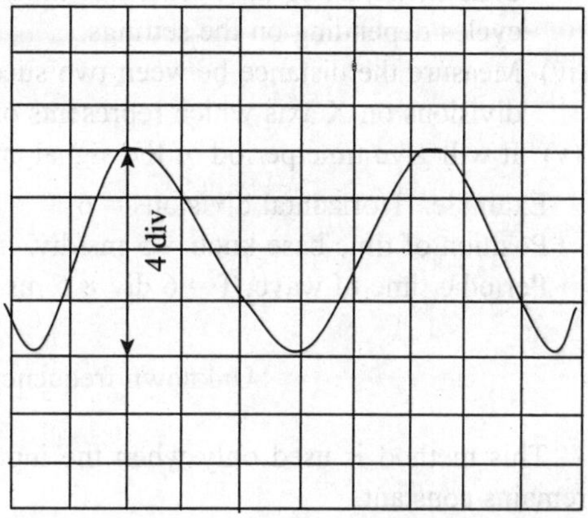

Fig. 2.15b AC voltage measurement

For example:

$$\textit{Position of vertial amplifier, i.e. } \frac{V}{div} \textit{ knob} = 5V/\text{div}$$

$$\textit{Peak to peak voltage} = 4\,\text{div}$$

$$V_{pp} = 4\,div \times 5\,V/div = 20\,\text{V}$$

$$V_p = \frac{20V}{2} = 10\text{V}$$

$$V_{rms} = 10V \times 0.707 = 7.07\,\text{V}$$

$$V_{av} = 10V \times 0.636 = 6.36\,\text{V}$$

2.5.2 Frequency Measurement

A CRO is the most useful instrument for frequency measurement. There are two methods which can be employed for unknown frequency measurement.

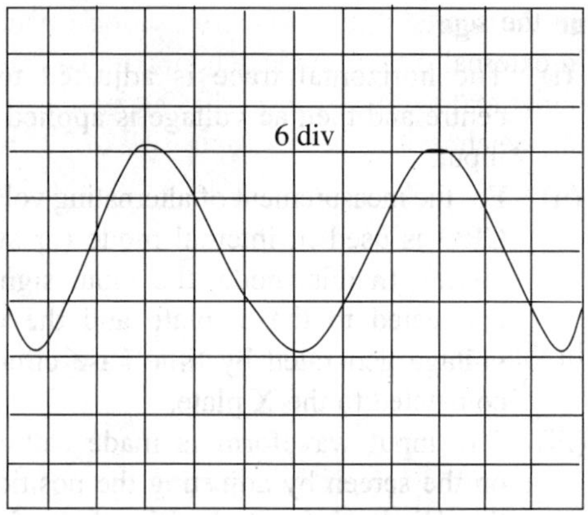

Fig. 2.16 Frequency measurement

Method I

(i) Connect unknown frequency voltage to Y input.

(ii) In this method for the measurement of unknown frequency of alternating voltage, CRO is used in internal mode (or sweep mode). In this mode the unknown frequency signal is connected to Y plate and the ramp voltage generated by time base circuit is connected to X plate.

(iii) Make the waveform stationary on the screen by synchronizing it with the time base voltage. This can be done by adjusting the position of time/div knob and level knob. The waveform may show one or two cycles depending on the settings.

(iv) Measure the distance between two successive peaks as shown in Fig. 2.16 in terms of the number of divisions on X axis which represents one cycle and multiply this by time/div knob setting.

(v) It will give time period of the signal and reciprocal of it is the frequency of input waveform.

Example: Horizontal divisions = 6

Position of time base knob = 5 ms/div.

Periodic time of wave, T = 6 div × 5 ms/div =30 ms

$$\text{Unknown frequency} = \frac{1}{T} = \frac{1}{30\text{ms}} = 0.033 \times 10^3 = 33 \text{ Hz}$$

This method is used only when the input signal comes from the standard source of which frequency remains constant.

Method II

If the unknown test signal has a slight variation in frequency, then Lissajous figures are created for the measurement of the unknown frequency. For producing Lissajous pattern, apart from unknown frequency source, a standard known sinusoidal frequency generator is required. In this method CRO is used in external mode (XY mode), that is, the time base voltage is delinked from X plate and the known and unknown inputs are applied simultaneously to X and Y plates. If the two sine waves oscillating in mutually perpendicular planes are combined, then different types of closed loop patterns are obtained. They are called *Lissajous patterns*. The form of the resultant pattern depends on the ratio of frequencies and the phase of the two waves.

In this method if the signal whose frequency is to be measured is applied to the Y plates and an accurate standard calibrated signal from oscillator is applied to the X plates, then the different waveforms can be obtained. The frequency of unknown signal is varied gradually to make the Lissajous figure steady

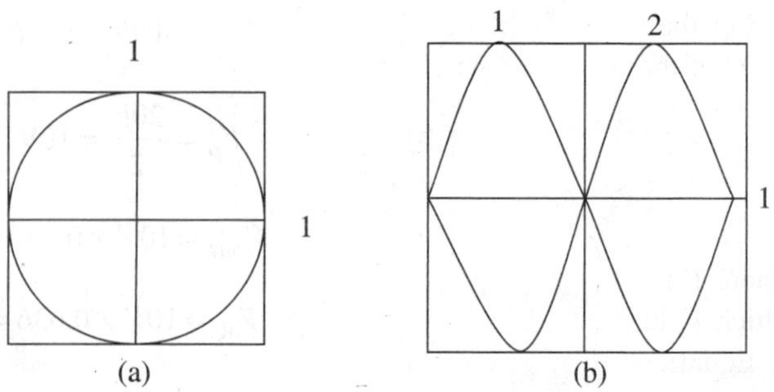

Fig. 2.17 Frequency measurement

on the screen. Thus, synchronization (matching of the frequencies) is to be also done when CRO is used in external or XY mode. The unknown frequency is calculated by measuring the number of tangencies on horizontal and vertical lines as shown in Fig. 2.17 and the frequency of unknown signal is calculated using the synchronization condition

$$n_x f_x = n_y f_y \text{ or } \frac{f_x}{f_y} = \frac{n_y}{n_x} \tag{2.6}$$

In this case as the frequency given to Y plate is unknown, the formula can be rearranged as

$$f_y = f_x\left(\frac{n_x}{n_y}\right)$$

where n_x represents the number of contact points on the horizontal tangent and n_y is the number of contact points on the vertical tangent.

For example: From Fig. 2.17 (a), $n_x = n_y = 1$

$$\therefore \frac{f_x}{f_y} = \frac{1}{1}$$

$$\text{or } f_x = f_y$$

Therefore, unknown frequency is equal to the standard frequency.

From Fig. 2.17 (b), $n_x = 2,\ n_y = 1$

$$\therefore \frac{f_x}{f_y} = \frac{1}{2}$$

$$\text{or } f_y = 2 f_x$$

Therefore, the unknown frequency is equal to twice the standard frequency.

2.5.3 Phase Measurements

If two sine waves oscillating in mutually perpendicular planes are of the same frequency, then the Lissajous pattern takes the shape of an ellipse. The ellipse may be used to determine the phase relationship of the two identical sine waves.

Let the two sinusoidal voltages be of the same amplitude and frequency. Let V_x be the instantaneous value of the voltage applied to X plates and V_y be instantaneous value of the voltage applied to Y plates, then

$$V_x = \text{V} \sin \omega t \tag{2.7}$$

$$V_y = \text{V} \sin(\omega t + \varnothing) \tag{2.8}$$

where V is the maximum amplitude of the waves, ω is the angular frequency, and $\varnothing$ is the phase angle by which V_y leads V_x.

Equation (2.8) can be expanded as

$$V_y = V(\sin \omega t \cos\varnothing + \cos\omega t \sin\varnothing) \tag{2.9}$$

Using Eq. (2.7) we substitute for $\sin \omega t$ and $\cos \omega t$ in Eq. (2.9), we get

$$V_y = V\left[\left(\frac{V_x}{V}\right)\cos\varnothing + \sqrt{\left(1-\left(\frac{V_x}{V}\right)^2\right)} sin\varnothing\right]$$

$$V_y = V_x \cos\varnothing + V\sqrt{\left(1-\left(\frac{V_x}{V}\right)^2\right)} sin\varnothing$$

$$V_y - V_x \cos\varnothing = V\sqrt{\left(1-\left(\frac{V_x}{V}\right)^2\right)} sin\varnothing$$

Squaring both sides of the above equation, we get

$$V_y^2 + V_x^2 \cos^2\varnothing - 2V_xV_y cos\varnothing = V^2 \sin^2\varnothing\left[1-\left(\frac{V_x^2}{V^2}\right)\right]$$

$$V_y^2 + V_x^2 \cos^2\varnothing - 2V_xV_y cos\varnothing = V^2 \sin^2\varnothing - V_x^2 \sin^2\varnothing$$

Rearranging the terms in the above equation, we get

$$V_x^2 + V_y^2 - 2V_xV_y cos\varnothing = V^2 \sin^2\varnothing \tag{2.10}$$

Equation (2.10) represents an ellipse. Therefore, the oscillogram will be an ellipse as shown in Fig. 2.18. When t = 0, from Eqs (2.7) and (2.8) we get

$$V_x = V\sin \omega t = 0 \text{ and } V_y = V\sin\varnothing$$

$$\therefore \sin\varnothing = \frac{V_y}{V} \tag{2.11}$$

Thus, theoretically phase difference $\varnothing$ between two waves can be estimated by using Eq. (2.11). Practically, it can be found by carrying out measurements on the ellipse as given below. From Fig. 2.18, the distance $2Y_1$ is given by

$$2Y_1 = 2V \sin\varnothing \tag{2.12}$$

The length $2Y_2$ obviously represents the peak-to-peak voltage $V_{p\text{-}p}$ which is 2V.

$$2Y_2 = 2V \tag{2.13}$$

Dividing Eq. (2.12) by Eq. (2.13), we get

$$\therefore \sin\varnothing = \frac{2Y_1}{2Y_2}$$

Thus, the phase angle $\varnothing$ is given by

$$\varnothing = \sin^{-1}\left(\frac{2Y_1}{2Y_2}\right) \tag{2.14}$$

By measuring the length $2Y_1$ and $2Y_2$ on the ellipse and using them in Eq. (2.14) the phase angle ∅ is evaluated.

Thus, to find the phase difference one sine wave is applied to X plates and the other to Y plates. If both have the same frequency, then an ellipse is seen on the CRO screen. It is adjusted at the centre of the screen as shown in Fig. 2.18 and the lengths $2Y_1$ and $2Y_2$ are measured. By using Eq. (2.14) we can find the phase difference between the waves.

For example: If $2Y_1/2Y_2 = 0.5$, then $\varnothing = 30^0$ or 330^0

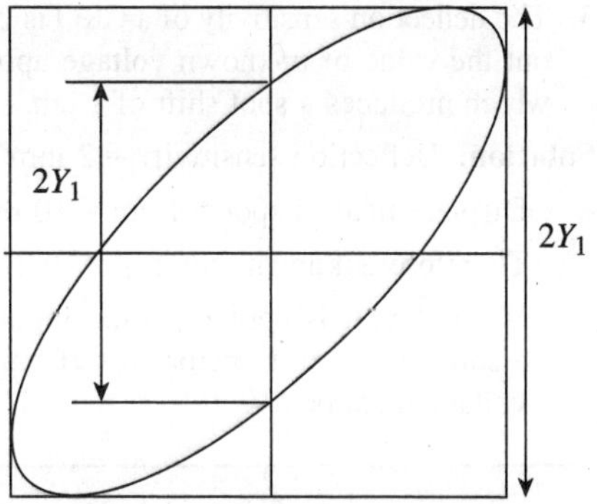

Fig. 2.18 Phase measurement

2.5.4 Typical Lissajous Figures

Figure 2.19 shows some typical Lissajous figures for various frequency ratios verses phase difference between two XY signals.

The frequency ratio is the ratio between the vertical and the horizontal frequency F_y/F_x.

To one set of plates the signal of unknown frequency is given and to the other set a signal of known frequency is given from a variable frequency oscillator. The frequency of the oscillator is varied till a suitable Lissajous pattern is obtained.

As the frequency of the oscillator signal is known, the unknown frequency can be found.

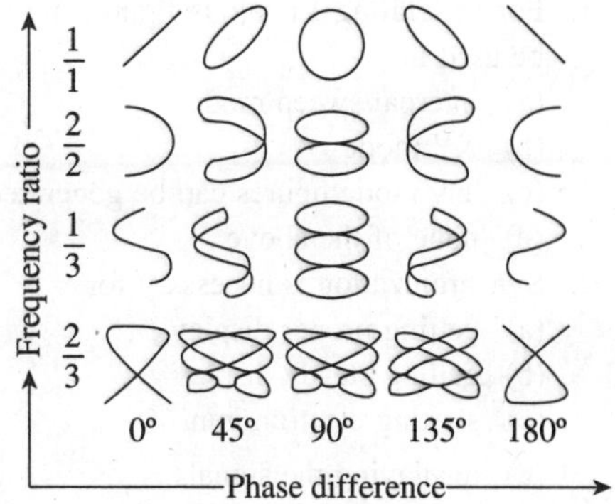

Fig. 2.19 Lissajous figures

Solved Numericals

1. Electrons accelerated by a potential of 200 V enter a region of different electric field at an angle of incidence of 40° and get refracted through an angle of 20°. Find the potential difference between the two regions.

Solution: $V_1 = 200$ V, θ_1=40°, θ_2=20°. By using Bethe's law

$$\frac{\sin\theta_1}{\sin\theta_2} = \sqrt{\frac{V_2}{V_1}}$$

$$V_2 = V_1\left(\frac{\sin\theta_1}{\sin\theta_2}\right)^2 = 200\times\left(\frac{\sin 40}{\sin 20}\right)^2$$

$$V_2 = 706.64 \text{ V}$$

Hence, the potential difference is

$(V_2 - V_1) = (706.64 - 200)$ V $= 506.64$ V

2. The horizontal deflection sensitivity of CRT is 0.04 mm/V. When a voltage is applied across the horizontal plates, the spot shifts 7.3 mm horizontally. Calculate the value of the applied voltage.

Solution:

Deflection sensitivity = 0.04 mm/V

Shift in the electron spot = 7.3 mm

Total shift = Deflection sensitivity × Applied voltage

$$\text{Applied voltage} = \frac{\text{Total shift}}{\text{Deflection sensitivity}}$$

$$= \frac{7.3 \text{ mm}}{0.04 \text{ mm/V}}$$

$$= 182.5 \text{ V}$$

3. The deflection sensitivity of a CRO is 2 mm/V. Find out the value of unknown voltage applied to input which produces a spot shift of 1 cm.

Solution: Deflection sensitivity = 2 mm/V

Displacement of spot = 1 cm = 10 mm

Therefore unknown voltage = 10/2 = 5V

When 1 volt is applied to the input it produces a shift of 2 mm; therefore for 10 mm shift input voltage must be 5V.

4. An unknown frequency applied to vertical input takes 4 cm for one cycle on time axis. Find out the unknown frequency if time/division knob is on 2 ms/cm.

Solution: Distance on X axis = 4 cm

Time base position = 2 ms/cm

Therefore period $T = 4 \times 2 = 8$ ms

Therefore, unknown frequency $= 1/T = 1/(8 \times 10^{-3})$ $= 125$ Hz

Checkpoint 3

1. For generating Lissajous figure, the CRO should be used in
   (a) internal/sweep mode
   (b) XY mode
   (c) any mode figures can be generated
   (d) none of the above
2. Synchronization is necessary for
   (a) getting proper display
   (b) getting steady display
   (c) starting electron gun
   (d) displaying the signals
3. The frequency of known sinusoidal voltage given to X plate is 2000 Hz. The horizontal and vertical tangencies of Lissajous figure are 3 and 4 respectively. The unknown frequency of applied voltage applied to Y plate is
   (a) 2000 Hz
   (b) 500 Hz
   (c) 1500 Hz
   (d) 2667 Hz
4. Ramp voltage waveform is a
   (a) triangular waveform
   (b) square waveform
   (c) sinusoidal waveform
   (d) sawtooth waveform
5. Ramp voltage controls
   (a) intensity of electron beam
   (b) vertical movement of electron beam
   (c) horizontal movement of beam
   (d) focuses the beam on CRO screen
6. Acquadag coating is a
   (a) non-conducting aqueous graphite coating
   (b) conducting aqueous graphite coating
   (c) coating to prevent leakage of electrons
   (d) coating on the screen to see movement of electron
7. In a CRO the control knob for changing the amplitude of the displayed signal is
   (a) time/div (c) level
   (b) focus (d) volt/div
8. Is it essential to synchronize the signals if CRO is used in XY mode?
   (a) No (c) Yes
   (b) Not possible (d) Sometimes
9. When phase difference between two sinusoidal signals of same frequency is 90°, then Lissajous figure is a
   (a) straight line (c) circle
   (b) ellipse (d) none of these
10. During the retrace time the electron beam is not visible as
    (a) temperature of cathode decreases.
    (b) electrons are absorbed by screen.
    (c) extreme negative potential is applied to grid.
    (d) extreme positive potential is applied to acquadag coating.
11. Synchronization of two signals refers to
    (a) making their amplitudes equal
    (b) making them start at the same time
    (c) making their frequencies equal
    (d) making their phases equal
12. An acquadag coating is used in a CRO to collect
    (a) primary electrons
    (b) secondary emission electrons
    (c) both primary and secondary emission electrons
    (d) none of these

13. By using CRO one can measure physical parameters like
    (a) temperature
    (b) pressure & strain only
    (c) none of these
    (d) both 1 & 2

14. Time taken by the sweep voltage to dip from its positive maximum to negative maximum value is
    (a) sweep time
    (b) retrace time
    (c) trace time
    (d) none of the above

Summary

- Bethe's law explains the motion of electron in a non-uniform electric field.
- Electron travels along a curved path in a non-uniform electric field.
- For converging or diverging electron beam, non-uniform electric field is used.
- CRTis a vacuum tube device that consists of electron gun, deflection system, and fluorescent screen.
- Electron gun produces, accelerates, and focuses electron beam.
- Two-dimensional control of the electron beam on the screen isobtained by using a deflection system.
- Fluorescent screen makes the electron beam visible.
- CRO is a display device, which can display electrical and non-electrical signals.
- CRO can measure amplitude and frequency of the alternating periodic signals. It can also measure the phase difference between two sinusoidal signals of same frequency.

Exercises

1. An electron beam enters from a region of potential 75 V into a region of potential 100 V making an angle of 45° with the direction of electric field. Find the angle through which the beam refracts. (Ans. $\theta r = 37.75°$)
2. The Lissajous pattern on a CRO is stationary. It has 5 horizontal and 2 vertical tangencies. The frequency of horizontal input is 1000 Hz. Determine the frequency of vertical input. (Ans. $F_y = 2500$ Hz)

Self-study Questions

1. Discuss the refraction of electron beam across an equipotential surface. Hence explain the working of electrostatic lens.
2. What is Bethe's law? In what way does it resemble the Snell's law and differs from it.
3. Draw a schematic of an electrostatic CRT. Describe the role of
   (a) Electron gun
   (b) Deflection system
   (c) Fluorescent screen
   (d) Acquadag coating
4. What is the need of the acquadag coating inside the glass bulb in a CRT? Which voltage controls the intensity of electron beam on the screen in CRT? What is the function of voltage given to the anode cylinders in it?
5. Explain the deflection systems used in CRT. What is deflection sensitivity?
6. Draw the block diagram of CRO and explain the function of time base circuit, delay line, and trigger circuit.
7. Give the graphical construction of Lissajous pattern you expect when the ratio of vertical frequency to horizontal frequency is:
   (i) 1: 2 with 90° phase difference and
   (ii) 1: 1 with 45° phase difference
8. Show using a graphical method that the resultant of sinusoidal voltage applied to the Y input and time

base voltage applied to the X input is a sine wave. Which figure will be observed on the CRO screen if sinusoidal voltage to the Y input is replaced by

(i) Square wave voltage

(ii) Sawtooth voltage

9. Explain how the actual waveform of the signal can be traced on the CRO screen with the help of time base generator by drawing the necessary sketches.

10. What is synchronization and how is it achieved in CRO?

11. Explain three controls of CRO.

12. Explain the use of CRO to find frequency, phase, and amplitude.

Answers

Checkpoint 1

1(b), 2(c), 3(c), 4(a)

Checkpoint 2

1(c), 2(d), 3(b), 4(c), 5(d), 6(b), 7(d), 8(d),

Checkpoint 3

1(b), 2(b), 3(c), 4(d), 5(c), 6(b), 7(d), 8(c), 9(c), 10(c), 11(c), 12(c), 13(d), 14(b)

Quantum Mechanics

3

Learning Objectives

After going through this chapter, you will be able to:

- Learn the basic concepts of quantum mechanics viz: particle nature of quanta, its properties, wave-particle duality principle, and uncertainty principle
- Explain Schrödinger wave equations and their application to a simple problem of particle trapped in one-dimensional infinite and finite potential well

3.0 Introduction

From 17th until 19th centuries, Newtonian mechanics, Maxwell's electromagnetic theory, and thermodynamics, regarded as Classical Theories, were successful in explaining all observed scientific phenomena. In classical theories, matter and fields were treated as separate and totally independent entities, where physical parameters like energy and position can vary continuously, that is, they can take any random values. Classical theory is valid for systems of large dimensions, whereas when a microscopic description of physical systems is necessary, the principles of Quantum Physics are invoked. Some physical phenomena such as interference and diffraction can be explained by considering wave nature (i.e., by using Wave Theory of light) and some viz. photoelectric effect, the Compton effect, the anomalous Zeeman effect, ultraviolet catastrophe, radioactivity, the short wavelength limit of the continuous X-rays spectrum, etc. by particle nature of radiation (i.e. by using Plank's quantum theory).

Quantum mechanics has been described as one of the most outstanding achievements of the 20th century as it answered many questions viz. how do atoms absorb or emit light?; how are elements placed in the periodic table?; how do microelectronic devices work? and the like. Quantum physics or mechanics deals with the behaviour of matter and energy at atomic and subatomic scale. In 1900, Max Planck gave a path-breaking idea, which symbolized the beginning of the field of quantum mechanics. The fundamental concept of this theory is that of the quantum—the idea that at the smallest of scales everything is discrete. Quantum theory was developed with the contributions from distinguished physicists including Neils Bohr, Einstein, Heisenberg, Schrodinger, Davisson and Germer, Dirac, and Max Born among others. Some important applications of quantum mechanical theory includes light-emitting diodes, solid state laser, superconducting magnets, medical and research imaging such as magnetic resonance imaging (MRI), nanotechnology, and electron microscopy, and explanations for many biological and physical phenomena.

3.1 Planck's Quantum Hypothesis

Max Planck, a German physicist, succeeded in developing a theory that correctly predicted the frequency distribution of thermal radiation. According to the classical wave theory, accelerated charged particles produce electromagnetic waves. These electromagnetic waves spread out and thus the energy of the wave is uniformly distributed. Further, the energy of the wave was related to the amplitude of the wave instead of its frequency. Planck's theory was a fundamental departure from the classical theory. According to his theory:

(i) The atomic oscillators in a body cannot have any random amount of energy. They could have only discrete units of energy given by

$$E_n = nh\nu$$

where n is any positive integer ($n = 1, 2, 3, \ldots$); ν is the frequency of oscillation, and h is the Planck's constant and has the value

$$h = 6.62\times10^{-34}\ Js$$

(ii) The atomic oscillators cannot absorb or emit any random amount of energy. They absorb or emit energy in undividable discrete units. The amount of radiated energy in each unit is called the *quantum* of energy. Each quantum carries an energy,

$$E \quad h\nu$$

Each quantum has a minimum amount of energy that cannot be further subdivided. It is the smallest quantity of energy of radiation of that frequency. Thus, the energy of an oscillator can change only by $h\nu$ or an integral multiple of $h\nu$. The hypothesis that radiation energy is emitted or absorbed in only discontinuous manner and in the form of quantum is called the Planck's quantum hypothesis. It marked the birth of quantum physics. The integer 'n' is called the quantum number. The energies of the atoms are said to be quantized and the allowed energy states are called quantum states. The quantum states of an atomic oscillator are shown in Fig. 3.1.

Fig. 3.1 Discrete energy levels

An atom absorbs or emits energy by jumping from one quantum state to another quantum state. The equation $E = h\nu$, indicates that the energy content of a wave is related to the frequency of the wave rather than to the square of its amplitude as postulated in the classical wave theory.

3.2 Einstein's Extension of Planck's Theory

Einstein refined Max Planck's hypothesis of discontinuous emission and absorption by bodies. Einstein also rechristened the energy quanta as *photons*. Einstein's proposals are as follows.

1. Einstein proposed that when the energy is emitted, it is not distributed evenly over the entire expanding wavefront, but rather remains concentrated in discrete photons. For example, a light beam is regarded as a stream of photons travelling with a velocity 'c'.
2. The energy of a photon depends on the frequency of radiation. Higher the frequency 'ν' of the radiation, higher is the energy content of each photon. Hence, the $\gamma-ray$ and $x-ray$ photons are far more energetic compared to optical photons, whereas the photons of radio frequencies are of the least energy.
3. Einstein further proposed that the intensity of a monochromatic light beam is related to the concentration of photons present in the beam. Thus, the intensity is given by $I = N h\nu$, whereas according to the classical theory, the intensity of light is given by $I = A^2$ where A is the amplitude of the electromagnetic wave.

4. During the interaction between energy and matter, it is an all or none process. This means that when a photon encounters matter, it is either completely absorbed, that is, it imparts all its energy to the particles of matter and vanishes or the photon continues as it is.

Thus, Einstein's proposals strengthened Planck's quantum theory. The quantum of energy or 'photon' is considered to have particle properties instead of wave properties. Most of the observed effects are due to the interaction between energy and matter. As energy is quantized and contains streams of quanta or photons, let us investigate first the properties of photon before finding the theoretical explanation of the observed effects of interaction between energy and matter.

3.3 Properties of Photons

As energy is considered to be made up of spatially localized quanta or photons, it is reasonable to attribute particle properties to photons.

Energy The energy of a photon is determined by the frequency of radiation and is given by Eq. (3.1)

$$E = h\nu \tag{3.1}$$

Here ν is the linear frequency of radiation and h is Planck's constant. Using $2\pi\nu = \omega$, where ω is the angular frequency and writing $\hbar = h/2\pi$, the above equation may be rewritten as

$$E = \hbar\omega \tag{3.2}$$

Velocity Photons always travel with the velocity of light c = 3×10^8 m/s.

Mass A photon can never be at rest. If it exists then it will be moving with velocity of light 'c'. Hence, photon is a relativistic particle and the rest mass of a photon is zero as it cannot be at rest. Thus, $m_0 = 0$.

However, as it travels with the velocity 'c' it will have relativistic mass. It is given by the mass–energy equivalence relation as $E = mc^2$.

Therefore, using Eqn (3.2), $m = \frac{E}{c^2} = \frac{\hbar\omega}{c^2}$

Thus, mass of photon can also be written as

$$m = \frac{\hbar\omega}{c^2} \tag{3.3}$$

Linear momentum The linear momentum associated with a photon can be written as

$$p = mc = \frac{\hbar\omega}{c}$$

It can be expressed as

$$p = \frac{\hbar\omega}{c} = \hbar k \tag{3.4}$$

where $k = \frac{\omega}{c} = \frac{2\pi\nu}{c} = \frac{2\pi}{\lambda}$ and k is called *propagation constant.*

Angular momentum Angular momentum is also known as *spin momentum,* which is an intrinsic characteristic of all microparticles. Photon has a spin of one unit, hence its spin angular momentum is given by

$$s = 1\hbar$$

Electrical nature Photons are electrically neutral and cannot be influenced by electric and magnetic fields. They cannot ionize matter.

In many atomic problems, it is essential to calculate energy and momentum of the atomic particle trapped in a potential field. There is a simple relation between energy and momentum which helps us to find the momentum, if energy is known and vice versa. Let us find the relation between energy and momentum for non-relativistic particles (particles moving with moderate speeds) and for relativistic particles (particles moving with speeds comparable to speed of light).

3.4 Energy–Momentum Relationship

For a non-relativistic particle, a simple relation as given below relates energy and momentum.

$$E = \frac{1}{2}mv^2 \text{ and } p = mv$$

or,

$$p = \sqrt{2mE}$$

For a relativistic atomic particle of mass m, the relation between energy and momentum is obtained as follows.

$$m = \frac{m_0}{\sqrt{1 - \frac{v^2}{c^2}}}$$

Here, m_0 is the rest mass of the particle. Squaring both sides and then rearranging the terms leads to the equation

$$m^2c^2 = m^2v^2 + m_0^2c^2$$

Multiplying the above equation by c^2 on both sides, we get

$$m^2c^4 = m^2v^2c^2 + m_0^2c^4$$

As $E = m^2c^2$ and $p = mv$, we get

$$E^2 = p^2c^2 + m_0^2c^4$$

Therefore,

$$E = \sqrt{p^2c^2 + m_0^2c^4} = c\sqrt{p^2 + m_0^2c^2} \tag{3.5}$$

Thus, there exists a simple relation between magnitude of momentum and energy of the particle which is used frequently in theoretical concepts and while solving numericals.

3.5 Compton Effect

There are many observed effects due to the interaction between photon and matter. When a photon interacts with matter, it transfers all its energy and momentum to matter. Let us investigate Compton Effect, which supported the view that both energy and momentum are transferred via photon. In an experiment conducted by A.H. Compton, X-rays were incident on a target viz. block of graphite. The intensities of scattered X-rays were measured at different angles as a function of wavelength as shown in Fig. 3.2. When a monochromatic beam of X-rays is scattered by a target, the scattered X-rays were found to contain radiation having the same wavelength as the incident radiation and radiation of longer wavelength. Thus, the scattered radiation contained two wavelengths—one, which is same as the incident radiation, called the unmodified radiation λ and the other with higher wavelength called the modified radiation λ'. The difference in the modified and unmodified wavelength is called the Compton shift denoted by $\Delta\lambda$. The shift in the wavelength depends only on the angle of scattering and is independent of the nature of the scatterer and the incident wavelength. The existence of unmodified radiation was explained by the classical theory, but it could not explain the presence of the modified radiation.

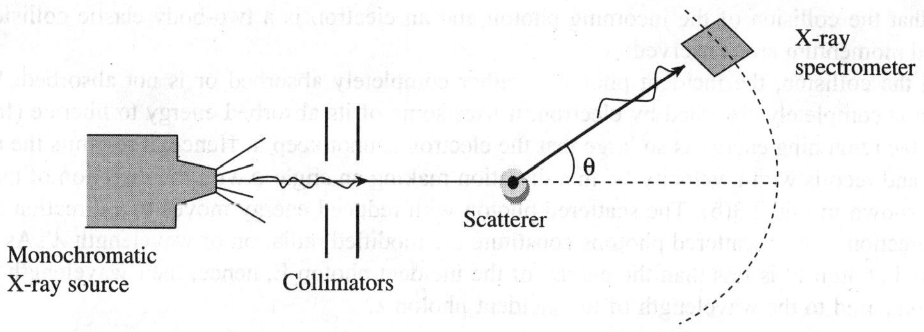

Fig. 3.2 Experimental arrangement of Compton effect

On the basis of quantum theory, Compton gave a satisfactory explanation for the existence of the modified radiation. Thus, the phenomenon that gives the modified radiation is called the 'Compton effect'. Compton considered that the scattering of X-rays from the scatterer takes place in terms of transfer of energy and momentum via photon, between the incident X-ray and the loosely bound electron from the scatterer. Let us first consider the conceptual (qualitative) aspect and then the quantitative aspect to understand how quantum mechanical principles explain all the observed features of Compton Effect.

3.5.1 Qualitative Explanation of Existence of λ'

Compton used the quantum theory to explain the presence of modified wavelength in the scattered radiation. The primary X-ray beam is made up of photons of energy $E = h\nu$, where 'h' is Planck's constant and ν is the frequency of the incident X-rays. These photons travel with the speed of light c and possess momentum. Compton assumed that during the scattering process the incident photon collides with a valance electron that is relatively free as compared to the inner shell electrons as shown in Fig. 3.3(a). He also assumed this free electron to be initially at rest in the scattering material as the energy that the colliding electron acquires after collision is tremendous. Hence, the assumption that the valance electron is at rest is valid. He further

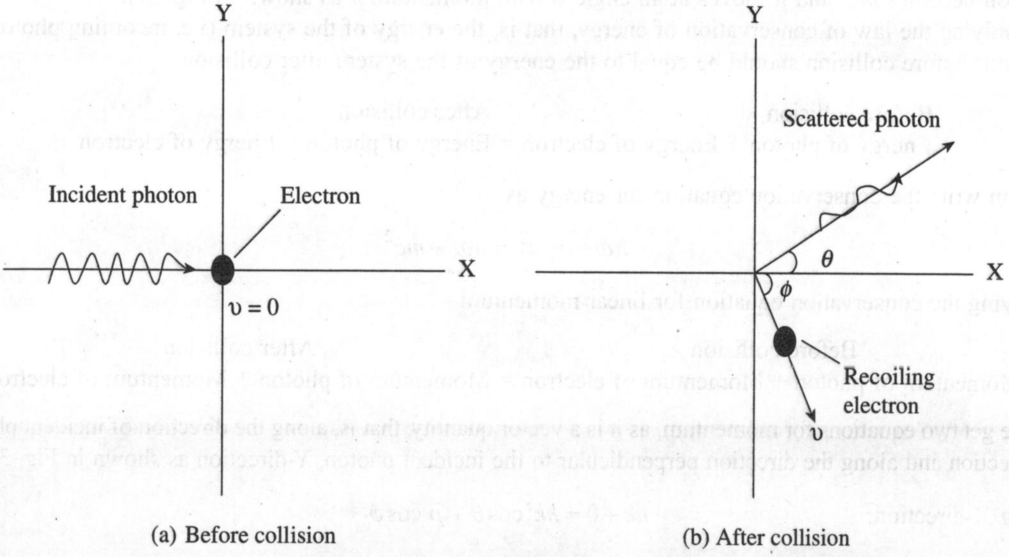

Fig. 3.3 Incident photon and free electron (a) before collision (b) after collision

assumed that the collision of the incoming photon and an electron is a two-body elastic collision, where energy and momentum are conserved.

During the collision, the incident photon is either completely absorbed or is not absorbed. Whenever the photon is completely absorbed by electron, it uses some of its absorbed energy to liberate (free) itself. However, the remaining energy is so large that the electron cannot keep it. Hence, it re-emits the remaining energy E' and recoils with a velocity 'v' in a direction making an angle ϕ with the direction of the incident photon as shown in Fig. 3.3(b). The scattered photon with reduced energy moves in a direction θ with the original direction. These scattered photons constitute the modified radiation of wavelength λ'. As energy of the scattered photon E' is less than the energy of the incident photon E, hence, their wavelength λ' will be more as compared to the wavelength of the incident photon λ.

3.5.2 Quantitative Explanation for Existence of λ'

Let us apply the laws of conservation of energy and momentum to the collision between a photon and an electron. The energy and momentum of the incident photon are respectively $\hbar\omega$ and $\hbar k$ as discussed in Section 3.3. The energy and momentum of the scattered photon are denoted respectively as $\hbar\omega'$ and $\hbar k'$. Here ω' is the frequency of the scattered radiation. Since the recoiling electron may have speed v, which is comparable with that of light, we must use the relativistic expressions for the energy and momentum of the electron. The mass of the stationary electron in the target is assumed to be m_0. Thus, the rest energy of the electron before collision is m_0c^2, but its momentum is zero as it is at rest. After collision the energy of the electron becomes mc^2 and it moves at an angle ϕ with momentum p as shown in Fig. 3.4.

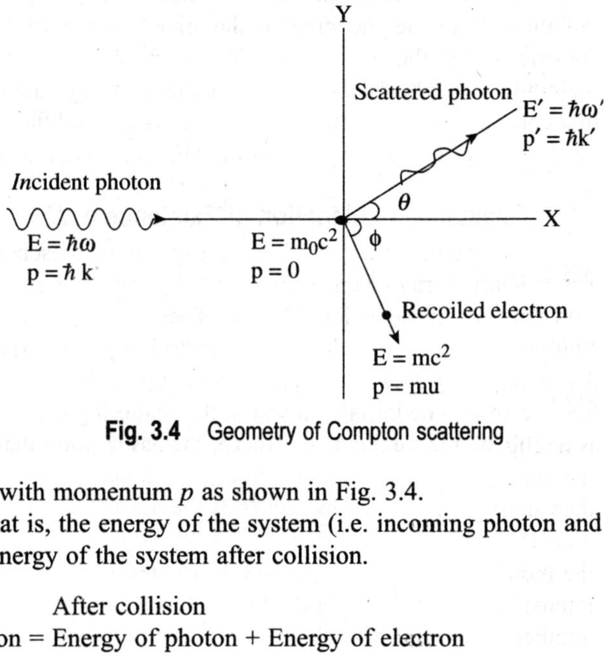

Fig. 3.4 Geometry of Compton scattering

Applying the law of conservation of energy, that is, the energy of the system (i.e. incoming photon and electron) before collision should be equal to the energy of the system after collision.

Before collision		After collision
Energy of photon + Energy of electron	=	Energy of photon + Energy of electron

We can write the conservation equation for energy as

$$\hbar\omega + m_0c^2 = \hbar\omega' + mc^2 \tag{3.6}$$

Applying the conservation equation for linear momentum

Before collision		After collision
Momentum of photon + Momentum of electron	=	Momentum of photon + Momentum of electron

We get two equations for momentum, as it is a vector quantity, that is, along the direction of incident photon, X-direction and along the direction perpendicular to the incident photon, Y-direction as shown in Fig. 3.4.

Along X-direction:

$$\hbar k + 0 = \hbar k' \cos\theta + p \cos\phi \tag{3.7}$$

Along Y-direction:

$$0 + 0 = \hbar k' \sin\theta - p \sin\phi \tag{3.8}$$

By using the energy and momentum conservation equations given by Eqs (3.6), (3.7), and (3.8) and solving them, the expression for the Compton shift can be obtained as

$$\Delta\lambda = \lambda' - \lambda = \frac{h}{m_0 c}(1 - \cos\theta) \tag{3.9}$$

3.5.3 Compton Wavelength

The quantity h/m_0c in Eq. (3.9) has dimensions of length, and hence is known as the Compton wavelength of the electron, λ_c. Its magnitude is

$$\lambda_c = \frac{h}{m_0 c} = \frac{6.63\times10^{-34}\,\text{Js}}{\left(9.11\times10^{-31}\,\text{kg}\right)\times(3.0\times10^{8}\,\text{m/s})} = 0.0243\times10^{-10}\,m = 0.0243\ \text{A}^\circ$$

This shows that the change in wavelength (Compton shift) depends neither on the incident wavelength nor on the nature of the scattering material, but depends only on the angle of scattering. Thus, Compton successfully explained the existence of the modified wavelength in the scattered radiation. The Compton equation (Eq. 3.9) can be written as

$$\Delta\lambda = \lambda' - \lambda = 0.0243\ (1 - \cos\theta) \tag{3.10}$$

Thus from Eq. (3.10), $\Delta\lambda = 0$ for scattering angle $\theta = 0$ and it increases with θ. For $\theta = 90^\circ$, $\Delta\lambda = \lambda_c = 0.0243$ Å. For $\theta = 180^\circ$, $\Delta\lambda = 2\lambda_c = 0.0486$ Å. This result is in close agreement with the experimental findings. The Compton effect offers one of the strongest evidences in support of the Planck's quantum theory of radiation.

3.5.4 Presence of the Unmodified Radiation

The above theory holds for the 'free' or 'loosely-bound' electrons of the atom. If the photon strikes an electron tightly bound to the atom, then the whole atom and not simply the electron, recoils. Since the mass of the atom 'M' is much larger than the mass of electron 'm_o', the Compton shift $\Delta\lambda$ becomes $\frac{h}{Mc}(1-\cos\theta)$, which is negligible. This accounts for the unmodified radiation. Thus, whenever the incoming photons collide with the valence electrons, then it results in modified radiation and when it collides with the inner shell electrons, then unmodified radiation occurs in the scattered radiation. In heavier elements (such as chromium, iron, nickel) most of the electrons are tightly bound; hence, the intensity of unmodified radiation is strong and the modified radiation is very weak. In case of lighter elements (such as carbon, sodium, magnesium) the intensity of the modified radiation is strong as compared to unmodified radiation as there is relatively more number of valence (free) electrons as compared to the inner shell electrons.

Now both photoelectric and Compton effects arise due to the action of photons on electrons but the two effects are not same. Let us consider a Compton collision between a photon and a free electron (initially at rest). The photon has energy hv and momentum hv/c. Suppose the photon transfers its entire energy (and momentum) to the electron that acquires a velocity v. Then, by considering the conservation of energy and momentum, we have

$$hv = \frac{1}{2}mv^2 \text{ ......Conservation equation for energy}$$

$$\frac{hv}{c} = mv \text{ ....Conservation equation for momentum}$$

where m is the mass of the electron. Eliminating hv between these two equations we find that the velocity of the electron comes out to be $v = 2c$

that is, twice the velocity of light. This is impossible according to the relativistic mechanics. Hence we conclude that a 'free' electron cannot absorb 'all' the energy of a photon.

In photoelectric effect, an electron absorbs the incident photon completely before it leaves the photosensitive surface as a photoelectron. Obviously, this electron cannot be free, that is, it must be bound to an atom; otherwise, it will not absorb the photon completely. Hence, we can conclude that the photoelectric effect cannot occur for completely free electrons; the electrons must be bound to an atom. The conservation of momentum occurs between the incoming photon and the atom as a whole.

Thus, there is an important difference between the photoelectric effect and the Compton effect. In the photoelectric effect, the photon completely disappears and all of its energy is given to the photoelectron. In Compton effect, there is still a photon after the collision but its energy is less than that of the incident photon. Only a part of the energy of the incident photon is given to the Compton electron.

3.5.5 Absence of Compton Effect for Visible Light

The change in wavelength due to the Compton effect is given by

$$\Delta\lambda = \frac{h}{m_0 c}(1-\cos\theta) = 0.0243(1-\cos\theta)\ \text{Å}$$

where θ is the scattering angle. Obviously, the change $\Delta\lambda$ is independent of the wavelength of the radiation suffering Compton effect, and depends only on the scattering angle θ. The highest value of $(1 - \cos\theta)$ is 2 when $\theta = 180°$; that is, when the photon recoils back. Thus only in the direction of incident radiation maximum wavelength-change is possible which is 0.048 Å or roughly 0.05 Å. This means that Compton effect can be detected only for those radiations whose wavelength is not greater than a few Angstroms. Now with the detector kept in the direction of the incident photon let us calculate the percentage change in the wavelength, that is, $\frac{\Delta\lambda}{\lambda}\times 100\%$ for different regions in the spectrum. For example, for $\lambda = 5$ Å, the maximum percentage change in wavelength is 1% while for $\lambda = 1$ Å there is a 5% change. For visible light $(\lambda \approx 5000\,\text{Å})$ the maximum percentage wavelength change is only about 0.001% of the initial wavelength that is undetectable. Hence, Compton effect occurs for the visible range but due to a very small shift in the wavelength the detector cannot detect it.

Checkpoint 1

1. Quantum mechanics differs from classical mechanics in the context of application to
   (a) atomic size particles
   (b) waves
   (c) particles of macro dimensions
   (d) waves and atomic particles
2. The magnitude of energy and momentum of a photon is always
   (a) discrete
   (b) continuous
   (c) both discrete & continuous
   (d) cannot be determined
3. The assumptions involved in Compton effect are
   (a) electron is moving & free
   (b) electron is free & at rest
   (c) electron is bound & at rest
   (d) electron is moving while bound to nucleus.
4. In Compton effect the intensity of the unmodified radiation goes on increasing with
   (a) increasing atomic number
   (b) decreasing atomic number
   (c) increasing frequency of incoming radiation
   (d) increasing dimensions of target
5. In Compton effect, the Compton shift when incident photon recoils back, is
   (a) 0
   (b) 0.024 Å
   (c) 0.012 Å
   (d) 0.048 Å

6. From the Compton equation, the Compton shift depends on
   (a) wavelength of incident radiation
   (b) scattering angle
   (c) nature of scattering material
   (d) all of the above
7. In Compton effect the unmodified radiation results when incoming photon interacts with
   (a) valance electron
   (b) bound electron
   (c) nucleus
   (d) does not interact with atom
8. The Compton effect is not observed in the visible region because of
   (a) very small percentage change in the wavelength
   (b) inadequacy of the measuring instrument
   (c) smallness of Compton shift
   (d) All of the above
9. Scattering angle for maximum Compton Shift is
   (a) 90°
   (b) 0
   (c) 180°
   (d) 360°
10. The modified radiation occurs due to the collision between incident photon and
   (a) free electron
   (b) an atom
   (c) bound electron
   (d) none of the above

3.6 De Broglie's Theory of Matter Waves

From the discussion in the previous sections, it is clear that light possesses a dual character, as it behaves as a wave as well as a particle (photon). This dual nature of light cannot be readily accepted because of the apparently contradictory aspects of the two natures. A wave is specified by ν or λ, velocity and amplitude. A wave is spread out and occupies a relatively large region of space. A particle on the other hand is specified by mass m, velocity v, momentum p, and energy E. Its position is well defined as it is confined to a very small region of space. It is very difficult to reconcile or harmonize with the two conflicting ideas. However, the countless experiments give convincing evidence that light possesses dual nature.

Louis de Broglie realized that all things in nature tend to be symmetrical. Out of the two fundamental forms in which nature manifests itself, energy, that is, radiations have been shown to possess a dual nature and therefore matter should inherit the same dual nature. Hence, he argued that like radiation, matter should also exhibit dual nature. Thus, he put forward the view that the wave character is also associated with all matter particles in motion. He proposed that, material particles such as electrons and protons in motion must be accompanied by waves and they should give inference effect. These waves are referred to as *matter waves*.

3.6.1 De Broglie Wavelength of Matter Waves

A light wave of frequency ν is associated with a photon of energy E given by

$$E = h\nu$$

According to Einstein theory of relativity, a mass m is equivalent to an amount of energy E according to the relation

$$E = mc^2$$

$$m = \frac{E}{c^2}$$

where c is the velocity of light in free space.

For a photon of energy $E = h\nu$, the mass of the photon becomes

$$m = \frac{h\nu}{c^2} \tag{3.11}$$

The photon travels with the velocity of light *c*. Therefore, the momentum of the photon is given by $p = mc$. Substituting the mass of photon from Eq. (3.11), the momentum of photon becomes

$$p = \frac{h\nu}{c^2} c = \frac{h\nu}{c}$$

But $\nu = c/\lambda$, where λ is the wavelength of light.

$$p = \frac{h}{\lambda} \quad \text{or} \quad \lambda = \frac{h}{p} \tag{3.12}$$

Thus the wavelength associated with a photon is h/p.

De Broglie postulated that the wavelength associated with a moving material particle is, in analogy with light, given by $\lambda = \frac{h}{p}$, where p is the momentum of the particle. A particle of (relativistic) mass m and moving with a velocity v has momentum mv. Therefore the wavelength associated with the particle is given by

$$\lambda = \frac{h}{p} = \frac{h}{mv}$$

This relation represents a connection between wave and particle. Thus matter also exhibits dual nature like light.

3.6.2 De Broglie Wavelength of Charged Particle (Non-relativistic Case)

Let us consider the case of any charged particle such as an electron or a proton, which has been accelerated from rest through a potential difference V. The velocity v acquired by the charged particle is given by

$$\frac{1}{2} m_0 v^2 = eV$$

where m_0, v, and e are the rest mass, velocity, and the charge of the particle respectively.

If V is expressed in volts, then velocity of the charged particle is

$$v = \sqrt{\left(\frac{2eV}{m_o}\right)}$$

Now, the de Broglie expression for matter waves is

$$\lambda = \frac{h}{mv}$$

If the relativistic variation of mass with velocity is to be ignored then

$$m \approx m_0$$

$$\lambda = \frac{h}{m_0 v}$$

Substituting the value of v from above, we get

$$\lambda = \frac{h}{m_0 \sqrt{\left(\frac{2eV}{m_0}\right)}}$$

$$\lambda = \frac{h}{\sqrt{2eV\, m_o}} \tag{3.13}$$

Here V is in volts. This is the expression for the wavelength of an accelerated charged particle in the non-relativistic case.

For most of the practical cases we come across electron as a charged particle. The wavelength of electron accelerated through a potential of 'V' volts can be calculated as follows.

Substituting $h = 6.625 \times 10^{-34}$ Js, $m_o = 9.1 \times 10^{-31}$ kg and $e = 1.6 \times 10^{-19}$ C, we get

$$\lambda = \frac{6.625\times10^{-34}}{\sqrt{2\times\left(9.1\times10^{-31}\right)\times\left(1.6\times10^{-19}\right)}} \times \frac{1}{\sqrt{V}} \text{ m}$$

$$\lambda = \sqrt{\left(\frac{150}{V}\right)}\text{Å} = \frac{12.3}{\sqrt{V}}\text{Å} \tag{3.14}$$

3.6.3 Davisson and Germer Experiment on Matter Waves

According to de Broglie a matter particle in motion is associated with a wave. These if exist should show the interference and diffraction effects. In the experiment performed by Davisson and Germer, accelerated electron beam was shown to demonstrate diffraction phenomenon. The apparatus is as shown in Fig. 3.5. Electron gun is used to produce an accelerated beam of electrons. The velocity of electrons in the beam can be varied by changing the potential in the electron gun. The accelerated electrons are made to fall normally on the target. The target used in this experiment is a Nickel crystal. The diffracted electrons are detected by using a detector (Faraday cylinder connected to a galvanometer). As it was speculated that electrons will be diffracted in all directions hence the arrangement was made in such a way that the detector can be moved along a circular scale with target at the centre. As the crystal is mounted on a turntable, the angle at which the incident beam will fall on the crystal can also be changed.

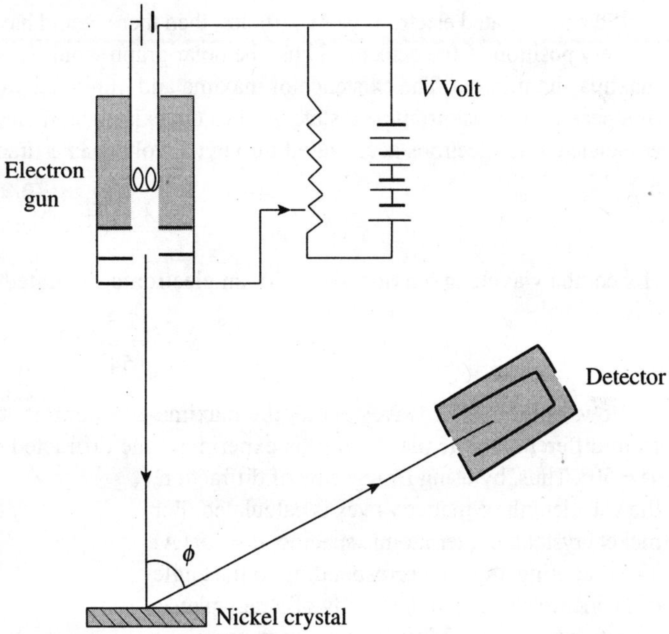

Fig. 3.5 Experimental arrangement in Davisson and Germer

During the experiment the accelerating potential V was changed gradually and the crystal for every value of potential was turned at any arbitrary angle. The intensity at every position was noted by rotating the detector on the circular scale. The intensity of the diffracted electron beam is plotted against the angle ϕ between the incident beam and the beam entering the detector. These plots are called the polar graphs shown in Fig. 3.6. From the polar graphs it is clear that the intensity of the diffracted beam shows distinct maxima and minima depending on the electron energy. The detector's position is fixed where the intensity is maximum and then the accelerating potential was gradually changed. From the different plots as shown in the Fig. 3.6, it is seen that a peak begins to appear in the curve at an angle $\phi = 50°$ for accelerating potential equal to 44 V. With detector position fixed at the same angle, with increasing accelerating potential the peak becomes most prominent when the accelerating potential reaches 54 V. Further increase in the accelerating potential reduces the intensity of the peak and it gradually disappears.

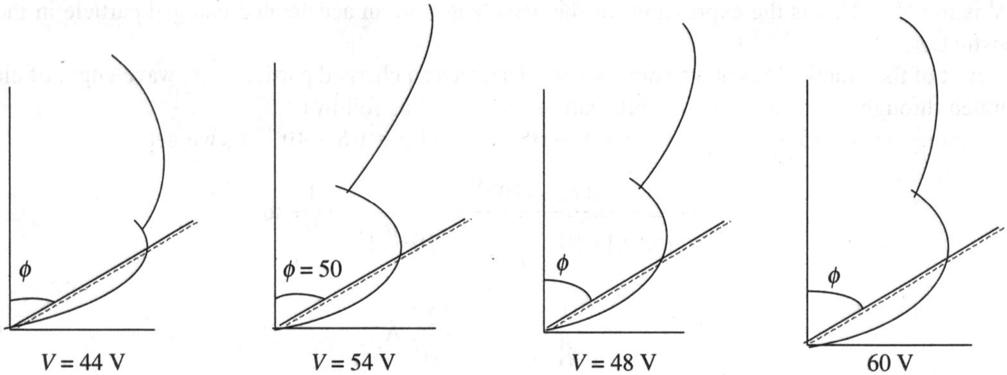

Fig. 3.6 Polar graphs- Intensity variation with scattering angle

If the accelerated electrons were particles, then there would have been equal probability of detecting electrons at every position of the detector. Thus, the polar graph would have been a circle. But the polar graph exhibited maxima and minima. The existence of maxima and minima is the characteristics of waves. Thus, the presence of a peak in its most prominent state verifies the existence of electron waves. The wavelength of matter waves associated with electrons accelerated through V volts was estimated by using de Broglie hypothesis as

$$\lambda = \frac{12.3}{\sqrt{V}} \text{ Å}$$

Hence, the wavelength associated with an electron accelerated through 54 V is

$$\lambda = \frac{12.3}{\sqrt{(54)}} = 1.67 \text{ Å} \tag{3.15}$$

Now, as the matter waves depict the maxima and minima, they must have been diffracted by the atoms from different atomic planes. In this experiment the diffracted electron beam was received only at an angle $\phi = 50°$. Thus, by using Braggs law of diffraction, the wavelength of matter waves is calculated. For nickel crystal the inter-atomic spacing, d = 0.91 Å. The scattering angle θ corresponding to the angle of incidence relative to the family of Bragg planes by referring to Fig. 3.7 is calculated as

$$\theta = \frac{180° - 50°}{2} = 65°$$

Hence using the Bragg's equation (taking the order $n = 1$), we get

$$\lambda = 2d \sin\theta = 2(0.91\text{Å})\sin 65 = 1.65\text{Å} \quad (3.16)$$

The wavelength estimated by using de Broglie concept (Eq. 3.15) and Braggs law (Eq. 3.16) are in close agreement. Hence, the experiment proved the existence of matter waves. Thus, accelerated electrons behave like a wave.

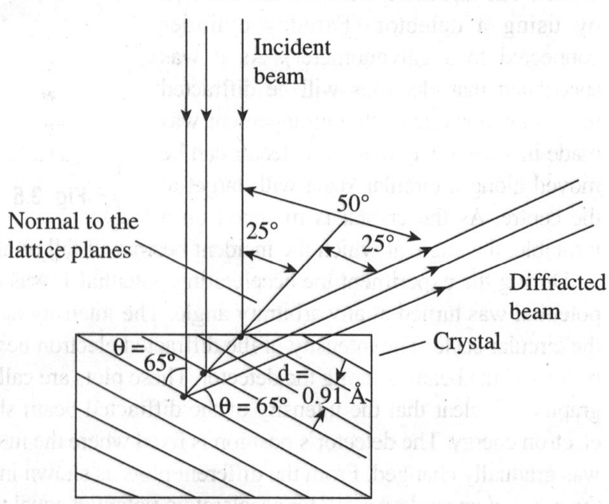

Fig. 3.7 Bragg's reflection of electron beam at lattice plane of nickel crystal

3.6.4 Concept of Matter Waves Under Trivial or Ordinary Situation

The wave nature of matter is not apparent to our daily observation. The reason is that in our daily life we come across macroscopic objects for which the de Broglie wavelength is extremely small. For example, the de Broglie wavelength of a bullet of mass 1 gm moving with a velocity of 10^5 m/s is given by

$$\lambda = \frac{h}{mv} = \frac{6.625 \times 10^{-34}\ Js}{10^{-3}\,\text{Kg} \times 10^{5}\ m/s} = 6.625 \times 10^{-32}\,\text{m}$$

The wavelength is too small to show any effect.

3.6.5 Application of de Broglie Concept

Theoretical proof of Bohr's quantization condition is provided by the de Broglie's concept of matter waves.

Bohr's quantization condition states that in an atom only those orbits are permitted for which the angular momentum of electron is an integral multiple of ħ. Bohr intuitively stated this postulate without any proof. De-Broglie gave the theoretical proof of Bohr's postulate. According to de-Broglie concept, an electron of mass m moving with velocity v has associated with it a wave whose wavelength is $\lambda = h/mv$ where the Planck's constant is h. This wave-character of electron can account for the limited number of permissible orbits in an atom and also enables us to deduce the quantized angular momentum of an electron in an atom.

We know that according to Bohr, the electron in an atom revolves in circular non-radiating orbits around the nucleus. Let us fit this idea with the wave-nature of the electron. Since the electron does not radiate energy while moving in its orbit, the wave associated with it must be a 'stationary' wave (in which there is no transfer of energy). Further, the electron wave can be adjusted around an orbit only when the circumference of the orbit is equal to an integral multiple of the wavelength as shown by Fig. 3.8.

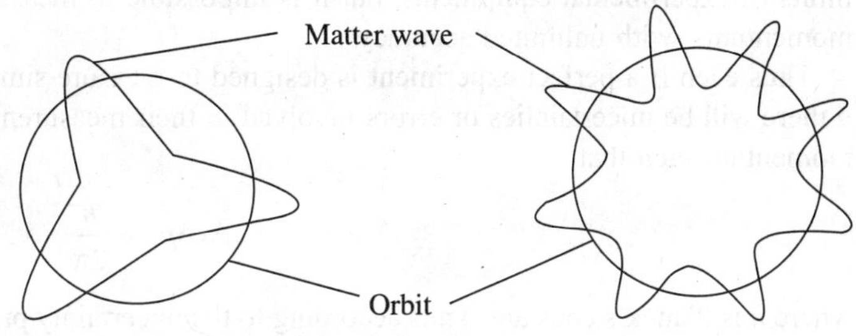

Fig. 3.8

This means that only those orbits are permissible for which the radius r is given by

$$2\pi r = n\lambda = n\frac{h}{mv}$$

Hence, the angular momentum of the electron in its orbit must be

$$L = mvr = \frac{nh}{2\pi}$$

This is the Bohr's quantization condition.

Checkpoint 2

1. The hypothesis of matter waves was experimentally verified by
   (a) Davisson–Germer
   (b) de-Broglie
   (c) Heisenberg
   (d) Schrödinger
2. The diffraction peak in Davisson - Germer experiment occurs at an angle and voltage equal to
   (a) $\theta = 54°$, V = 50 V
   (b) $\theta = 25°$, V = 50 V
   (c) $\theta = 50°$, V = 54 V
   (d) $\theta = 65°$, V = 54 V

3. The wavelength of a wave associated with a car (m = 100 kg) moving with the speed of 100 km/hr is
   (a) 216.72×10^{-29} m (c) 216.72×10^{-19} m
   (b) 216.72×10^{-29} Å (d) none
4. The wavelength of matter waves associated with electrons accelerated through potential difference of 100 V is
   (a) 100 Å (b) 100 m
   (c) 12.3 Å (d) 1.23 Å

3.7 Heisenberg's Uncertainty Principle

Heisenberg realized that as soon as we associate wave with the particle, then its momentum is well defined but its position becomes ambiguous. If we want to measure either position or momentum, we can do so with 100% accuracy. However, if we want to measure both the parameters viz. momentum and position simultaneously, then the measuring operation perturbs the particle in such a way that some uncertainties are introduced in the measurements. Thus we can determine both the parameters simultaneously but not with 100% accuracy. The uncertainty or errors introduced during simultaneous measurements are not due to the inability of the measuring instrument, but due to the constraint put by the nature that we cannot see both wave and particle characteristics simultaneously. Thus, the uncertainty principle states that we can measure either the position or the momentum of a particle with any desired degree of accuracy (within the limits of experimental equipment), but it is impossible to measure both simultaneously—the position and momentum—with unlimited accuracy.

Thus even if a perfect experiment is designed to measure simultaneously the position x and momentum p there will be uncertainties or errors involved in their measurement, viz. Δx in the position and Δp_x in the momentum such that

$$\Delta x \Delta p_x \geq \frac{h}{2\pi} \tag{3.17}$$

where h is Planck's constant. Thus according to the uncertainty principle the product of the uncertainty Δx in the position of a body at some instant and the uncertainty Δp_x in its momentum at the same instant is equal to or greater that $h/2\pi$. If effort is made to measure accurately one quantity, then the measurement disturbs the particle in such a way that the other quantity becomes completely uncertain, and vice versa. From the uncertainty principle it also follows that if there is maximum uncertainty involved in the measurement of position Δx_{max}, then there is minimum uncertainty involved in the measurement of momentum $\Delta p_{x\,min}$ and vice versa, as the product of uncertainties has to be constant. Let us illustrate the principle by simple experiments.

3.7.1 Diffraction of Electron Beam at a Narrow Slit

Let us consider that a beam of mono-velocity electrons is moving along the X-axis. The electrons can be anywhere as the beam has a lateral spread. Hence, we know the velocity of electrons but its position is completely unknown. To find the position of electrons we allow them to move through a slit. Thus, the position of electrons is now uncertain over the width of the slit only, that is, Δy. However,

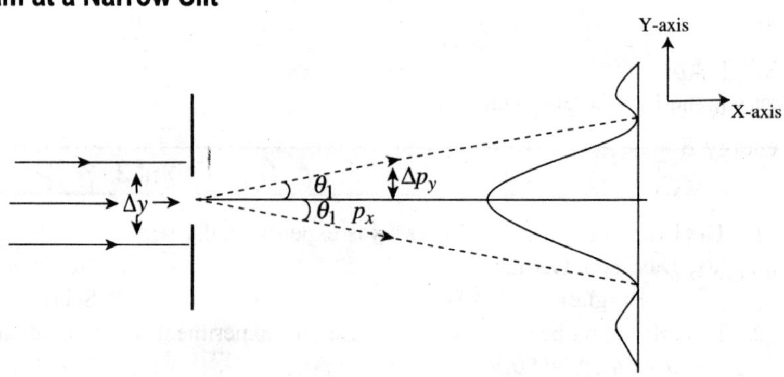

Fig. 3.8 Diffraction of electrons at a single slit

when the beam of electrons passes through the slit, the matter waves associated with it produce a diffraction pattern on a screen as shown in Fig. 3.8.

In Fraunhofer's diffraction (when diffraction pattern is viewed at a long distance from the diffracting object) at a single slit, the direction of the first minimum is given by

$$\sin\theta = \frac{\lambda}{\Delta y}$$

Here, Δy is the width of the slit. Electrons are diffracted at angles greater than θ_1 also, but most of them lie within the angular range $+\theta_1$ to $-\theta_1$. As θ_1 is very small, $\sin\theta_1 \cong \theta_1$.

Therefore,
$$\theta = \frac{\lambda}{\Delta y} \tag{3.18}$$

The fact that a diffraction pattern appears indicates that each electron has passed through the slit, but the position in the slit is indefinite. Thus, the Y co-ordinate of the position of the electron has an uncertainty Δy.

Initially the electrons moving along the X-axis have zero momentum along the Y direction. But at the slit, they deviate from their initial path to form the pattern and have component of momentum in the Y-direction also. The momentum of the electron is given by de Broglie relation, $px = h/\lambda$. Hence its Y-component can lie anywhere between $(+h/\lambda)\sin\theta_1$ and $(-h/\lambda)\sin\theta_1$. This means that, the uncertainty in the Y component of the momentum can be found by using the geometry as

$$\tan\theta_1 = \frac{\Delta p_y}{p_x}$$

Therefore,
$$\theta_1 = \frac{\Delta p_y}{p_x} \text{ as } \theta_1 \text{ is very small.} \tag{3.19}$$

Equating the two equations, namely Eqs (3.18) and (3.19) we get

$$\frac{\lambda}{\Delta y} = \frac{\Delta p_y}{p_x}$$

Therefore,
$$\Delta y \Delta p_y = \lambda p_x = \frac{h}{p_x} p_x = h > \frac{h}{2\pi} \tag{3.20}$$

This is in agreement with the uncertainty principle. If we try to improve the accuracy of the position by decreasing the width of the slit, the diffraction pattern will be widened. This means that the uncertainty in momentum will increase. Thus, both the parameters can be measured simultaneously but with some inaccuracies in it.

3.7.2 Application of Uncertainty Principle for Correcting Concept of Bohr Orbits

The uncertainty relation $\Delta x \Delta p_y \geq h/2\pi$ can also be written in terms of the conjugate pair of quantities, energy E and time t. We know that the kinetic energy of a particle is

$$E = \frac{1}{2}mv^2 = \frac{p^2}{2m}$$

where p is the linear momentum ($p = mv$). Thus

$$\Delta E = \frac{p\Delta p}{m}$$

But
$$p = mv = m\frac{\Delta x}{\Delta t}$$

Therefore,
$$\Delta E = \frac{\Delta x}{\Delta t}\Delta p$$
$$\Delta E \Delta t = \Delta x \Delta p$$

Since $\Delta x \Delta p \geq h/2\pi$, we can write
$$\Delta E \Delta t \geq h/2\pi \tag{3.21}$$

According to the Bohr's concept of orbit, an electron in an atom revolves in one of the quantized orbits, each having a sharply defined energy, with zero uncertainty in energy. This means that if $\Delta E = 0$, then according to the uncertainty relation Δt must be infinite. Thus, that all energy states of the atom must have an infinite lifetime. However, we know that the excited states of the atom from which radiation occurs do not last indefinitely and the lifetime is of the order of 10^{-8} to 10^{-10} s. Thus, the concept of Bohr orbits violates the uncertainty principle.

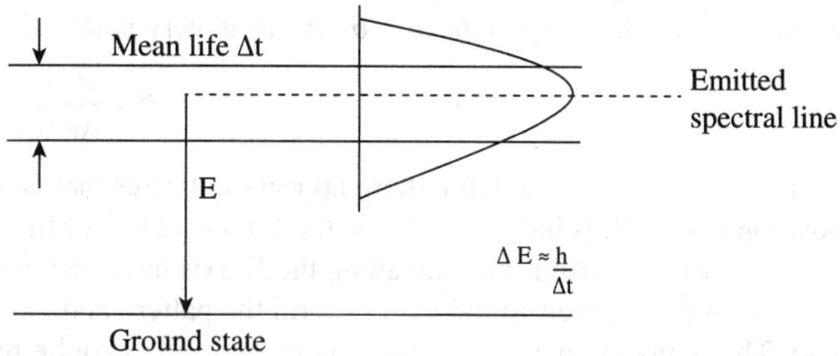

Fig. 3.9 Finite width of discrete energy levels

The finite lifetime Δt gives a finite width (uncertainty) to the energy levels, $\Delta E = \dfrac{h}{2\pi\Delta t}$ as shown in Fig. 3.9. This means that excited energy levels have finite energy spread so that the radiation emitted when atoms fall back to the ground state from a particular level is not truly monochromatic. In other words, the spectral lines can never be infinitely sharp, but must have a natural finite width (natural broadening). Thus, well-defined or single valued energy levels do not exist. The energy levels will have a finite energy spread and hence the width.

The uncertainty principle is applicable to all material particles, from atomic particles like electrons to large bodies. In case of large bodies, however the uncertainties are negligibly small compared to the ordinary experimental errors as explained below.

3.7.3 Uncertainty Principle with Macroscopic Objects

The uncertainty principle does not reveal itself when dealing with macroscopic objects. This is due to the extremely small value of $h\ (= 6.63\times10^{-34}\ \text{Js})$. In the case of macroscopic objects, the experimental errors involved in simultaneous measurements of position and momentum, viz. Δx and Δp are always such that $\Delta x \Delta p >> h/2\pi$. Hence, the fundamental limit $\Delta x \Delta p = h/2\pi$ is never reached due to our inability of measurement to this accuracy.

Let us consider a 1-gm bullet moving at the speed of 10^5 m/s. Its position at any instant might be known within an experimental accuracy of $\Delta x = \pm 10^{-3}$ m and its momentum to within $\Delta p = \pm 10^{-3}$ kg m/s. Then
$$\Delta x \Delta p = \pm 10^{-6}\ \text{Js}$$
which is very much greater that $\dfrac{h}{2\pi} = 1.04\times10^{-34}$ Js.

3.7.4 Wave and Group Velocity

Wave velocity In subatomic world according to de Broglie, a moving particle is represented by a matter wave as the wave nature predominates over particle properties. The velocity of propagation of the matter wave is called *wave velocity*. If we consider that only single matter wave is associated with a moving particle, then its velocity of advancement in the medium is called the wave velocity. For example, if we represent the matter wave by a plane harmonic wave travelling along the + X-direction as given by $Y = a \sin(\omega t - kx)$, where 'a' is the amplitude, ω ($=2\pi v$) is the angular frequency, k ($=2\pi/\lambda$) is the propagation constant of the wave and $(\omega t - kx)$ is the phase of the wave motion. If we want to find wave velocity, then we will have to fix a point on the wave. The point fixed on the wave represents the phase at that point. The wave velocity then can be defined as the velocity with which the plane of constant phase advances in the medium. Hence, wave velocity is also called *phase velocity* v_p. The planes of constant phase are defined by $(\omega t - kx)$ = constant. Differentiating with respect to time, we get

$$\omega - k\frac{dx}{dt} = 0 \text{ or } \frac{dx}{dt} = \frac{\omega}{k}$$

Thus the wave velocity v_p by definition is the ratio of angular frequency ω to the propagation constant k.

$$v_p = \frac{\omega}{k} \tag{3.22}$$

Group velocity In practice a single matter wave does not exist which can be mathematically proved as follows. Consider a particle is moving with velocity v and the matter wave associated with it is moving with velocity v_p. Since velocity is related to wavelength and frequency through the equation, $c = v\lambda$, the velocity of matter wave is given by

$$v_p = v\lambda$$

As frequency $v = \frac{E}{h}$, where $E = mc^2$ and wavelength of matter wave is $\lambda = h/p$.

Therefore,
$$v_p = \frac{mc^2}{h} \times \frac{h}{mv} = \frac{c^2}{v} \tag{3.23}$$

Thus,
$$v_p >> c.$$

This means that the matter wave moves much faster than light and thus leaves the particle far behind. Hence, we cannot represent a moving particle with a single de Broglie/matter wave. Therefore, a moving particle can be represented by numerous matter waves moving with slightly different wave velocities. As waves with spread in velocity cannot move independently in the same direction, they superpose on each other forming a wave packet or a wave group. *Wave packet* is defined as the packet that contains a group of matter waves moving with slightly different wavelength or frequency (i.e. spread in wave velocity) such that they interfere constructively over a very small region and destructively elsewhere. The wave packet travels in a medium with a velocity called group velocity v_g. *Group velocity* is the velocity with which the maximum amplitude of the group advances in the medium. Thus, the energy in the wave packet travels with the group velocity. The individual waves inside the wave packet travel with their phase velocity.

As the group contains a number of matter waves with spread in their velocity, the group velocity is written as $v_g = \frac{d\omega}{dk}$. Thus, a wave packet best represents a moving particle, as the particle velocity is same as the group velocity. The proof is as follows:

$$v_g = \frac{d\omega}{dk} \tag{3.24}$$

As
$$\omega = \frac{E}{\hbar} \quad \text{and} \quad k = \frac{p}{\hbar}$$

Therefore,
$$v_g = \frac{dE}{dp} \qquad \text{since } E = \frac{p2}{2m}$$

Thus
$$v_g = \frac{p}{m}$$

$$\therefore v_g = v \qquad (3.25)$$

Thus, a wave packet contains the complete information about the moving particle. Wave packet gives information about the position of the particle as it has a large amplitude only over a very small region in the space and zero elsewhere. Therefore, the particle is most likely to be found where the amplitude of the wave packet is largest and its velocity is same as that of the particle. The amplitude of wave packet is denoted by a complex mathematical function called *wave function* given by $\psi(x, t)$ (in one dimension). In analogy with the simple harmonic wave travelling along the + X-direction (one dimension is considered for simplicity) is given by Y = a sin ($\omega t - kx$), the amplitude of the wave packet which is modulated with respect to space and time coordinates can be expressed as $\Psi(x,t) = Ae^{-i\,(kx-wt)}$.

For mathematical convenience, a complex mathematical function is considered to represent the amplitude of wave packet. The next section will answer some of the questions viz. 'How to find wave function' and 'What is the physical meaning of ψ?'

Now, to describe the motion of a particle in a subatomic surrounding, an equation of motion should be set up for the wave packet. E. Schrödinger developed this equation of motion for atomic particles confined in atomic surroundings which is explained in the next section.

Checkpoint 3

1. If the uncertainty in locating a particle is 0.02%, the uncertainty in momentum is
   (a) 0.527×10^{-34} kg m/s
   (b) 1.054×10^{-30} kg m/s
   (c) 0.527×10^{-30} kg m/s
   (d) none of the above
2. According to the statement of uncertainty principle the product of uncertainty in position and momentum is
   (a) always less than or equal to Planck's constant, h
   (b) always equal to h
   (c) always equal to zero
   (d) always more than equal to h
3. The relation between the matter wave velocity v_p and the velocity of the particle v is
   (a) $v_p = \frac{c^2}{v}$
   (b) $v_p = \frac{c}{v}$
   (c) $v_p = v$
   (d) $v_p = v \times c$

3.8 Schrödinger Equations

In classical mechanics, in order to determine the position of a given particle under some conditions we use Newton's equations. Similarly, in quantum mechanics we make use of Schrodinger equations to understand the behaviour of subatomic particles.

For a non-relativistic particle of mass *m* moving in a potential region 'V', time-dependent Schrodinger equation in one dimension is as follows.

$$i\hbar\frac{\partial\psi(x,t)}{\partial t} = \frac{-\hbar^2}{2m}\frac{\partial^2\psi(x,t)}{\partial^2 x} + V\psi(x,t) \qquad (3.26)$$

Here $\psi(x, t)$ is the wave function as discussed in Section 3.7.4 which represents the amplitude of the wave packet, modulated with respect to space (one-dimension) and time expressed as a complex mathematical function as $\psi(x,t) = Ae^{-i\,(kx-wt)}$

Expressing the wave function in terms of observables by using equations p= ħK and E= ħω as

$$\psi(x,t) = Ae^{\frac{i}{\hbar}(px-Et)} \tag{3.27}$$

Equation (3.27) is the one-dimensional form of Schrödinger equation. A three-dimensional form can be written as

$$i\hbar\frac{\partial\psi(x,y,z,t)}{\partial t} = \frac{-\hbar^2}{2m}\left(\frac{\partial^2}{\partial x^2}+\frac{\partial^2}{\partial y^2}+\frac{\partial^2}{\partial z^2}\right)\psi(x,y,z,t)+V(x,y,z,t)\psi(x,y,z,t) \tag{3.28}$$

We can use Laplacian notation viz.

$$\frac{\partial^2}{\partial x^2}+\frac{\partial^2}{\partial y^2}+\frac{\partial^2}{\partial z^2} = \nabla^2$$

and r for (x, y, z).

Equation (3.28) then takes a compact form as

$$i\hbar\frac{\partial\psi(r,t)}{\partial t} = \frac{-\hbar^2}{2m}\nabla^2\psi(r,t)+V(r,t)\psi(r,t) \tag{3.29}$$

In most of the situations, the atomic particles move in a time-independent potential region. In order to find allowed energies and positions of such particles, we require Schrödinger's time-independent equation. By using the method of separation of variables, we can obtain Schrödinger's time-independent equation from dependent equation as follows. Assuming that wave function $\psi(x, t)$ can be separated into two parts—one varying with position and the other with time—it can be written as

$$\psi(x,t) = \phi(x)\theta(t) \tag{3.30}$$

By using the method of separation of variables, substituting Eq. (3.30) in Eq. (3.26), we get

$$\frac{-\hbar^2}{2m}\theta(t)\frac{d^2\phi(x)}{dx^2}+V(x)\phi(x)\theta(t) = i\hbar\phi(x)\frac{d\theta(t)}{dt}$$

Dividing the above equation throughout by $\phi(x)\ \theta(t)$, we get

$$\frac{-\hbar^2}{2m\phi(x)}\frac{d^2\phi(x)}{dx^2}+V(x) = \frac{i\hbar}{\theta(t)}\frac{d\theta(t)}{dt} \tag{3.31}$$

Note that in Eq. (3.31) the left hand side is only position dependent and the right hand side is only time dependent. When a particle is moving in potential that is independent of time, then the time part of the wave equation becomes constant. Therefore, we find the value of the right side of Eq. (3.31). The time part of the wave equation, Eq.(3.27) is given by,

$$\theta(t) = Ae^{\frac{-i}{\hbar}Et}$$

Differentiating the above equation with respect to time

$$\frac{d\theta(t)}{dt} = \left(\frac{-iE}{\hbar}\right)\theta(t)$$

Therefore,

$$\frac{i\hbar}{\theta(t)}\frac{d\theta(t)}{dt} = E$$

Substituting this value of right side of Eq. (3.31), we get

$$\frac{-\hbar^2}{2m\phi(x)}\frac{d^2\phi(x)}{dx^2}+V(x)=E$$

Multiplying both sides by $\phi(x)$

$$\therefore \frac{-\hbar^2}{2m}\frac{d^2\phi(x)}{dx^2}+V(x)\phi(x)=E\phi(x) \tag{3.32}$$

Equation (3.32) is one-dimensional, time-independent Schrödinger equation.

For three dimensions we can write $\psi(r,t)=\psi(r)\phi(t)$

Also, time-independent three-dimensional Schrödinger equation becomes

$$\frac{-\hbar^2}{2m}\nabla^2\psi(r)+V(r)\psi(r)=E\psi(r) \tag{3.33}$$

To determine the allowed energy states of a particle in one dimension it is necessary to solve differential equation Eq. (3.32) for the known form of potential 'V'. The solution of the Schrödinger equation will be a mathematical function representing the variation of the amplitude of the wave packet in space represented by wave function ψ(r).

Now, for a given problem, there may be a number of mathematical functions which satisfy Eq. (3.32), but any mathematical function cannot be treated as solution of the Schrödinger equation. The acceptable wave function should satisfy the physical boundary conditions. Only well-behaved wave functions ψ can become the solution of Schrödinger equation. The wave function is called well behaved if it satisfies the following conditions.

1. ψ(r) must be finite and single valued at all points in coordinate space.
2. It is necessary that ψ(r) is a continuous function at all points in coordinate space.
3. First derivatives of ψ viz. $\frac{\partial\psi}{\partial x}$, $\frac{\partial\psi}{\partial y}$ and $\frac{\partial\psi}{\partial z}$ also need to be finite, single valued, and continuous functions in coordinate space.

Thus, only well-behaved functions can become the solution of Schrödinger's equation. Now, let us find an answer for the question 'What is the physical significance of the wave function ψ'?

The wave function ψ characterizes the state of the micro particle. But as it is a complex quantity it is not an observable. Thus, the wave function does not have any physical significance. Max Born suggested that the square of the absolute value can be calculated by multiplying ψ by ψ^*, that is, $|\psi\psi^*|$. The value of $|\psi|^2$ is interpreted as the probability density of the particle, that is, the probability of finding the particle at that point at the given instant. The interpretation of ψ provides a statistical connection between the wave and the associated particle. It tells us where the particle is likely to be and not where the particle is. Thus, the probability of finding the particle in a given volume dV by given by $\int|\psi|^2 dV$.

For example, if a complex waveform of ψ is expressed as

$\psi = u + iv$ with u and v as real functions.

Then $\psi^* = u - iv$ is the complex conjugate of ψ.

Therefore, $$\psi\psi^* = u^2 + v^2 \text{ (as } i^2 = -1)$$

Thus, the physical interpretation of de Broglie waves is given as follows.

1. The wave function $\psi(r, t)$ is useful to determine the probability of a moving particle.
2. $\psi(r, t)$ may be complex.
3. $|\psi(r, t)|^2$ is the probability of finding the particle at time t at a point r and $r + dr$.
4. As the particle should be found somewhere in space, the total probability is of course unity. Thus, wave function ψ can be normalization by using the condition $\int|\psi(r, t)|^2 dV = 1$. Normalization condition is used to find the value of constants involved in the wave function ψ.

The following section discusses the applications of Schrödinger's mechanics.

3.8.1 Electron Confinement

We will now apply quantum theory (Schrödinger's mechanics) to a simple and unreal situation to understand the basic principles of quantum physics of an atom. To begin with, we will first examine the wave function representing matter waves associated with a non-relativistic electron confined to a limited region of the space. We will first find what positions and energy values are permissible for an electron trapped in a one-dimensional infinite box. Confinement of matter waves leads to quantization. For the nano-structured materials, at least one of its dimensions should be in the range of nanoscale. Thus, the motion of electron in nanomaterial is confined or restricted. Due to the confinement, the nano material shows drastic change in its various properties including physical, chemical, electrical, magnetic and optical properties as compared to bulk material. The change in properties of materials can be understood by using quantum mechanics. Quantum mechanics is also used to understand the behaviour of bulk material as the bulk properties are ultimately determined by the properties of atoms in molecules. Thus, quantum mechanics helps us in understanding the properties of the subatomic particles in atomic surroundings. The second application discussed is a realistic case where an atomic particle with less energy encounters a potential barrier of finite height.

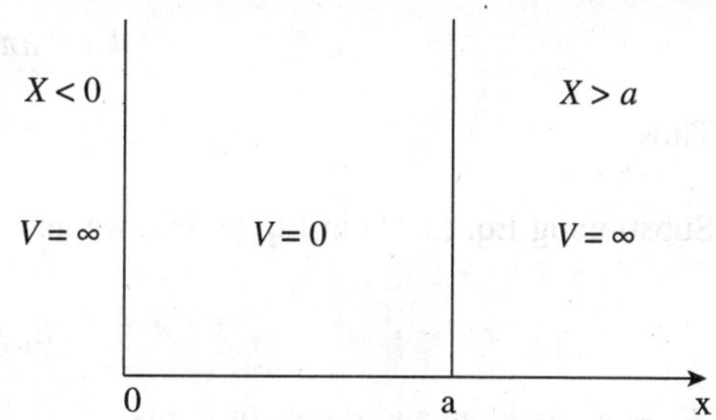

Fig. 3.10 One-dimensional box of infinite height

Particle in an Infinite One-dimensional Box

Consider a one-dimensional box of width 'a' and infinite height such that potential $V = 0$ if $0 < x < a$ and $V = \infty$ if $x < 0$ or $x > a$ as illustrated in Fig. 3.10.

Energy states of the particle of mass m inside the box can be obtained using time-independent Schrodinger equation for one dimension as

$$\frac{-\hbar^2}{2m}\frac{\partial^2\psi(x)}{\partial x^2} + V(x)\psi(x) = E\psi(x) \tag{3.34}$$

As the potential inside the box is zero, the equation becomes

$$\frac{-\hbar^2}{2m}\frac{\partial^2\psi(x)}{\partial x^2} = E\psi(x)$$

Rearranging the equation, we get

$$\frac{\partial^2\psi(x)}{\partial x^2} + \frac{2mE}{\hbar^2}\psi(x) = 0 \tag{3.35}$$

Substituting
$$k^2 = \frac{2mE}{\hbar^2} \tag{3.36}$$

We can rewrite Eq. (3.35) as
$$\frac{\partial^2 \psi(x)}{\partial x^2} + k^2 \psi(x) = 0 \tag{3.37}$$

Equation (3.37) is a standard differential equation, the solution of which is given by
$$\psi(x) = A \sin kx + B \cos kx \tag{3.38}$$

As the particle exists only inside the box, the wave function should not exist outside the box and should be zero at the boundaries. Now boundary conditions are superimposed on the solution.

At $x = 0$, $\psi(x) = 0$. Substitution of this boundary condition in Eq. (3.38) leads to B = 0. Therefore, the solution reduces to
$$\psi(x) = A \sin kx \tag{3.39}$$

Applying the second boundary condition as follows.
At $x = a$, $\psi(a) = 0$. Equation (3.39) becomes
$$0 = A \sin ka \tag{3.40}$$

However, 'A' cannot be zero, therefore
$$ka = n\pi, \text{ where } n = 1, 2, 3, \ldots$$

Thus
$$k = \frac{n\pi}{a} \tag{3.41}$$

Substituting Eq. (3.41) in Eq. (3.36), we get
$$E_n = \frac{n^2 \hbar^2 \pi^2}{2ma^2} = \frac{n^2 h^2}{8ma^2} \tag{3.42}$$

$E_n \propto n^2$ as h, m, and a are constants.
Thus the energy of the particle in a box depends only on quantum number n. As n can take only discrete values, hence energy of the particle is quantized.
Substituting Eq. (3.41) in Eq. (3.39), we get
$$\psi_n = A \sin\left(\frac{n\pi}{a}\right)x \tag{3.43}$$

Although n can take any integer value, in practice if $n = 0$, then $\psi_n = 0$. Hence, n cannot take value 0 as if allowed, $|\psi|^2$ would be 0 and probability of finding the particle inside the box would be zero. Another explanation is that if $n = 0$, then $E_n = 0$. As $E_n = \frac{p^2}{2m}$ therefore $p = 0$. Since $p = \frac{h}{\lambda}$ therefore $\lambda = \infty$, which means the particle can exist outside the box which is incorrect. Therefore n takes the values 1, 2, 3, ... This shows that the energy of particle in a one-dimensional potential box is quantized.

For macroscopic objects, quantization is not noticed. This is explained in the following section.

Quantization does not Exist in Macroscopic World

Equation (3.42) suggests that the quantization of energy is dependent on the width of the box a and the quantum number n. Let us study the effect of a on the quantization of electron energy. If we consider two adjacent energy levels by E_{n+1} and E_n the separation between the levels ΔE is given by

$$\Delta E = E_{n+1} - E_n = \frac{h^2}{8ma^2}(2n+1)$$

$$\Delta E = \frac{h^2 n}{4ma^2} \text{ as } 2n \gg 1, \text{ we can neglect } 1$$

Now for two extreme values of a, let us find the separation between successive energy levels.

(i) Taking a = 1 cm = 10^{-2} m, we get $\Delta E \approx n \times 10^{-15} eV$

This energy difference between successive levels is so insignificant that an electron possessing thermal energy $kT \approx 10^{-3}\, eV$ at room temperature can easily move from one level to the next level without the need of external energy.

Thus, energy levels are quasi-continuous for an atomic particle trapped in atomic surrounding as shown in Fig. 3.11 (a). Hence, quantization is not applicable for atomic particles trapped in macroscopic surroundings. When an electron moves through large apertures, it acts therefore essentially as a classical particle without revealing the quantum aspects.

(ii) On the other hand, if a = 10 Å, $\Delta E \approx 0.75(n)\, eV$.

This energy difference between the adjacent levels is very large as shown in Fig. 3.11(b) and the electron cannot move into higher levels without an energy input from outside source. Thus, quantization of energy comes into effect in the realm of atomic world.

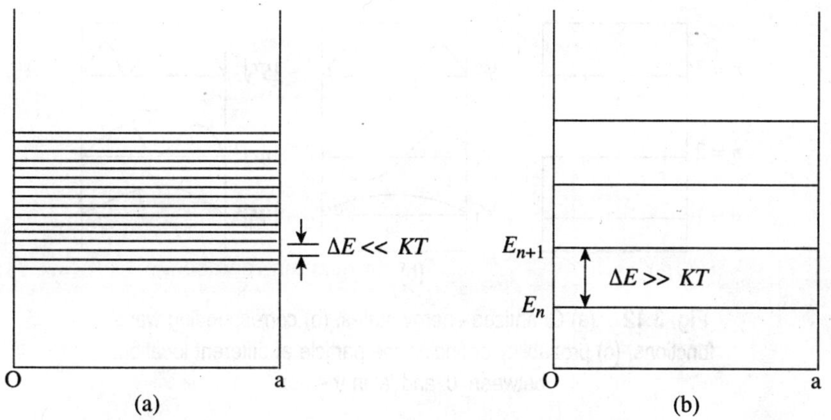

Fig. 3.11 The effect of 'a' on energy levels: (a) when a is large, the electron energy levels form a quasi-continuous spectrum: (b) when a is small, the energy levels are discrete

Wave function

The wave function corresponding to the above-allowed discrete energy levels could be obtained as follows. From Eq. (3.43)

$$\psi_n = A \sin\left(\frac{n\pi}{a}\right)x \tag{3.44}$$

Applying normalization condition, $\int_0^a |\psi(x)|^2 dx = 1$,

Substituting the wave function, we get $\int_{O}^{a} |A|^2 \sin^2\left(\frac{n\pi}{a}\right) x\, dx = 1$

Therefore,
$$A^2 \int_0^a \left[\frac{1 - \cos 2\left(\frac{n\pi}{a}\right) x}{2} \right] dx = 1$$

Solving the above equation we get
$$A = \sqrt{\frac{2}{a}} \tag{3.45}$$

Inserting the value of A in Eq. (3.44), we get
$$\psi(x) = \sqrt{\frac{2}{a}} \sin \frac{n\pi}{a} x \quad \text{and} \quad |\psi(x)|^2 = \frac{2}{a} \sin^2 \frac{n\pi}{a} x \tag{3.46}$$

The corresponding wave functions and probabilities of finding a particle in the box qualitatively would look like those in Fig. (3.12 (b) and (c)) respectively.

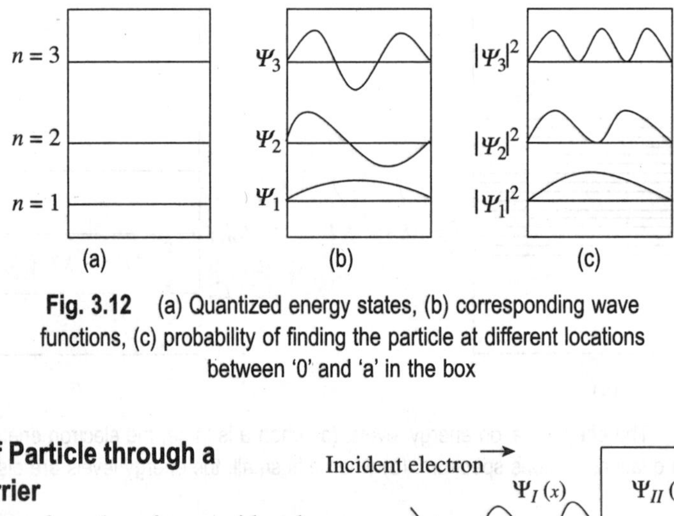

Fig. 3.12 (a) Quantized energy states, (b) corresponding wave functions, (c) probability of finding the particle at different locations between '0' and 'a' in the box

3.8.2 Tunnelling of Particle through a Potential Barrier

A particle trapped in a box (can be considered as potential well) of infinite height is an ideal or unrealistic situation. Figure 3.13 shows a realizable barrier potential on which a particle with kinetic energy less than the barrier potential is incident. For mathematical convenience, we have considered a one-dimensional potential barrier. In region I, the potential $V = 0$ and the energy of the particle is purely kinetic energy. Classically, it is impossible

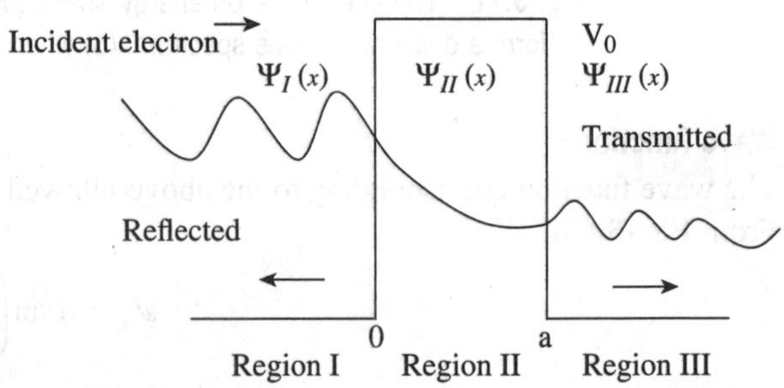

Fig. 3.13 Particle tunnelling through a potential barrier of height V_0.

for a particle with energy less than the barrier height to surmount it. However, by using Schrödinger equation we can show that for a particle with less energy than the potential energy in region I, there exists a possibility that the particle can tunnel through region II and gets transmitted in region III and propagate as long as barrier potential is not infinite.

The phenomenon of tunnelling in quantum mechanics can be understood by considering three different wave functions $\psi_I(x)$, $\psi_{II}(x)$, and $\psi_{III}(x)$ in regions I, II, and III respectively. The Schrodinger's equations for the three regions can be written as

$$\frac{\partial^2 \psi_I(x)}{\partial x^2}+\frac{2mE}{\hbar^2}\psi_I(x)=0 \tag{3.47}$$

$$\frac{\partial^2 \psi_{II}(x)}{\partial x^2}+\frac{2mE}{\hbar^2}\psi_{II}(x)=0 \tag{3.48}$$

$$\frac{\partial^2 \psi_{III}(x)}{\partial x^2}+\frac{2mE}{\hbar^2}\psi_{III}(x)=0 \tag{3.49}$$

The solution of these equations will consist of the superposition of waves that will move to the right (denoted by e^{ikx} and $e^{\alpha x}$) and left (denoted by e^{-ikx} and $e^{-\alpha x}$) as given below.

$$\psi_I(x)=Ae^{ikx}+Be^{-ikx} \tag{3.50}$$

$$\psi_{II}(x)=Ce^{\alpha x}+De^{-\alpha x} \tag{3.51}$$

$$\psi_{III}(x)=Ee^{ikx}+Fe^{-ikx} \tag{3.52}$$

Here k and α are positive and are given by Eqs (3.53) and (3.54). In all these wave functions, the reflected wave is Be^{-ikx} and transmitted wave is E^{ikx}. In regions I and III, waves are travelling as values of k and α are positive; however, in region II there is only a decaying part.

$$k=\sqrt{\frac{2mE}{\hbar^2}} \tag{3.53}$$

$$\alpha=\sqrt{\frac{2m(V_0-E)}{\hbar^2}} \tag{3.54}$$

The particle with kinetic energy $E < V_0$ is shown in Fig. 3.13. The particle is incident from left side in region I and by tunnelling through region II, escapes to region III with some probability of transmission. The transmission or tunnelling probability 'T' for the particle to tunnel through the barrier is the ratio of the number of particles that emerge from the barrier to those which reach the barrier. After solving the equations, we get tunnelling probability T defined as

$$T=\frac{|\psi_{III}(x)|^2}{|\psi_I(\text{incident})|^2}=\frac{F^2}{A^2} \tag{3.55}$$

Thus, there is always a non-zero probability of finding the particle on the other side of a potential barrier even though its kinetic energy is less than the height of the potential barrier. This result is very useful in understanding many phenomena observed for subatomic particles that cannot be explained by classical mechanics.

One can also show that reflection probability of wave in region I is

$$R = \frac{|\psi_I(x)|^2}{|\psi_I(\text{incident})|^2} = \frac{B^2}{A^2} = 1 - T \tag{3.56}$$

Further it should be remembered that even if $E > V_0$ there is always a probability of reflection at each barrier.

Thus according to quantum mechanics it is possible that the particle with relatively less energy than the barrier potential can tunnel through the potential barrier. Thus, there exists a finite probability for the particle to be on the other side of the barrier. This concept finds application in solid-state lasers, light emitting diodes, tunnel diode, nucleons inside nucleus, quantum dots, quantum computing, scanning tunnelling microscope, and so on.

Checkpoint 4

1. Normalization of wave function means
   (a) wave function is equated to 1.
   (b) probability of finding particle in the given volume is equated to 1.
   (c) probability of finding particle in the given volume is equated to 0.
   (d) wave function is equated to 0.
2. The wave function is said to be well behaved if it is
   (a) finite, discontinuous & single valued
   (b) finite, continuous & multiple valued
   (c) infinite, continuous & single valued
   (d) finite, continuous & single valued
3. The width of the infinite potential well is such that it accommodates
   (a) odd multiples of half wavelength
   (b) integral multiples of half wavelength
   (c) integral multiples of full wavelength
   (d) even multiples of half wavelengths
4. The zero point energy of any particle in quantum mechanics is
   (a) greater than zero
   (b) equal to zero
   (c) negative
   (d) does not exist

Solved Numericals

1. The wavelength of the incident photon in Compton effect is 3.0×10^{-10} m. Find the wavelength of the photon scattered at an angle of 60°. ($h = 6.62 \times 10^{-34}$ Js, $c = 3.0 \times 10^8$ m/s, $m_0 = 9.1 \times 10^{-31}$ kg).

Solution: The Compton shift (increase in wavelength) is given by

$$\Delta\lambda = \frac{h}{m_0 c}(1 - \cos\theta)$$

Substituting the given values:

$$\Delta\lambda = \frac{6.62\times10^{-34}}{(9.1\times10^{-31})\times(3.0\times10^{8})}(1-\cos 60°)$$

$$= 0.024\times10^{-10}\left(1-\frac{1}{2}\right)$$

$$= 0.012\times10^{-10}\,\text{m}$$

The incident wavelength is $\lambda = 3.0\times10^{-10}$ m. Therefore, the scattered wavelength will be

$$\lambda' = \lambda + \Delta\lambda$$

$$= (3.0\times10^{-10}\,\text{m}) + (0.012\times10^{-10}\,\text{m})$$

$$= 3.012\times10^{-10}\,\text{m} = 3.012\,\text{Å}$$

2. X-rays of wavelength 1Å are scattered at such an angle that the recoil electron has the maximum kinetic energy. Calculate the wavelength of the scattered ray.

Solution: The recoil electron would acquire maximum kinetic energy when the Compton shift is maximum.

The Compton shift is given by $\Delta\lambda = 0.024(1-\cos\theta)$

It is maximum when $\theta = 180°$ *i.e.*, $\cos\theta = -1$. Thus

$$\Delta\lambda = 0.048\,Å$$

The incident wavelength is $\lambda = 1\,Å$. Therefore, the scattered wavelength will be

$$\lambda' = \lambda + \Delta\lambda$$

$$= 1 + 0.048 = 1.048\,Å$$

3. An X-ray photon is found to have its wavelength doubled on being scattered through 90°. Find the wavelength and energy of the incident photon (m_0 of electron = 9.1×10^{-31} kg).

Solution: The Compton change in wavelength is given by

$$\Delta\lambda = \frac{h}{m_0 c}(1-\cos\theta)$$

When θ = 90°, $\Delta\lambda = \lambda$ (wavelength is doubled). Thus

$$\lambda = \frac{h}{m_0 c} = 0.024\,Å$$

This is the wavelength of the incident photon. Its energy would be

$$E = h\nu = \frac{hc}{\lambda}$$

$$= \frac{hc}{h/m_0 c} = m_0 c^2$$

$$= \left(9.1\times10^{-31}\right)\times(3.0\times10^{8})^2$$

$$= 8.1\times10^{-14}\,\text{J}$$

4. X-rays of 1.0 Å are scattered from a carbon block. Find (a) the wavelength of the scattered beam in a direction making 90° with the incident beam, (b) kinetic energy imparted to the recoiling electron, (c) direction of the recoiling electron.

($h = 6.63\times10^{-34}$ Js, $c = 3.0\times10^{8}$ m/s and $1eV = 1.6\times10^{-19}$ J)

Solution: (a) The Compton shift is

$$\Delta\lambda = 0.024(1-\cos\theta)\,Å$$

At $\theta = 90°$, we have

$$\Delta\lambda = 0.024\,(1-\cos 90°) = 0.024\,Å$$

Hence the wavelength of the scattered beam is

$$\lambda' = \lambda + \Delta\lambda = 1.0 + 0.024 = 1.024\,Å$$

(b) The energy of the incident X-rays photon is $h\nu(= hc/\lambda)$ and that of the scattered photon is $h\nu'(= hc/\lambda')$. The balance has been imparted to the recoiling electron as its kinetic energy. Then

$$\text{Kinetic energy of recoiled electron} = \frac{12400}{\lambda} - \frac{12400}{\lambda'}$$

$$= \frac{12400}{1} - \frac{12400}{1.024}$$

$$= 290\,\text{ev}$$

(c) $$\tan\varnothing = \frac{\lambda}{\lambda'} = \frac{1.0}{1.024} = 0.98$$

$$\varnothing = \tan^{-1}(0.98) = 44.42°$$

5. Find the de Broglie wavelength of

(a) an electron accelerated through a potential difference of 182 V and

(b) a 1 kg object moving with a speed 1 m/s. Comparing the results explain why the wave nature of matter is not more apparent in daily observations.

Solution: Formulae to be used are (a) $\lambda_e = \dfrac{h}{\sqrt{2emV}}$

(b) $\lambda_m = \dfrac{h}{mv}$

(a) $$\lambda_e = \frac{6.626\times10^{-34}}{\sqrt{2\left(1.602\times10^{-19}\right)\left(9.11\times10^{-31}\right)182}}$$

$$= \frac{6.626\times10^{-34}}{7.29\times10^{-24}} = 0.91\times10^{-10}$$

$$= 9.1\times10^{-11}\,\text{m} = 0.91\,Å$$

(b) $$\lambda_m = \frac{6.626\times10^{-34}}{1\,\text{kg}\times1\,\text{m/s}} = 6.6\times10^{-34}\,\text{m}$$

It is seen from the above that the wavelength of the accelerated electron is about 10^4 times larger than its own size (size of electron $\approx 10^{-15}$m) and is therefore significant. On the other hand, the wavelength associated with the macroscopic object is negligibly small and is hidden due to the sizeable dimension of the object. Thus, wave nature is not apparent for macroscopic objects.

6 (a) The uncertainty in the location of a particle is equal to its de Broglie wavelength. Show that the uncertainty in its velocity is equal to its velocity.

(b) Apply uncertainty principle to investigate whether electrons are present in atomic nuclei. The nuclei radius is about 10^{-14} m and the energy of a β-particle emitted from a nucleus is at most of the order of 4 MeV.

Ans. (a) Using the uncertainty principle

$$\Delta x \Delta p \approx h$$

$$\Delta p \approx \frac{h}{\Delta x}$$

But it is given that $\Delta x \approx \frac{h}{p}$ (de Broglie wavelength). Substituting this value of Δx in the above equation, we obtain

$$\Delta p \approx p$$

Substituting $p = mv$

$$\Delta(mv) \approx mv$$

$$m\Delta v \approx mv$$

$$\Delta v \approx v$$

(b) Electron in nuclei Let us suppose that an electron is present in an atomic nucleus whose radius is of the order of 10^{-14} m. For an electron to be confined within a nucleus, using the uncertainty relation $(\Delta x \Delta p \geq h/2\pi)$, the corresponding uncertainty in the electron's momentum is

$$\Delta p \geq \frac{h}{2\pi\Delta x}, \; [\Delta x = 2r = 2 \times 10^{-14} \text{ m}]$$

$$\geq \frac{6.63\times10^{-34}}{2\times3.14\times10^{-14}\times2}$$

$$\geq 0.52\times10^{-20} \text{ kg m/s}$$

If this is the uncertainty in the electron's momentum, then the momentum itself must be comparable to it, (i.e. $p \approx 0.52 \times 10^{-20}$ kg m/s). Thus, we consider that the minimum momentum possessed by the electron should be at least equal to 0.52×10^{-20} kg m/s, if it is the constituent of nucleus. An electron having this momentum must have a velocity comparable to the velocity of light. Hence, let us apply the relativistic formula for calculating the minimum energy of the electron. The total energy, that is the sum of kinetic energy and rest mass energy of the electron with a rest mass m_0 and momentum p is given by

$$E = \sqrt{\left(p^2c^2 + m_0^2c^4\right)}$$

The value of $pc = 0.52 \times 10^{-20} \times (3 \times 10^8)$

$$\therefore pc = 1.56\times10^{-12} \text{ J}$$

$$pc = \frac{1.56\times10^{-12}}{1.6\times10^{-19}} \cong 10\times10^6 \text{ eV} \cong 10 \text{ MeV}$$

The rest mass energy of the electron $E_o = m_0c^2$ is 0.51 MeV, which is negligible as compared to the above value of pc. Hence, it is neglected in the above calculations.

Thus, the electron must have a minimum kinetic energy of the order of 10 MeV if it is to be a nuclear constituent. However, since the energy of a particle (electron) emitted from a nucleus is found to have maximum energy not exceeding 4 MeV, therefore, we conclude that the electrons cannot be present within nuclei. Thus, this theoretical proof for showing non-existence of electron in nucleus is an application of the uncertainty principle.

7. An electron and a 100 gm baseball are traveling at 200 m/s measured to an accuracy of 0.05%. Calculate and compare the uncertainty in position of each of the bodies.

Solution: The inaccuracy in the velocity measurement is

$$\Delta v = v\times0.05\% = 200\times\frac{0.05}{100} = 0.1 \text{ m/s}$$

(a) For baseball

$$\Delta x = \frac{\hbar}{m\Delta v}$$

$$\Delta x_b = \frac{1.05\times10^{-34}}{100\times10^{-3}\times0.1}$$

$$\Delta x_b = 1.05\times10^{-32} \text{ m}$$

(b) For electron

$$\Delta x_e = \frac{1.05\times10^{-34}}{9.11\times10^{-31}\times0.1}$$

$$\Delta x_e = 0.12\times10^{-2} = 1.2 \text{ mm}$$

The uncertainty in measuring the position of baseball is extremely small. In fact, it is insignificant as the errors involved in the measurement of the position will be larger than this. However, for electron the uncertainty or inaccuracy in the measurement of position is appreciable. Hence, the uncertainty principle is applicable to only atomic particles and not the macroscopic objects.

8. Compare the lowest three energy states for (a) an electron confined in the infinite potential well of width 1 Å and (b) a grain of dust (m = 10^{-4} gm) moving with a speed of 10^5 m/s in an infinite potential well of width 0.1 mm. What can you conclude from this comparison?

Solution: $E_n = \dfrac{n^2h^2}{8mL^2}$; $n = 1, 2, 3, \ldots$

(a) For electron, L = 1Å = 10^{-10} m and mass of electron is 9.11×10^{-31} kg

$$E_1 = \frac{(1)^2(6.626\times10^{-34})^2}{8(9.11\times10^{-31}\text{kg})(10^{-10})^2}$$

$$= 6.03\times10^{-17}\,\text{J} = 37\text{eV}$$

$$E_2 = 4\times6.03\times10^{-17}\,\text{J} = 24\times10^{-17}\,\text{J} = 148\,\text{eV}$$

$$E_3 = 9\times6.03\times10^{-17}\,\text{J} = 54\times10^{-17}\,\text{J} = 333\,\text{eV}$$

(b) For grain of dust, m = 10^{-4} g = 10^{-7} kg and $L = 10^{-4}$ m

$$E_1 = \frac{(1)^2(6.626\times10^{-34})^2}{8(10^{-7})(10^{-4})^2}$$

$$E_2 = 4\times5.49\times10^{-53} = 22\times10^{-53}\,\text{J}$$

$$E_3 = 9\times5.49\times10^{-53} = 48.7\times10^{-53}\,\text{J}$$

Now the kinetic energy of the grain of dust is

$$K.E. = \frac{1}{2}mv^2 = \frac{1}{2}\times10^{-9}\times(10^5)^2 = 50\,\text{J}$$

The following may be inferred from the above calculations:

(i) For electron the energy levels in an infinite potential well are spaced apart and an electron cannot jump from one energy level to another only due to the available thermal energy which is very small(approximately 0.026 eV). Thus, energy levels are discrete and an electron can hop from lower to higher energy level only when energy is provided from outside. Thus, quantization exists for an electron trapped in an atomic sized well.

(ii) For a grain of dust the energy levels are very closely spaced, as the difference between successive energy levels is very small thus forming a continuum. The kinetic energy of grain is 50 J, which is large as compared to the difference between energy levels. Hence, the grain of dust can easily go from one energy level to another due to available energy. It can move through all these energy levels without an external supply of energy. Thus quantization of energy level is not at all significant in case of microscopic bodies.

Summary

- Quantum mechanics deals with the behaviour of matter and energy at atomic and subatomic scale.
- The quantum of energy or 'photon' is considered to have particle properties instead of wave properties. Most of the observed effects are due to the interaction between energy and matter.
- In Compton effect when a monochromatic beam of X-rays is scattered by a target, the scattered X-rays were found to contain two wavelengths—unmodified radiation, which is same as the incident radiation and modified radiation with higher wavelength. The difference between these two wavelengths called the Compton shift, depends only on the angle of scattering and is independent of the nature of the scatterer and the incident wavelength.
- When the incoming photons collide with the valence electrons, then it results in modified radiation and when it collides with the inner shell electrons, then unmodified radiation occurs in the scattered radiation.
- In the photoelectric effect, the interaction takes place between the incident photon and the bound electron whereas in Compton effect photon interacts with the free electron.

- De broglie proposed that material particles such as electrons and protons in motion must be accompanied by waves called matter waves. Davisson and Germer experimentally verified the existence of matter waves.
- The theoretical proof of Bohr's quantization condition, which states that in an atom only those orbits are permitted for which the angular momentum of electron is an integral multiple of ħ, is provided by the de Broglie's concept of matter waves.
- The uncertainty principle states that we can measure either the position or the momentum of a particle with any desired degree of accuracy (within the limits of experimental equipment), but it is impossible to measure both simultaneously—the position and momentum—with unlimited accuracy.
- If there is maximum uncertainty involved in the measurement of position Δxmax, then there is minimum uncertainty involved in the measurement of momentum Δpxmin and vice versa, as the product of uncertainties has to be a constant.
- Wave packet consists of groups of matter waves moving with slightly different wavelength or frequency (i.e. spread in wave velocity) such that they interfere constructively over a very small region and destructively elsewhere. The wave packet travels in a medium with a velocity called group velocity *vg*. The amplitude of wave packet is denoted by a complex mathematical function called wave function given by y(x, t) (in one dimension).
- In quantum mechanics, Schrodinger equations are used to find the behaviour of the subatomic particles trapped in atomic surroundings. Only well-behaved mathematical functions can become the solution of Schrodinger equation.
- Energy and position of an electron trapped in one-dimensional infinite potential well is discrete.
- According to quantum mechanics, it is possible that a particle with relatively less energy than the barrier potential can tunnel through the potential barrier. This phenomena is called tunnelling which finds application in solid-state lasers, light emitting diodes, tunnel diode, nucleons inside nucleus, tunnel diodes, quantum dots, quantum computing, and scanning tunnelling microscope.

Self-study Questions

1. Explain Compton Effect. Explain how wave theory of light failed to explain Compton Effect.
2. Write down the equations of energy and momentum conservation in Compton effect.
3. What is Compton shift? Mention clearly the assumptions made as well as discuss the limitations of Comption shift.
4. On the basis of quantum theory, explain the existence of modified and unmodified radiations in Compton Effect.
5. Why is Compton shift detectable only in X-rays region and not in visible range of light?
6. Why is it impossible for a free electron to completely absorb photon?
7. In Compton effect what happens when:
   (a) A photon collides with a free electron in the scattering block? Write down the equations of energy and momentum conservation.
   (b) photon collides with a bound electron?
8. Calculate the wavelengths of electron using (i) Bragg's law and (ii) De-Broglie's hypothesis using the data as obtained in the Davisson and Germer experiment. Explain the importance of the results obtained.
9. Describe in detail, the experiment that provides the validity of de-Broglie's hypothesis regarding the wave nature of matter.
10. What are matter waves? Why does a single monochromatic wave not represent a localized particle?
11. Explain the synthesis of wave packet.
12. Explain how de Broglie's hypothesis regarding wave nature of matter leads to Bohr's quantization condition for angular momentum of electron.
13. Why is the wave nature of matter not noticeable in our daily observations?
14. Write and explain Heisenberg's uncertainty principle. Why is it significant only for subatomic particles and not for heavy bodies?
15. Discuss the significance of uncertainty principle. Describe the thought experiment to show how it validates the uncertainty principle.

16. The uncertainty in the location of a particle is equal to its de Broglie wavelength. Show that the uncertainty in the velocity is equal to its velocity.
17. An electron moving along X-axis passes through a narrow slit. It is observed to develop Y-component of velocity. Explain how it leads to Heisenberg's uncertainty principle?
18. How does Heisenberg's uncertainty principle correct Bohr's First Postulate?
19. Define the terms: (i) phase velocity (ii) group velocity (iii) wave packet. Show that group velocity is equal to the velocity of particle.
20. What is wave function? When is ψ called well-behaved function? Explain.
21. Give the physical significance of wave function ψ. Explain in brief the mathematical conditions imposed on ψ.
22. Why is normalization of ψ required?
23. Using Schrödinger's time-dependent equation, obtain Schrödinger's time-independent equation.
24. Show that the energy of an electron confined in a 1D potential well of length 'L' and infinite depth is quantized. Is the electron trapped in a potential well allowed to take zero energy? Why?
25. Write down Schrödinger's time-dependent and time-independent wave equations for matter waves. Explain, why. (i) Wave function ψ must be single valued and a continuous function of position (ii) The integral of $|\psi^2|$ overall space must be equal to unity.
26. Why is quantization not noticed for macroscopic particles?
27. Show that the wave function for a particle confined in an infinite one-dimensional potential well of length l is given by $\psi n(x) = \sqrt{2/l \sin(n\pi x/l)}$. Hence, discuss the energy levels and their discreteness.
28. What is tunnelling? How does quantum mechanics explain it?
29. Explain in short the phenomenon of tunnelling that occurs when a beam of particles is incident on a potential barrier of finite width.
30. Write briefly on the following:
    (a) Quantum nature of radiation
    (b) Dual nature of matter
    (c) Heisenberg's uncertainty principle
    (d) Physical significance of wave function, ψ

Numerical Exercises

1. Suppose that an X-ray with a wavelength of 3.75 pm is fired at a stationary electron such that the scattering angle is 85°. What is the resulting energy of the electron?
   (Ans. 1.96×10^{-14} J or 1.23×10^5 eV)
2. During Compton collision with an electron, a photon of violet light ($\lambda = 400$ nm) is backward scattered through an angle 180°. How much energy is transferred to the electron in this collision?
   (Ans. 1.904×10^{-21} J or 1.19×10^{-2} eV)
3. Incident X-rays have a wavelength of 0.31 nm and are scattered by the free electrons in graphite. The scattering angle in is $\theta = 135.0°$. What is the magnitude of the momentum of (a) the incident photon and (b) the scattered photon?
   (Ans. h = 6.626×10^{-34}J.s and c = 2.998×10^8 m/s.) [(a) 21.23×10^{-25} kg.m/s (b) 21.11×10^{-25} kg. m/s]
4. A beam of gamma radiations having energy 500 keV is incident on an aluminium foil. Calculate (a) wavelength of scattered radiation at 60°. (b) energy of recoiled electron.
   [Ans. (a) 0.0372 A (b) 166.67 KeV])
5. A photon collides on the graphite target at right angles. Find the percentage change in the energy of the photon when (a) Gamma ray photon with wavelength 0.013 Å is used (b) visible photon of wavelength 6000 Å is used.
   [Ans. (a) 65.14 % (b) 4.05×10^{-4} %])
6. A bacterium (mass = 2×10^{-15} kg) in the blood is moving at 0.33 m/s. What is the de Broglie wavelength of this bacterium? (Ans. 10.04×10^{-19} m)
7. How fast does a proton have to move in order to have the same de Broglie wavelength as an electron moving at the speed of 4.5×10^6 m/s?
   (Ans. 2.45×10^3 m/s)
8. The de Broglie wavelength of a proton in a particle accelerator is 1.3×10^{-14} m. Determine the kinetic energy (in joules) of the proton.
   (Ans. 7.78×10^{-13} J)

9. The kinetic energy of a particle is equal to the energy of a photon. The particle moves at 5 % of the speed of light. Find the ratio of the photon wavelength to the de Broglie wavelength of the particle. (Ans. 20)
10. A particle has a speed of 1.2×10^6 m/s. Its de Broglie wavelength is 8.4×10^{-14} m. What is the mass of the particle? (Ans. 6.57×10^{-27} Kg)
11. A proton is accelerated through a potential difference of 1000 V. What is its de Broglie wavelength and kinetic energy? (Ans. 9.06×10^{-3} A and Ep = 1.6×10^{-16} J)
12. For electron and photon each having de Broglie wavelength as 1 A, compare their momentum and total energy.

    (Ans. Momentum will be the same, i.e., 6.63×10^{-34} kg m/s, $E_e = 2.42 \times 10^{-17}$ J, E_p = 12400 eV)
13. An electron is trapped within a sphere that has a diameter 10^{-15} m (about the size of the nucleus of an oxygen atom). What is the minimum uncertainty in the electron's momentum?

    (Ans. 1.04×10^{-19} kg m/s)
14. In the lungs there are tiny sacs of air, which are called alveoli. The average diameter of one of these sacs is 0.25 mm. Consider an oxygen molecule (mass = 5.3×10^{-26} kg) trapped within a sac. What is the minimum uncertainty in the velocity of this oxygen molecule? (Ans. 7.92×10^{-6} m/s)
15. Consider a line that is 2.5 m long. A moving object is somewhere along this line, but its position is not known. (a) Find the minimum uncertainty in the momentum of the object. Find the minimum uncertainty in the object's velocity, assuming that the object is (b) a golf ball (mass = 0.045 kg) and (c) an electron.

    [Ans. (a) 4.2×10^{-35} kg.m/s (b) 9.3×10^{-34} m/s, (c) 4.61×10^{-5} m/s)
16. Compare uncertainties in the velocities of an electron and proton confined in a box of equal dimensions. (Ans. 1.84×10^3)
17. Find the first three energy levels of (a) a marble of 10 g in a box of width 1 cm and (b) an electron in a box of width 1 A.

    [Ans. (a) 5.47×10^{-62} J, 2.18×10^{-61} J, 4.92×10^{-61} J (b) 6.01×10^{-18} J, 24.04×10^{-18} J, 54.36×10^{-18} J)
18. Find the first three energy levels of (a) a feather of 1 μg in a box of width 100 cm and (b) an electron in a box of width 0.1 A.

    [Ans. (a) 5.47×10^{-59} J, 2.18×10^{-58} J, 4.92×10^{-58} J (b) 3.75 keV, 15 keV, 33.75 keV]
19. An electron is confined to move in one-dimensional potential box of 0.01 nm side. If the eigen value of the excited state of electron is 20000 eV, find the order of the excited state.(2)
20. An electron of least energy 37.7 eV is confined in an infinite potential well. Calculate the width of the barrier through which it continuously moves along X-axis. (1 A)
21. A particle is moving in a one-dimensional infinite potential box of width 30Å. Calculate the probability of finding the particle within a small interval of 0.05Å at the centre of the box when it is in its state of least energy.

    (Ans. 3.33×10^{-3})
22. If an electron is confined in a one-dimensional box of 0.1 nm length, calculate the energy required to excite it from ground state to fifth excited state. (Ans. 905 eV)
23. The lowest energy of the particle in an infinite potential well of width 10A is 0.065 eV. Calculate the mass of the particle.

    (Ans. 52.83×10^{-31} kg)

Answers

Checkpoint 1
1(a), 2(a), 3(b), 4(a) 5(d), 6(b), 7(b), 8(a), 9 (c), 10(a)

Checkpoint 2
1(a), 2(c), 3(b), 4(d)

Checkpoint 3
1(c), 2(d), 3(a)

Checkpoint 4
1(b), 2(d), 3(b), 4(a)

Laser

4

Learning Objectives

After going through this chapter, you will be able to:

- Explain the characteristics of laser
- Understand how laser is different from ordinary source of light
- List three quantum processes which occur when energy interacts with matter
- Learn the basic requirements of laser, namely population inversion and optical resonator
- Understand the construction and working details of ruby laser, He–Ne laser, and semiconductor laser
- Highlight the applications of laser, viz. holography

4.0 Introduction

Laser is the acronym for '**L**ight **a**mplification by **s**timulated **e**mission of **r**adiation'. Lasers and laser systems find wide applications in many fields of science and technology due to their remarkable properties. Lasers are different from ordinary sources such as incandescent lamps or arc lamps, as laser produces radiation with a highly regular light field with exceptionally high degree of coherence, monochromaticity, and directionality. These special properties of laser are discussed in Section 4.1. The physics of processes resulting in coherent optical radiation in laser is discussed in Section 4.2.

The two requirements for obtaining laser are *population inversion* and *optical resonator*; these are discussed at length in Section 4.3. Some of the different types of lasers are explained in Section 4.4; and Section 4.5 explains one interesting application of laser.

4.1 Properties of Laser

The most remarkable features of laser are its directionality, high intensity, extraordinary monochromaticity, and high degree of coherence. Let us examine these characteristics in the following sections.

4.1.1 Directionality

If we want to produce a beam of light in a particular direction using a conventional light source, we put an aperture in front of the source. Lasers on the other hand, emit light in only one direction. The directionality of laser beams is usually expressed in terms of full angle beam divergence. This angle is twice the angle that the outer edge of the beam makes with the axis of the beam. The outer edge of the beam is defined as the point at which the strength of the beam has dropped to $1/e$ times its value at the centre 'I_0'. The beam usually propagates as a parallel beam for a distance of about d^2/λ which is called Rayleigh's range (Fig. 4.1).

After this distance, the beam begins to spread linearly with the distance because of the inherent diffraction phenomenon. The angular spread $\Delta\theta$ of the far field beam is related to the aperture diameter 'd' by

$$\Delta\theta = \frac{\lambda}{d}$$

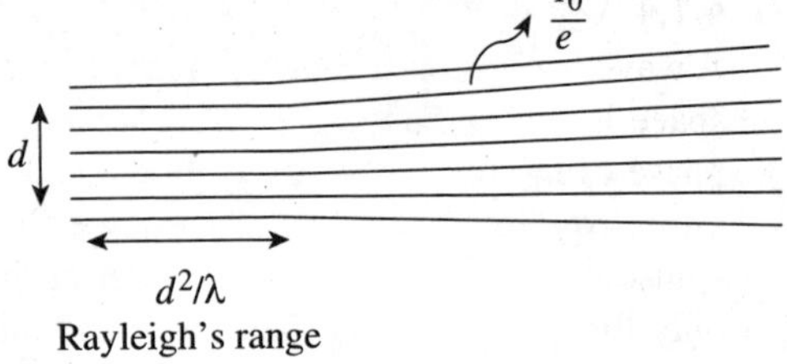

Fig. 4.1 Schematic showing directionality of laser beam

The typical values of '$\Delta\theta$' for the laser are usually less than 0.01 milliradian. This means that the beam spreads less than 0.01 mm for every metre it travels. Thus, as the spread is very less, the laser beam is highly directional; whereas for ordinary light the value of '$\Delta\theta$' is one kilo radians, that is, the beam will spread one kilometre laterally for every kilometre it travels. Thus, laser is a highly directional beam as compared to the ordinary light beam.

4.1.2 Intensity

An ordinary light source such as the light emitted by the sun, an incandescent bulb, a CFL (compact fluorescent lamp) or a candle emits energy in all directions. Hence, the energy is uniformly distributed in all directions. On the contrary, laser light forms a narrow beam and its energy is concentrated in a small region. As the energy is confined to a very small area it accounts for the strong intensity of lasers. Thus, even a low wattage laser, viz. 0.5 milliwatt would appear many thousand times more intense than a 1000 watt ordinary lamp. Thus, the high intensity of laser beam is a direct consequence of its directionality property.

4.1.3 Monochromaticity

When the light emitted from a source has only one frequency of oscillation, then such light is said to be purely monochromatic and the source is called a perfectly monochromatic source. However, in reality no light source, not even laser can produce absolute monochromatic light. Usually light emitted by any source consists of a band of frequencies closely spaced around a central frequency, ν_0 (Fig. 4.2). The band of frequencies $\Delta\nu$ is called the *bandwidth* and the spread in wavelength $\Delta\lambda$ is called the *line width*. The degree of monochromaticity of light can be estimated by the degree of non-monochromaticity of light, which is defined by

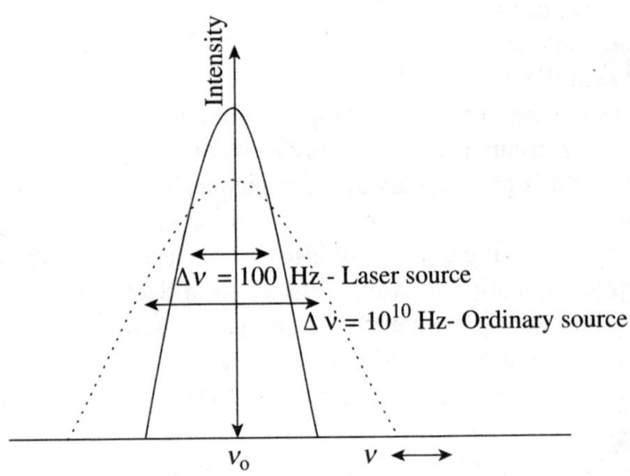

Fig. 4.2 Normal and laser bandwidth

$$\xi = \frac{\Delta\nu}{\nu_0}$$

Absolute monochromaticity is not possible as for it the bandwidth, $\Delta\nu$ should be zero. The light from conventional sources (shown by dotted curve in Fig. 4.2) has large bandwidths of the order of 10^{10} Hz or more; whereas light from lasers is almost a perfect monochromatic source with a very small bandwidth about 100 Hz. Thus, we can consider that the laser light has a high degree of monochromaticity and hence is called spectrally pure source as it emits at lower wavelengths.

4.1.4 Coherence

A wave that appears to be a pure sine wave for an infinitely large period of time or in an infinitely extended space is said to be a *perfectly coherent wave*. In such a wave, there is a definite relationship between the phase of the wave at a given time and at a certain time later, or at a given point and at a certain distance away. No real light source emits a perfectly coherent wave, that is, the waves which continue to be pure sinusoidal for infinite time and over an infinite distance. In practice, light waves are pure sine waves but only for a limited period of time or in limited space. Such waves are called *partially coherent waves*. There are two criteria of coherence—criteria of time and criteria of space. This gives rise to temporal coherence and spatial coherence.

Temporal Coherence

In this type of coherence the correlation between the fields at a point as shown in Fig. 4.3, at two different instances of time is considered; that is, the relation between $E(x, y, z, t_1)$ and $E(x, y, z, t_2)$ is considered. Now during the period normally covered by observations which is of the order of a few microseconds, if the phase difference between these two fields is constant, then the wave is said to have temporal coherence. Temporal coherence is also termed as *longitudinal coherence*. If the phase difference changes many times and in an irregular way during the shortest period of observation, the wave is said to be incoherent.

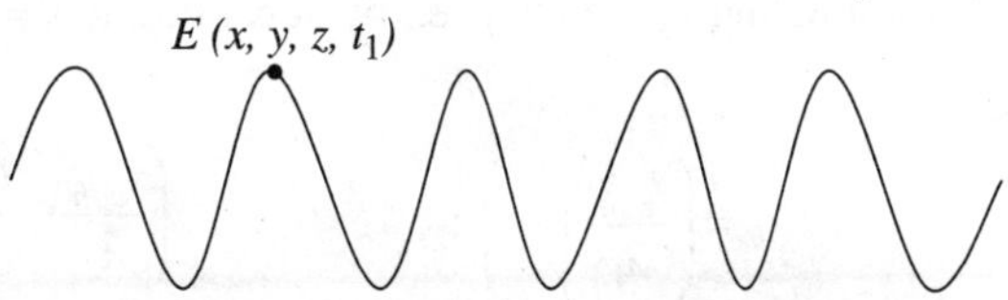

Fig. 4.3 Illustration of temporal coherence

The property of temporal coherence is closely related to monochromaticity. The actual sources of light are not perfectly monochromatic. On the other hand, they consist of a sequence of harmonic wave trains of finite length, each separated from the other by a discontinuous change in phase. The discontinuities in phase of the two adjacent wave trains reflect the randomness of the emission of light waves by the different atoms. Accordingly, the light wave trains emitted from each source can be characterized by an average time τ, during which the waves remain sinusoidal. This time period corresponds to the lifetime of an excited state of atoms and is called *coherent time*. The distance l_{coh} for which the field remains sinusoidal is given by

$$l_{coh} = \tau c \tag{4.1}$$

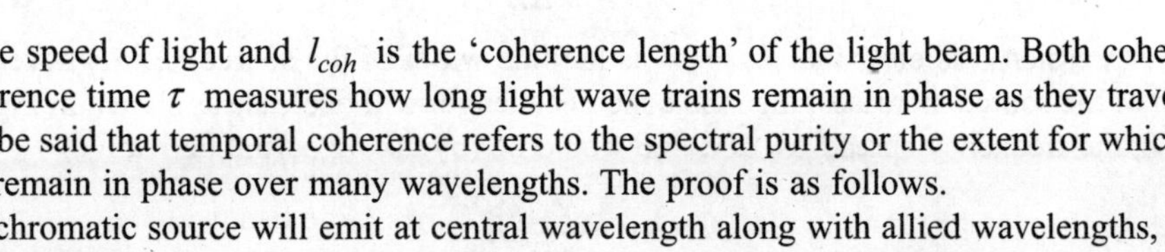

Fig. 4.4 Schematic of incoherent wave

Here, 'c' is the speed of light and l_{coh} is the 'coherence length' of the light beam. Both coherence length l_{coh} and coherence time τ measures how long light wave trains remain in phase as they travel in space.

It can also be said that temporal coherence refers to the spectral purity or the extent for which two waves of laser light remain in phase over many wavelengths. The proof is as follows.

Any monochromatic source will emit at central wavelength along with allied wavelengths, that is, there will be a finite spectral width. This means that it corresponds to a continuous distribution of wavelength in some narrow interval between λ and $\lambda + \Delta\lambda$. The interference pattern remains stable when all the waves $\lambda_1, \lambda_2, \lambda_3, \ldots$ emitted from the source produce the same interference pattern, which overlaps and ultimately gives a stable pattern. This is obvious from the fact that for every spectral line, the interference pattern will eventually disappear when the path difference between the interfering beams is gradually increased. Let us adopt the criteria that when the path difference between the interfering beams become equal to the coherence length l_{coh}, the interference pattern produced by these two closely spaced wavelengths λ_1 and λ_2 are completely out of step at the centre (i.e., bright band of λ_1 coincides with dark band of λ_2). Then we can write

$$l_{coh} = n\lambda_1 = (2n+1)\frac{\lambda_2}{2} \tag{4.2}$$

that is, for one it will be an even multiple of half wavelength $\lambda_1/2$ and for the other it will be an odd multiple of $\lambda_2/2$.

Eliminating *n*, we get (i.e., by substituting $n = \frac{l_{coh}}{\lambda_1}$)

$$l_{coh} = \left(\frac{l_{coh}}{\lambda_1} + \frac{1}{2}\right)\lambda_2 \quad \text{or} \quad \left(\frac{l_{coh}}{\lambda_2} - \frac{l_{coh}}{\lambda_1}\right) = \frac{1}{2} \tag{4.3}$$

$$\text{or} \quad l_{coh}\left(\frac{\lambda_1 - \lambda_2}{\lambda_1\lambda_2}\right) = \frac{1}{2} \quad \text{or} \quad l_{coh} = \frac{\lambda_1\lambda_2}{2(\lambda_1 - \lambda_2)} \tag{4.4}$$

If instead of two discrete wavelengths λ_1 and λ_2, the beam consists of all wavelengths lying between λ_1 and λ_2, then the pattern would disappear if

$$l_{coh} = \frac{\lambda_1\lambda_2}{(\lambda_1 - \lambda_2)}$$

Now by substituting mean wavelength λ for λ_1 and λ_2 and $\Delta\lambda$ for $(\lambda_1 - \lambda_2)$, we get

$$l_{coh} \cong \frac{\lambda^2}{\Delta\lambda} \quad \text{or} \quad \Delta\lambda \cong \frac{\lambda^2}{l_{coh}} \tag{4.5}$$

Thus, the fringes become indistinct when the path difference exceeds l_{coh}. So we can conclude that the spectral line (of mean wavelength λ) has a wavelength spread (width) given by

$$\Delta\lambda \cong \frac{\lambda^2}{l_{coh}}$$

The coherence length of the source can be estimated by knowing its line width. Therefore,

$$l_{coh} \cong \frac{\lambda^2}{\Delta\lambda}$$

From this expression it is clear that $l_{coh} \neq \infty$ as the line width of the source, $\Delta\lambda$ is never zero. Thus if the line width of the source increases, then the coherence length will decrease as shown in Fig. 4.5. Further, since $\nu = c/\lambda$, the frequency spread $\Delta\nu$ of a line would be (by taking partial derivative)

$$\Delta\nu = \frac{c}{\lambda^2}\Delta\lambda \tag{4.6}$$

By substituting Eq. (4.5) we get $\Delta\nu \cong \dfrac{c}{l_{coh}}$ (4.7)

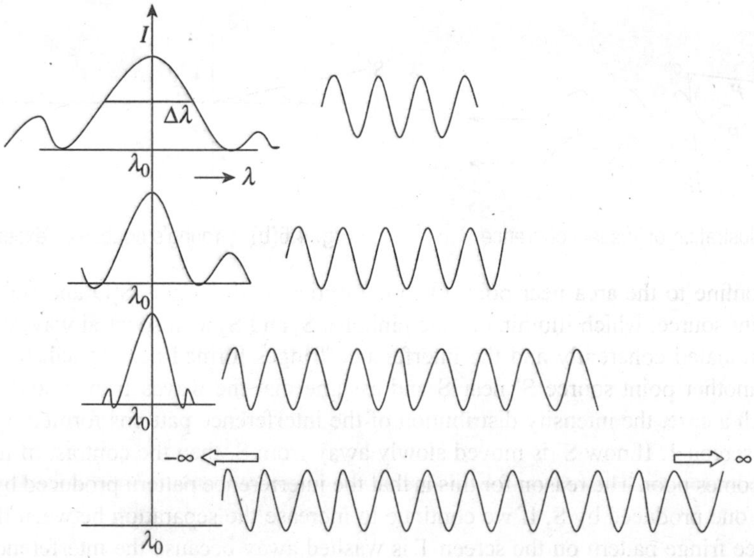

Fig. 4.5 Diagram showing dependence of wave train length on the linewidth

We know that $\tau \cong \frac{l_{coh}}{c}$, where τ is the coherence time. Therefore,

$$\Delta\nu \cong \frac{1}{\tau} \tag{4.8}$$

Thus, the frequency spread of the spectral line is of the order of the inverse of the coherence time. Subsequently, for a perfectly sharp monochromatic light the bandwidth $\Delta\nu = 0$; hence the coherence time should be infinite which is impossible. Thus, the monochromaticity of a light beam is a measure of its temporal coherence.

Spatial Coherence

A beam is said to possess spatial coherence, if the phase difference of the waves crossing the two points P and P' as shown in Fig. 4.6(a) on a plane perpendicular to the direction of propagation of the beam is independent of time. Hence, spatial coherence is also termed as *transverse coherence*. For understanding the concept of spatial coherence, let us consider two identical waves travelling along the same direction, but are at a distance from each other. Here '*abcd*' is a plane perpendicular to the waves and P and P' are two points on this wave as shown in Fig. 4.6(a). The beam is said to possess spatial coherence if the phase difference of the waves crossing P and P' at any instant of time is always constant or zero.

Let us consider that the phase difference between the waves crossing P and P' at any instant t_1 is ϕ_1 and at a later instant t_2, it is 'is ϕ_2'. If $\phi_1 = \phi_2$, then in such a case the beam possesses spatial coherence. Thus, spatial coherence measures the area over which there exists a definite phase relationship between the waves emitted by the source.

Another experiment which clearly explains the concept of spatial coherence, is Young's double slit experiment as shown in Fig. 4.6(b). In the figure, S represents a source placed in front of a screen with two holes S_1 and S_2 and the interference pattern between the waves emanating from S_1 and S_2 is observed on the screen T.

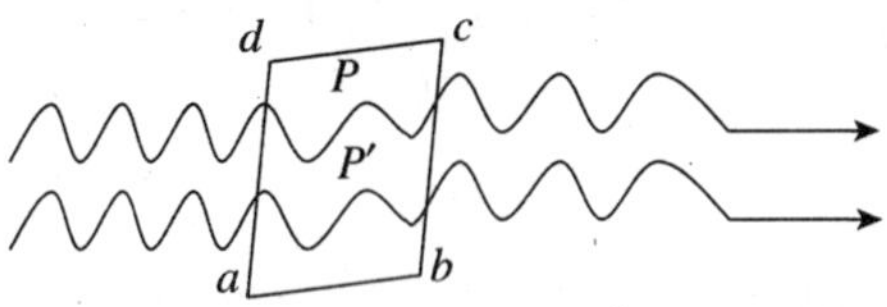

Fig. 4.6(a) Illustration of spatial coherence

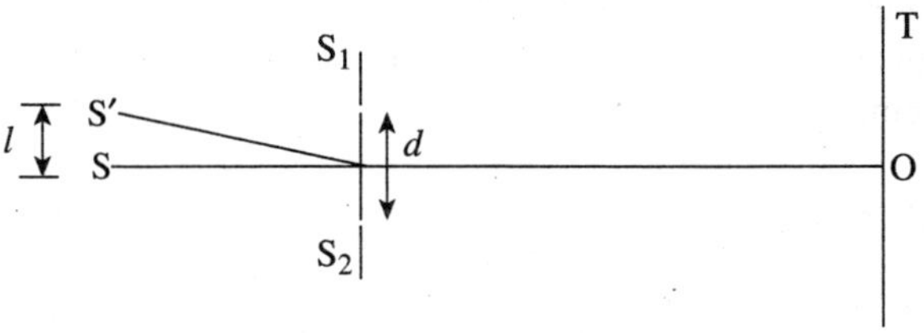

Fig. 4.6(b) Young's double-slit experiment

Here we will confine to the area near point O as the optical path lengths S_1O and S_2O are equal. If we consider S as a point source, which illuminates the pinholes S_1 and S_2 with spectral waves, then the holes S_1 and S_2 will be illuminated coherently and the interference fringes formed near O will be of good contrast. Now if we place another point source S′ near S and assume that the waves from S and S′ have no phase relationship: in such a case, the intensity distribution of the interference patterns formed by S and S′ will be superposed on the screen T. If now S′ is moved slowly away from S, then the contrast of the fringes formed on the screen T becomes poor. The reason for this is that the interference pattern produced by S′ shifts slightly with respect to the one produced by S. If we continue to increase the separation between these two sources, then the interference fringe pattern on the screen T is washed away because the interference maximum produced by S′ falls on the minimum by S.

Thus, if the source is a point source, it produces ideally spherical wavefronts as shown in Fig 4.6(c) and hence maintains high spatial coherence. Points S_1 and S_2 are spatially coherent.

But now if an extended source is used wherein different points in the source emit waves, the wavefronts are far from spherical and thus lead to *spatial incoherence* as shown in Fig. 4.6(d).

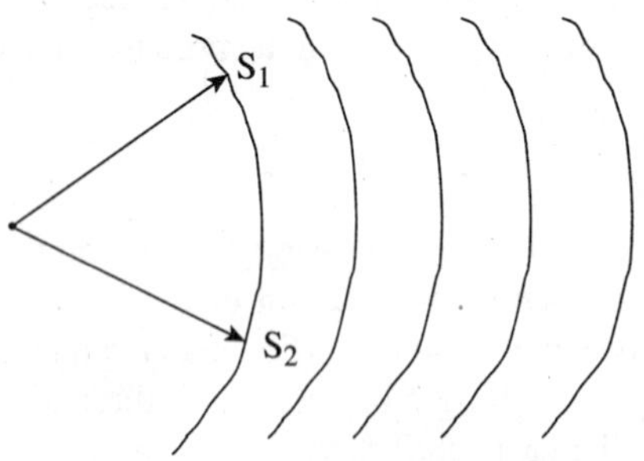

Fig. 4.6(c) Light waves from point source

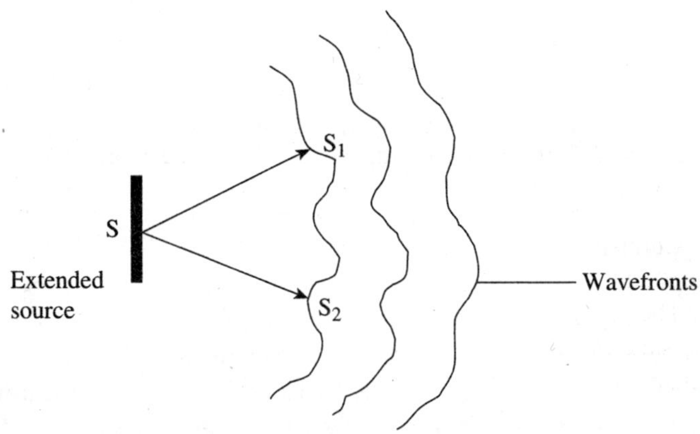

Fig. 4.6(d) Light waves from extended source

Thus, in Young's double-slit experiment, the area of the wavefront over which the pinhole can be moved so that the fringe pattern is still seen on the screen is known as *coherent area* of the light wave and is a measure of the spatial coherence of the wave. Therefore for an ordinary extended source where the light beam originates from different parts of the source, spatially coherent beam of light can be obtained only by passing light through the pinhole. In contrast, laser beam is highly spatially coherent.

Thus, the phenomenon of interference is a manifestation of coherence. The number of interference fringes will be larger, when the smaller the line width $(\Delta \nu)$ or longer the temporal coherence of the interfering waves. The degree of contrast of the interference fringes obtained is a measure of the degree of spatial coherence. Thus, higher the degree of spatial coherence of the source, smaller is the divergence of the light beam and consequently, accounts for its high directionality. Thus, light emerging from a laser source is both temporally and spatially coherent.

4.2 Three Quantum Processes

Let us consider a medium consisting of identical atoms capable of being excited from energy level 1 to energy level 2 by absorption of photons. Let the levels be denoted as E_1 and E_2 and their populations be N_1 and N_2 respectively. Let the atoms be in thermal equilibrium. In equilibrium conditions, the number of upward atomic transitions must be equal to the number of downward transitions. Thus, no net photons are generated or lost. However, when the atoms are subjected to an external light of photon density $\rho(\nu)$, the following three processes occur in the medium.

4.2.1 Stimulated Absorption

An atom can exist in a number of quantized energy states characterized by integral numbers. If an atom is in the lower state E_1, it can be raised to a higher state E_2 by absorption of photons of frequency ν such that

$$\nu = \frac{E_2 - E_1}{h}$$

This process is known as *stimulated* (or *induced*) *absorption* of radiation as depicted in Fig. 4.7.

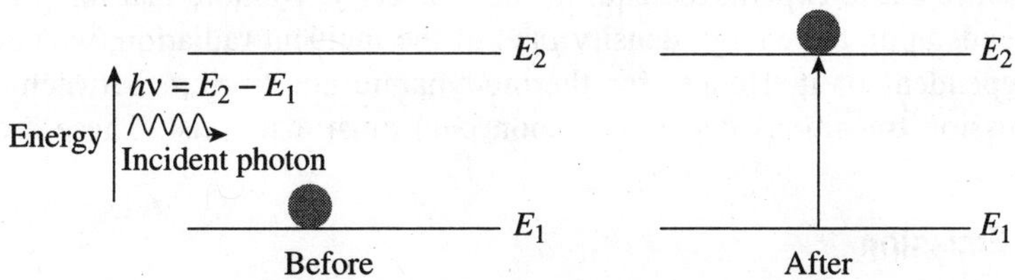

Fig. 4.7 Stimulated absorption

Corresponding to each absorption transition, one photon disappears from the incident light field and one atom adds to the population at the excited energy level E_2.

The probability of occurrence per unit time of this absorption transition depends upon the population of the state E_1 denoted as N_1 and the rate at which photons of frequency ν fall on the atom, that is, energy density $\rho(\nu)$ of the incident radiation. Thus, the probability of absorption is given by

$$P_{12} = B_{12}\, \rho(\nu) N_1 \tag{4.9}$$

Here B_{12} is the proportionality constant, which is known as *Einstein coefficient* for induced absorption.

4.2.2 Spontaneous Emission

Let us now consider the converse process of absorption. Suppose an atom is in a higher (excited state) state E_2. Observations show that its lifetime in this state is usually small ($\approx 10^{-8}$ s) and it, on its own accord jumps to a lower energy state E_1, emitting photons of frequency ν as shown in Fig. 4.8. This is 'spontaneous emission' of radiation. If there is an assembly of atoms in the excited state, then the radiation emitted spontaneously by each atom has a random direction and random phase and is therefore incoherent from one atom to another.

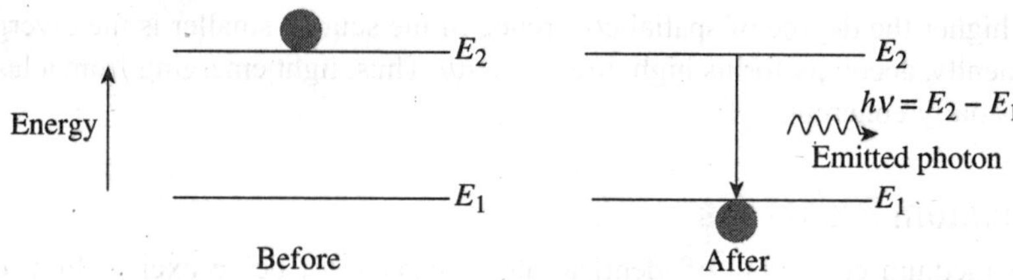

Fig. 4.8 Spontaneous emission

The probability of spontaneous emission 2 $\rightarrow$ 1 is determined only by the population of state E_2 denoted as N_2 which is given by

$$P_{21} = A_{21}N_2 \tag{4.10}$$

where A_{21} is Einstein's coefficient of spontaneous emission of radiation.

The important features of this process are:

1. It cannot be controlled from outside.
2. The light emitted due to this process is incoherent, that is, the emitted photons have different initial phases, planes of polarization, and directions of propagation.
3. The emitted light spreads in all directions around the source. The light intensity goes on decreasing rapidly with distance from the source.

Einstein studied these two processes—*stimulated absorption* and *spontaneous emission* and found that these are not sufficient to explain the equilibrium correctly. We note that the probability of absorption transition depends upon the energy density $\rho(\nu)$ of the incident radiation, whereas the spontaneous emissions are independent of it. Hence, for thermodynamic equilibrium between the atoms and the radiation field, emission transitions depending upon $\rho(\nu)$ must also exist. These are called *stimulated emission transitions*.

4.2.3 Stimulated Emission

According to Einstein, an atom in an excited state may, under the influence of the electromagnetic field of photons of frequency ν incident upon it, jump to a lower energy state, emitting an additional photon of the same frequency as shown in Fig 4.9. Thus now two photons instead of one, move on. This is known as *stimulated* (or *induced*) *emission of radiation*.

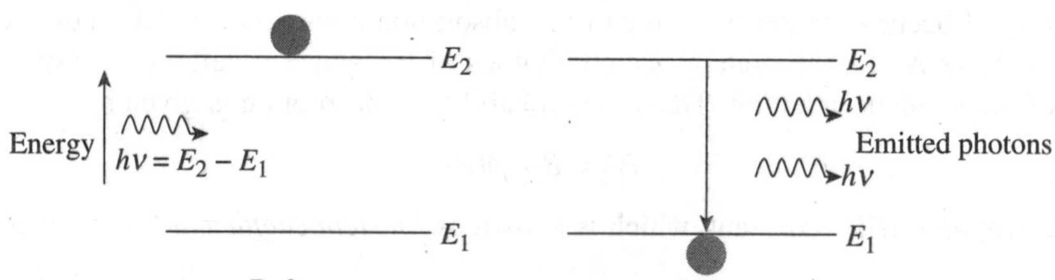

Fig. 4.9 Stimulated emission process

The direction of propagation, energy, phase, and state of polarization of the emitted photon is exactly the same as that of the stimulating photon. In other words, the stimulated radiation is completely coherent with the stimulating radiation. As a result of this process, radiation passing through an assembly of atoms is amplified; for example, for one photon hitting an atom in an excited state, two photons emerge that are exactly identical. These two photons stimulate two more excited atoms in their path and produce four photons, which are in phase and travel along the same direction. These photons can in turn stimulate four excited atoms and generate eight photons, and so on. Thus, this process results in the multiplication of photons, which are all identical giving an intense coherent beam as output.

The probability of stimulated emission transition $E_2 \rightarrow E_1$ is proportional to energy density and N_2, population of energy level E_2. Therefore

$$P_{21} = B_{21}\,\rho(\nu)\,N_2 \tag{4.11}$$

where B_{21} is Einstein's coefficient for stimulated emission.

The important features of simulated emission are:

1. This process can be controlled from outside the medium.
2. The photons emitted in any one direction have exactly the same frequency, phase, and plane of polarization as those of the incident photons.
3. The light emitted in this process along any one direction is highly directional, coherent, and monochromatic.

4.3 Principle of Laser

Laser is a strong, monochromatic, collimated, and highly coherent beam of light. When a substance is irradiated with light of frequency ν which coincides with one of the frequencies of the atom of the substance, three processes takes place:

(i) an (stimulated) absorption transition of atom from lower energy level E_1 to higher energy level E_2, for $\nu = (E_2 - E_1)/h$

(ii) spontaneous emission transition from E_2 to E_1, and

(iii) stimulated emission (triggered) transition from E_2 to E_1.

The key to laser action is the existence of stimulated emission. At thermal equilibrium, the ratio of the stimulated to spontaneous emission rate is generally very small and the contribution of stimulated emission is negligible.

$$\frac{\text{Stimulated emission rate}}{\text{Spontaneous emission rate}} = \frac{B_{21} N_2\,\rho(\nu)}{A_{21}\,N_2} = \frac{B_{21}}{A_{21}}\rho(\nu) \tag{4.12}$$

Here, N_2 and $\rho(\nu)$ are the population of energy level E_2 and photon density; B_{21} and A_{21} are the Einstein's coefficients. The above equation indicates that the way to enhance the stimulated emission over spontaneous emission is to have a very large photon field energy density $\rho(\nu)$. In laser, this is encouraged by providing an optical resonant cavity in which photon density can build up to a large value through multiple internal reflections at a certain frequency.

Similarly, to obtain more stimulated emission than absorption we must have $N_2 > N_1$, as

$$\frac{\text{Stimulated emission rate}}{\text{Stimulated absorption rate}} = \frac{B_{21} N_2\,\rho(\nu)}{B_{12}\,N_1\,\rho(\nu)} = \frac{B_{21}\,N_2}{B_{12}\,N_1} \tag{4.13}$$

Thus, if stimulated emission is to dominate over absorption of photons from the radiation field, we must have a way of maintaining more atoms in the upper level than in the lower level. But this is quite unnatural as

$$\frac{N_2}{N_1} = \exp\left[-\frac{(E_2 - E_1)}{kT}\right] \tag{4.14}$$

The negative exponent in Eq. (4.14) indicates that at equilibrium, $N_2 << N_1$, that is, most atoms are in lower energy levels as expected. The condition $N_2 > N_1$ is called *population inversion*. It is also referred to as a condition of negative temperature as the ratio N_2 / N_1 could be larger than unity only if the temperatures were negative.

In summary, Eqs (4.12) and (4.13) indicate that if the photon density is to be built up by predominating stimulated emission over spontaneous emission and absorptions, two requirements be met. We must provide (i) a means of obtaining population inversion and (ii) the optical resonant cavity to encourage photon field to build up. Before going into the details of these two, next we will learn about how to create a state of population inversion.

4.3.1 Population Inversion

Under ordinary conditions when a material is in thermal equilibrium, N_2, the number of atoms in higher energy state E_2, is considerably smaller than N_1, the number of atoms in lower energy state E_1, that is, $N_2 < N_1$. Therefore, there is very little stimulated emission compared to absorption. Now if, by some means a larger number of atoms is made available in the higher energy state, then we can increase the rate of stimulated emission. This situation in which there are more number of atoms in the higher energy state that in the lower state is known as *population inversion* (Fig. 4.10). When this is achieved, the photons in the light beam passing through the atoms go on multiplying by repeated stimulated emission. Hence, a strong coherent beam of light emerges from the material.

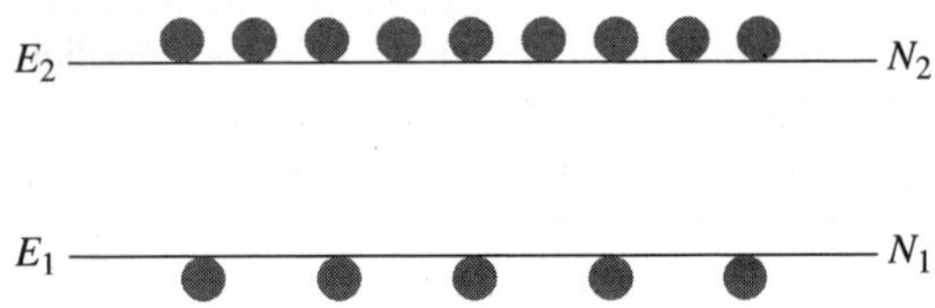

Fig. 4.10 Population inversion

In order to achieve and maintain a state of population inversion, it is necessary that atoms must be continuously raised from the lower energy level to the excited level. Hence, energy must be supplied somehow to the laser medium, so that the atoms from the lower level are promoted to the excited level, creating more population at the excited level than at the lower energy level. The process by which atoms are raised from the lower level to the upper level is called *pumping*.

Active Medium

A system in which population inversion is achieved is called an *active system* or *medium*. The active medium can be solid, liquid, or gas. The atoms or ions or molecules of active medium are called *active centres*. The active centres, that is, atoms in the excited states usually have a short lifetime. This means that they stay in the excited state only for a short span of time (10^{-9} s) and fall back to lower states through spontaneous transition. Thus, atoms do not stay at such excited states long enough to be stimulated to emit their energy. Hence, even though atoms are continuously pumped to the excited state, they come down immediately creating a state where the number of atoms in the lower energy levels always exceeds that in the upper energy levels. Due to this a state of population inversion cannot be established. For realizing or establishing a state of population inversion, the excited atoms should wait at the excited level for a little longer duration of time ($\approx 10^{-3}$ s), so that atoms start accumulating and population inversion can be obtained. The energy levels which have lifetime of about 10^{-6} to 10^{-3} s are called *metastable states*. Metastable states are a requisite for obtaining laser. Without metastable states population inversion cannot be realized and hence laser action is not possible. Thus, materials with metastable states are suitable for obtaining laser.

Pumping Methods

As explained earlier, the process of achieving population inversion is known as 'pumping' of atoms. There are various types of pumping processes such as optical, electrical, chemical pumping, and so on. The *optical pumping* method is employed usually for solids, as solids have broad energy bands to accommodate the atoms excited by a broad light source. Light sources such as an arc lamp, a flash discharge tube, or spark gaps are used as pump source.

Electrical pumping is used only in case of laser materials that can conduct electricity without destroying the laser activity. This is usually employed for gases. An electric current flowing through the gas excites the atom to the excited level creating a state of population inversion. *Chemical pumping* is used for liquids where exothermic reactions raise the active centres to the excited state. Some of the other methods of pumping include heat pumping and gas dynamics pumping.

Pumping Schemes

The medium in which we want to obtain laser has the active centres as the atoms, ions, or molecules. These active centres often exhibit a complicated system of energy levels. However, from all these complicated energy structures, the actual pumping schemes are obtained as a few rather simple diagrams, which correctly depict the pumping processes. These diagrams involve two, three, or four levels only. The three-level and four-level schemes are important and are widely employed. Let us discuss them.

Two-level pumping scheme The two-level pumping scheme is as shown in Fig. 4.11. For population inversion, the population of upper level should be built up faster than the rate at which it decays. Thus, pumping radiation is used to pump atoms from E_1 to E_2. But the pump radiation not only excites the ground state atoms, but it also works as trigger or stimulus for the excited atoms and induces a downward transition. This means that the pumping operation simultaneously populates and depopulates the upper level. Hence, population inversion cannot be attained in a two-level scheme. All that we may achieve is a system of equally populated levels. This is due to the overlap of pump transition levels and lasing transition levels.

Pump *transitions* are those which occur between the ground level and the upper level, whereas the transition between two levels that generate stimulated emission is called *lasing transition.* The lower lasing level and pump level is same which is ground state. Thus, here E_1 is the lower pump and lasing level, whereas E_2 is the upper lasing and pump level. Thus, such sharing of functions is quite disadvantageous in the sense that the lower lasing level should be less populated, but as it is the lower pump level and ground state, it is always densely populated. Hence, two-level pumping scheme is not suitable for producing laser.

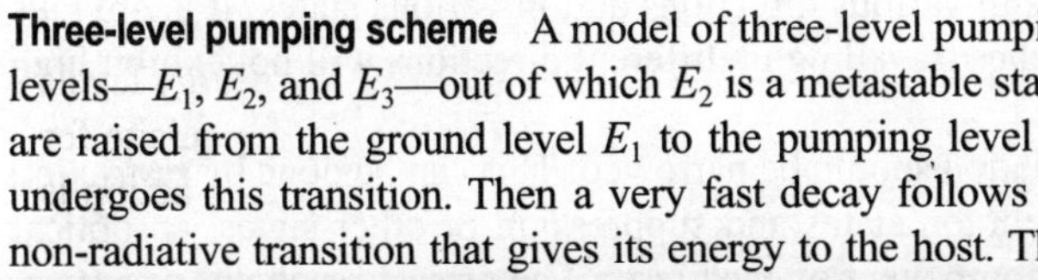

Fig. 4.11 Two-level pumping scheme

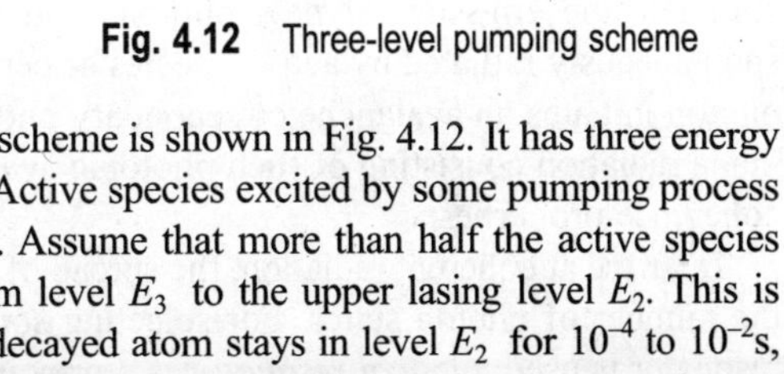

Fig. 4.12 Three-level pumping scheme

Three-level pumping scheme A model of three-level pumping scheme is shown in Fig. 4.12. It has three energy levels—E_1, E_2, and E_3—out of which E_2 is a metastable state. Active species excited by some pumping process are raised from the ground level E_1 to the pumping level E_3. Assume that more than half the active species undergoes this transition. Then a very fast decay follows from level E_3 to the upper lasing level E_2. This is non-radiative transition that gives its energy to the host. The decayed atom stays in level E_2 for 10^{-4} to 10^{-2}s, as this is a metastable state. Now as in this level there are more active centres as compared to the ground state, population inversion takes place. In this scheme, the ground state is simultaneously the lower lasing level. Such sharing of function is a disadvantage as ground state is always heavily populated instead of being depopulated or empty. To depopulate the ground state to about less than half its population, we need a very powerful pumping source. Thus, the disadvantage of the three-level pumping scheme is the requirement of a very intense pump source. But in this scheme also, the entire incident energy is not utilized in lifting atoms from the ground state to the upper level. Moreover, some energy is dissipated as heat, so cooling arrangement will also be required.

Thus, pump source is switched on for a short duration to enable removal of the heat dissipated as fast as possible; else the host material will melt off. The three-level scheme can produce light only in pulses as once the stimulated emission starts, it depopulates the upper lasing level very fast and the population of ground state increases quickly. As a result, population inversion ends and so some time is required to rebuild the population inversion. Hence, three-level lasers are operated in pulsed mode.

Four-level pumping scheme Consider the four-level pumping scheme as shown in Fig. 4.13. In this scheme, the terminal or lower lasing level E_2 is well above the ground state such that $(E_2 - E_1) >> kT$. It guarantees that at room temperature, the room temperature energy (kT) does not excite the atom from E_1 to E_2. Thus, this level (E_2) is virtually always empty. The pump energy elevates the atoms from E_1 to E_4. The atoms then drop spontaneously to a metastable upper laser level E_3. As the terminal lasing level E_2 is almost vacant, population inversion between the states E_3 and E_2 is quickly established.

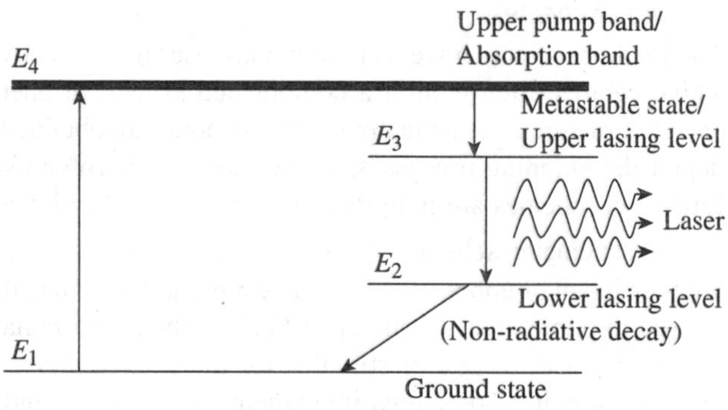

Fig. 4.13 Four-level pumping scheme

When atoms descend from E_3 to E_2 spontaneously, the emitted photons work as a trigger and initiate a chain of stimulated emissions resulting in laser. After emitting laser, atoms shift to level E_2. From there, the atoms finally go to the ground state E_1 by losing the remaining energy by a non-radiative transition. Thus, atoms are now available for excitation. In this scheme E_3 is a metastable state, which is an upper lasing level, whereas the lower lasing level is neither a metastable nor the ground state and so is virtually vacant. Hence, even if a few atoms are pumped to the upper lasing level, population inversion is achieved. Thus, to produce laser this pumping scheme requires less pumping energy as compared to the three-level pumping scheme. This is a major advantage of the four-level pumping scheme. The four-level pumping can be operated in a continuous wave (cw) mode as E_2 not being in metastable state is steadily depleted which helps sustain the population inversion.

4.3.2 Optical Resonator

We have learnt already that a light beam incident on an active medium can be amplified, because stimulated emission dominates over absorption in an inverted medium. In laser the role of primary photons, which stimulate the emission of new photons and thereby initiate laser oscillation, is played by the photons spontaneously radiated by active species or centres. Thus, in the active medium each spontaneously emitted photon initiates an avalanche of secondary photons in various directions and in various states. It is obvious that a radiation consisting of such photonic avalanches travelling in different directions will not exhibit high coherence properties.

To arrive at coherent radiation, the spread of radiation should be narrowed. This can be done by restricting the number of photon states. For selecting some photon states and suppression of other states, an optical resonator is used. *Optical resonator* is a principal component of each laser. The optical resonator or optical resonant cavity consists of a pair of mirrors set on an optic axis, which defines the direction of laser beam. The active material is placed in between these mirrors. One of the mirrors of resonant cavity is made semi-transparent and serves as an output element passing the light out of the resonator.

The successive stages of action of an optical resonator are depicted in Fig. 4.14. To begin with, the active centres in the laser medium are in the ground state as shown in Fig. 4.14(a).

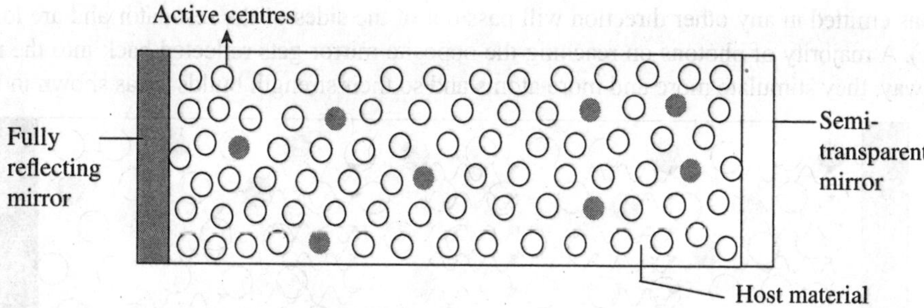

Fig. 4.14(a) Non-excited state

Through suitable pumping mechanism, the centres are taken into the state of population inversion as shown in Fig. 4.14(b). Now, some of the atoms in the excited state decay spontaneously, emitting photons in various directions. Each spontaneously emitted photon while travelling through the active medium triggers many stimulated transitions along the direction of its propagation.

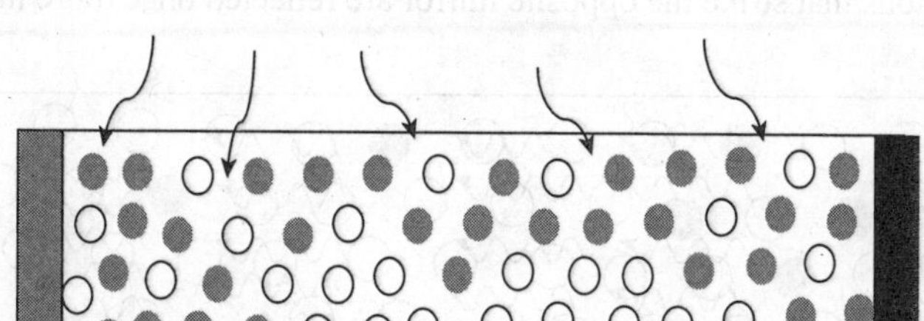

Fig. 4.14(b) Optical pumping

However, many of the photons leave the medium without reinforcing their strength. In the absence of end mirrors, the net effect would have been that incoherent light is produced without a specific direction. But in the presence of a mirror, the photon propagating along the optic axis of the pair of mirrors gets reflected and again travels through the medium and thus the number gets multiplied. Eventually the light propagating along the axis gets amplified.

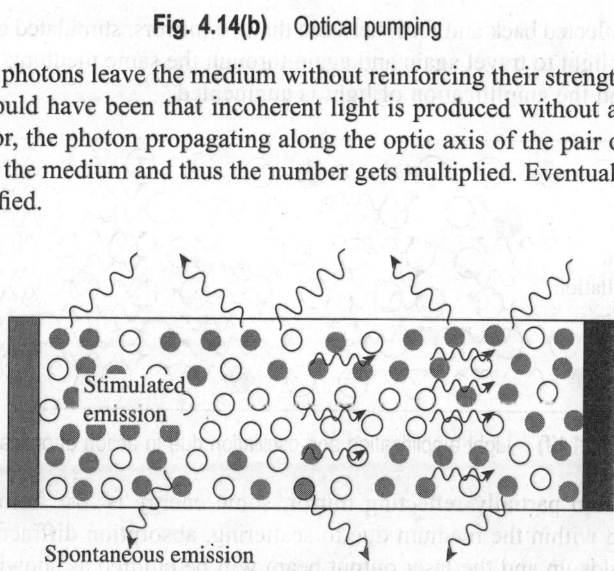

Fig. 4.14(c) Stimulated emission

The photons emitted in any other direction will pass out of the sides of the resonator and are lost, as shown in Fig. 4.14(c). A majority of photons on reaching the opposite mirror gets reflected back into the medium and thus on their way, they stimulate more and more atoms and so their strength builds up as shown in Fig. 4.14(d).

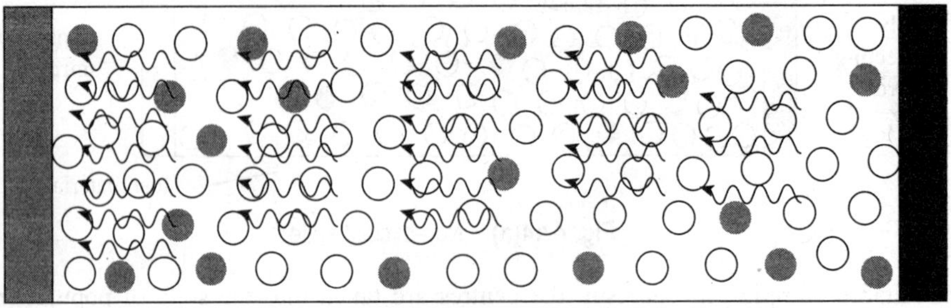

Fig. 4.14(d) Optical feedback

Thus, the mirrors provide positive feedback of light/photons in the medium, thereby enhancing the energy density of light. The photons that strike the opposite mirror are reflected once more into the medium as shown in Fig. 4.14(e).

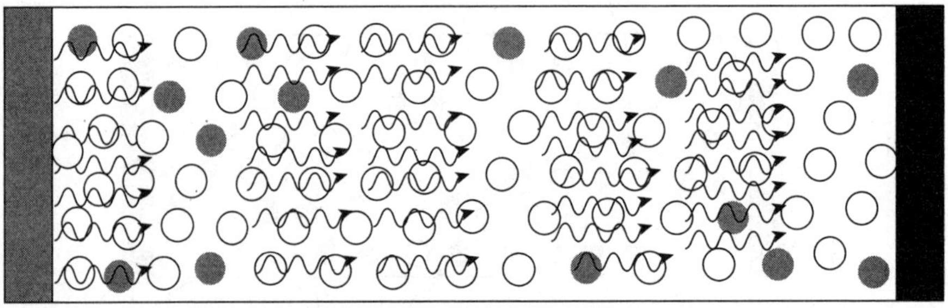

Fig. 4.14(e) Light amplification

As the photons are reflected back and forth between the two mirrors, stimulated emission sharply increases. Hence, by allowing the light to travel again and again through the same medium, the effective length of the medium is increased and the amplification of light is augmented.

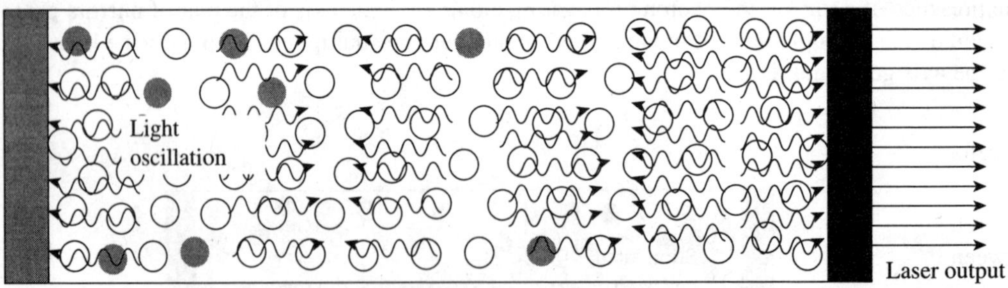

Fig. 4. 14(f) Light amplification and oscillation due to action of optical resonator

At each reflection from partially reflecting mirror, some energy is lost from the resonator. When the losses at the mirror and within the medium due to scattering, absorption diffraction, etc. balance the gain, the laser oscillation builds up and the laser output beam will be emitted as shown in Fig. 4.14(f). A highly collimated, intense, and coherent beam emerges from the cavity.

Condition for Steady State Oscillation

In the optical resonator, light amplification takes place. If we consider the wave nature of light then amplification means that there should be continuous and distinct increase in the amplitude of the light waves. Amplification of waves is possible only when the wave returning to some point in the medium after a number of reflections from the mirrors have the same phase as that of the original wave. This imposes a certain constraint on the relationship between the wavelength λ and the length of the optical resonator denoted by 'L'. Thus, it is required that the optical path length travelled by a wave between two consecutive reflections at the same end mirror should be an integral multiple of wavelength. The condition for amplification is

$$\text{Optical path} = 2\mu L = m\lambda_m \tag{4.15}$$

where μ is refractive index of the active medium and m = 1, 2, 3.....are integers.

$$L = m\left(\frac{\lambda}{2}\right) \quad \text{where } \mu \cong 1$$

Thus, only those wavelengths will be amplified which fit in the cavity. The cavity length supports only those light waves which can fit as an integral number of half wavelength (Fig. 4.15). Other wavelengths are not supported and hence are attenuated. Thus, the condition expressed in Eq. (4.15) is called the condition of resonance between the mirror cavity and the light waves.

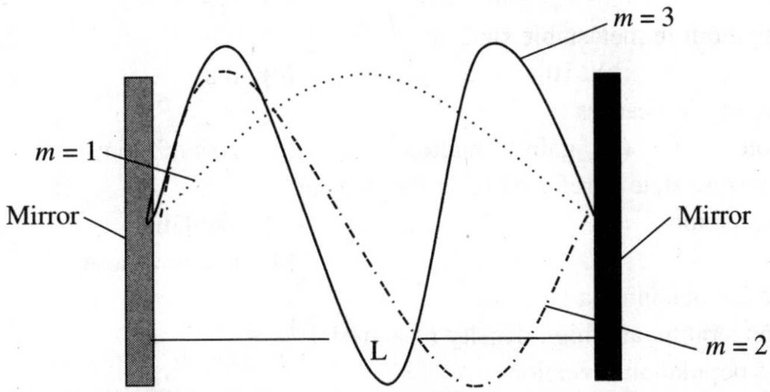

Fig. 4.15 Standing wave pattern and axial modes in optical resonator

Equation (4.15) can be rewritten in terms of frequency as follows

$$\lambda_m = \frac{2L\mu}{m} \because \nu = \frac{c}{\lambda}$$

Therefore,

$$\nu_m = \frac{cm}{2L\mu} \tag{4.16}$$

Theoretically the cavity can resonate at very large number of frequencies that satisfy Eq. (4.16). The spacing between the neighbourhood frequencies is constant and is given by

$$\Delta\nu = \nu_{m+1} - \nu_m = \frac{c}{2\mu L} \tag{4.17}$$

In terms of wavelength ($\Delta\lambda_m$) the mode separation can be written as follows. As we know

$$\nu_m = \frac{c}{\lambda_m}$$

Partially differentiating, we get

$$\Delta \nu_m = \frac{c}{\lambda^2}\Delta\lambda_m \quad \text{or} \quad \Delta\lambda_m = \frac{\lambda^2}{c}\Delta\nu_m$$

Using Eq. (4.17), we get $\Delta\lambda_m = \dfrac{\lambda^2}{2\mu L}$ (4.18)

However, it does not mean that the laser operates at all the frequencies given by Eq. (4.16). The laser will operate only at a select few frequencies for which the gain exceeds all the cavity losses.

Checkpoint 1

1. In population inversion there are
   (a) same number of atoms in upper and lower levels
   (b) more atoms in lower level than upper level
   (c) more atoms in upper than lower level
   (d) none of these
2. The optical cavity will resonate only when
   (a) L = m (λ/2) (b) l_{coh} = m (λ/2) (c) L = n (v/2) (d) L = $n\lambda$
3. The lifetime of an atom in metastable state is
   (a) infinite (b) 10 s (c) 0 s (d) none of these
4. The process of absorption causes
   (a) loss of photon (b) gain of photon (c) loss of atoms (d) gain of atoms
5. A negative temperature state is referred to as the state of
   (a) population inversion (b) pumping (c) excitation (d) none of these
6. The requirements for obtaining a laser are
   (a) minimum temperature and high density of material
   (b) pumping and population inversion
   (c) pumping and cavity
   (d) high input power and more active centres
7. Incoherent light is due to
   (a) stimulated emission (b) spontaneous emission (c) absorption (d) pumping
8. Negative temperature state can be represented by
   (a) $N_1 > N_2$ (b) $N_1 = N_2$ (c) $N_2 > N_1$ (d) $N_2 \neq N_1$

4.4 Types of Lasers

Lasers are normally classified by their active media and the means employed for excitation (pumping). This section is devoted to the study of the physical properties of media utilized as active materials and the method of pumping which allows laser action to occur in some specific laser systems, viz. solid, gas, liquid/chemical, dye and semiconductor. Therefore, only some of the important and practical lasers such as ruby laser, He–Ne laser, and semiconductor laser are described in this section.

4.4.1 Ruby Laser

The solid-state lasers are also called *doped insulator lasers*. It is built around active media prepared as insulating material (dielectric crystal or glass) doped with ions of impurity in its host structure. These lasers are rugged, simple to maintain, and capable of generating high peak powers. They employ only optical pumping. The first working laser was built in 1960 using a ruby crystal. Ruby crystal was shaped in the form of a rod. Ruby belongs to the family of gems consisting of Al_2O_3 doped with 0.05% chromium oxide (Cr_2O_3) by weight. The Cr^{3+} ions function as active centre in the ruby crystal.

Construction

The schematic of a ruby laser is shown in Fig. 4.16. Ruby rod is taken in the form of a cylindrical rod of 4 cm in length and 0.5 cm in diameter. Its ends are grounded and polished such that the end faces are exactly parallel and also perpendicular to the axis of the rod. One face is silvered to achieve 100% reflection while the other is silvered to give 10% transmission. The rod is surrounded by a helical photographic flash lamp filled with xenon.

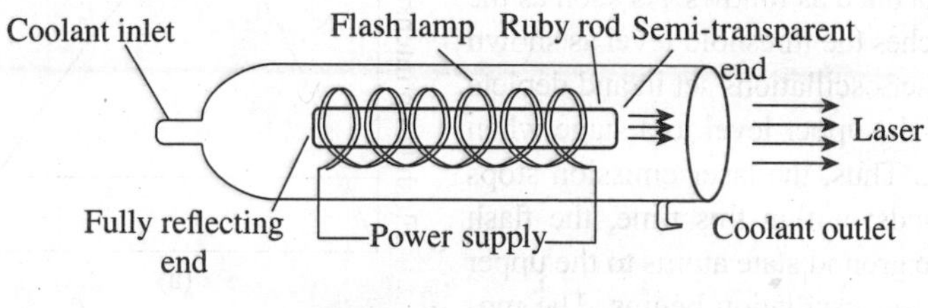

Fig. 4.16 Ruby laser

The lamp produces flashes of white light whenever activated by the power supply. The system is cooled with the help of a coolant circulating around the ruby rod.

Working

The energy levels of chromium ion are shown in Fig. 4.17. They form the three-level scheme. The states in the bands labelled E_1 and E_2 have a very small lifetime ($\leq 10^{-9}$ s), whereas the metastable state (labelled M) has a much longer lifetime (~ milliseconds). The flash lamp when switched on for a few milliseconds discharges a few thousand joules of energy. This results in a power output (from the flash lamp) of a few megawatts. A part of this energy is absorbed by the chromium ions which are in the ground state (labelled as G), resulting in their excitation to an energy level inside the bands E_1 and E_2. This transition corresponds to the absorption of light centred near green (5500 Å) and blue (4200 Å) regions of the white light respectively.

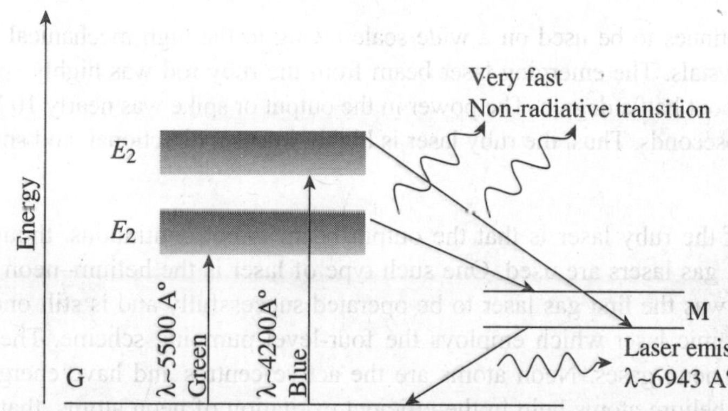

Fig. 4.17 Relevant energy levels of chromium ions in ruby

In order that the expended power of the source be very useful, it is necessary that the bandwidth of the absorption bands be sufficiently wide. One of the advantageous features of the solid-state systems is that they generally have a broad absorption band for the pump transition. In case of ruby crystal, each of the bands E_1 and E_2 is about $0.1\mu m$ wide, which makes optical pumping effective. Since most high-power optical sources have a broad spectral distribution in their radiant energy, a relatively high efficiency can be realized. The excited chromium ions make a very fast non-radiative transition from excited states E_1 and E_2 to the metastable state M and since this state has a long lifetime, the number of the atoms in this state keeps on increasing. This eventually leads to a state of population inversion between doublet M and G. Once the state of population inversion is achieved, lasing action is triggered by spontaneously emitted photons. The photons make many passes through the medium building up the stimulated emissions in a large way and the system gives out a laser beam, which is red in colour and corresponds to the wavelength 6943 Å. The photons travelling in any other direction would be lost after a few reflections.

The flash lamp operation leads to a pulsed output of the laser, which is explained as follows. As soon as the flash lamp power reaches the threshold level as shown in Fig. 4.18(a), the laser oscillations set in and deplete most of the atoms in the upper level to a stage when the laser action stops. Thus, the laser emission stops for a few microseconds; within this time, the flash lamp again pumps the ground state atoms to the upper level and again the laser oscillation begins. The moment the flash lamp stops operating, the population in the upper level is depleted very fast and lasing action stops until the arrival of the next flash from the lamp. Even in the short period of a few tens of microseconds in which the ruby is lasing, one finds that the emission is made up of spikes of high intensity emissions as shown in Fig. 4.18(b). This phenomenon is termed as *spiking of the laser.*

This process repeats itself until the flash lamp power falls below the threshold value and the laser action stops.

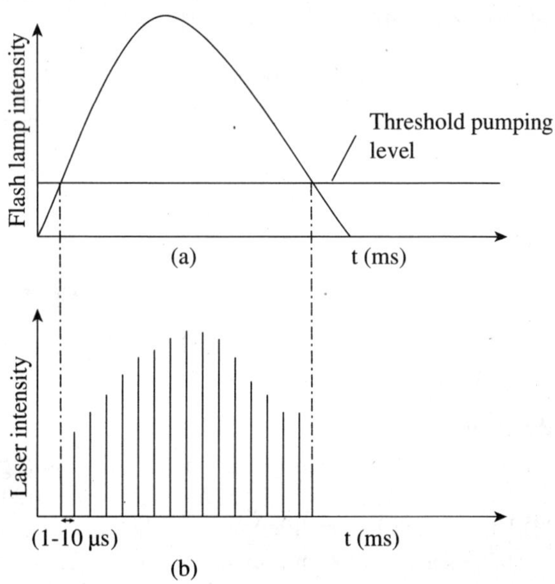

Fig. 4.18 Laser spikes in the output of a ruby Laser

Beam Characteristics

The ruby laser still continues to be used on a wide scale owing to the high mechanical strength and thermal conductivity of ruby crystals. The emerging laser beam from the ruby rod was highly spatially coherent. The beam divergence was about half a degree. The power in the output or spike was nearly 10 kilowatts, which lasts for about 0.1 to 1 microseconds. Thus, the ruby laser is highly intense, directional, and spatially coherent light.

4.4.2 He–Ne Laser

The main drawback of the ruby laser is that the output beam is not continuous, though very intense. For continuous laser beam, gas lasers are used. One such type of laser is the helium–neon (He–Ne) laser.

Helium–Neon laser was the first gas laser to be operated successfully and is still one of the most widely used lasers. It is an atomic laser which employs the four-level pumping scheme. The active medium is a mixture of helium and neon gases. Neon atoms are the active centres and have energy levels suitable for laser transitions, while helium atoms help in the efficient excitation of neon atoms, that is, they assist in the pumping process.

Construction

The construction of He–Ne laser is shown in Fig. 4.19. It consists of a long and narrow discharge tube of about 1 cm diameter and 80 cm length filled with 1 Torr of helium (Torr is a unit of pressure; 1 Torr = 1mm of Hg) and about 0.1 Torr of neon. Thus, the partial pressure of helium gas is 5 to 10 times more than that of neon. The gas mixture of He and Ne forms the lasing medium and is enclosed between a set of mirrors forming a resonant cavity. One mirror is completely reflecting and the other is partially reflecting so as to couple out the laser beam.

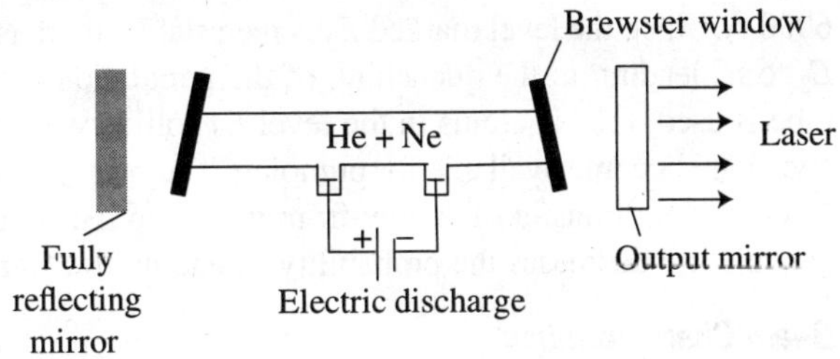

Fig. 4.19 He–Ne laser set-up

For laser systems employing gases, generally pumping is by electrical discharge and not by flash lamp, that is, optical pumping. This is because gas atoms are characterized by sharp energy levels compared to solids, which are characterized by the energy bands rather than energy levels. Hence, if optical pumping is used, only an insignificant fraction of the optical power will be utilized for the excitation of the active centre and the major part of it will be wasted in heating the gas. Hence, wideband optical pumping is ineffective for gaseous medium.

Working

When a discharge is passed through the gaseous mixture, electrons are accelerated down the tube. These accelerated electrons collide with helium atoms and excite them to higher energy levels. Figure 4.20 shows the relevant energy levels of helium and neon atoms. The helium atoms are efficiently excited to the levels F_2 and F_3. These levels happen to be metastable and hence the He atoms excited to these levels spend a sufficiently long time before getting de-excited. As shown in Fig. 4.20, some of the excited states of Ne, that is, E_4 and E_6 corresponds approximately to the excited energy levels F_2 and F_3 of helium. Thus, when He atoms in the levels F_2 and F_3 collide with Ne atoms in the ground level E_1, an energy exchange takes place. This results in the excitation of Ne atoms to the levels E_4 and E_6 and de-excitation of the He atoms to the ground level. Because of the long lifetimes of the levels F_2 and F_3 of helium, this process of energy transfer has a high probability. Thus, discharge through the gas mixture continuously populates the excited energy levels E_4 and E_6 of neon. This helps to create a state of population inversion between the levels E_4 (or E_6) and the lower lying energy levels E_5 and E_3. The various transitions lead to emission at wavelengths 33390 Å (3.39 μm), 11500 Å (1.15μm), and 6328 Å. The first two corresponds to the infrared region while the wavelength resulting from transition E_6 to E_3 corresponds to the well-known red light of He–Ne laser.

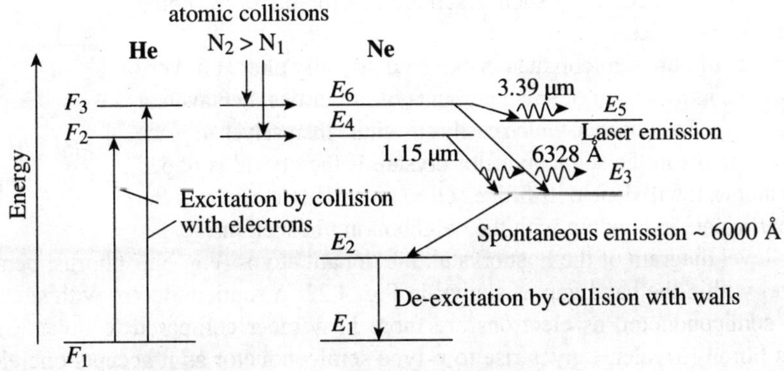

Fig. 4.20 Energy levels of He–Ne system

The excited Ne atoms drop down from level E_3 to E_2 by spontaneously emitting a photon around wavelength 6000 Å. Since the level marked E_2 is metastable, there is a finite probability of the excitation of Ne atoms from E_2 to E_3 leading to the quenching of the population inversion that exists between E_6 and E_3. When a narrow tube is used, the Ne atoms in the level E_2 collide with the walls of the tube and get de-excited to the level E_1 and again become available for pumping. The pressures of the two gases in the mixture are so chosen that there is an efficient transfer of energy from the He to the Ne atoms. The higher partial pressure of the helium in the gas mixture decreases the probability of energy transfer in the reverse direction, that is, from neon to helium.

Beam Characteristics

The light from gas lasers compared to that of solid state lasers is found to be more directional and much more monochromatic. He–Ne laser operates in cw (continuous wave) mode. The typical input power is 5 to 10 W, whereas the output power is 0.5 to 50 mW. The beam divergence is about one minute of arc. Thus, the He–Ne laser beam is highly collimated, spatially coherent, and extremely monochromatic. The use of Brewster's windows causes the laser output to be linearly polarized in the plane of Fig. 4.19.

4.4.3 Semiconductor Laser

A semiconductor laser is a specially fabricated p-n junction device that emits coherent light when it is forward biased. It is made from gallium arsenide (GaAs), which is operated at low temperatures and emits light in the near infrared region of the light spectrum.

Basics of Semiconductor

Semiconductors, as the name implies, are materials that have electrical conductivities lying somewhere between those of conductors and insulators. In a solid, the atoms lie very close to each other and the presence of one atom has an effect on the other atoms. As the atoms are brought close together to form a solid, the effect is that the atomic energy levels get perturbed, resulting in energy bands rather than energy levels. In an energy band, the energy levels are so close together that it is essentially continuous. The energy bands of intrinsic (pure) semiconductors are separated by regions of energy in which no energy level exists as shown in Fig. 4.21. The filled band represents the valance band, which comprises energy levels corresponding to the valance electrons. The next upper empty band is termed as *conduction band,* which comprises energy levels corresponding to conducting (free) electrons. If the temperature of the semiconductor is increased, then some electrons from the top of the valance band may gain enough energy to make a transition to the conduction band. This leaves behind a hole in the valance band. The excited electron and the hole can take part in conduction. In such a semiconductor there are equal number of electrons and holes.

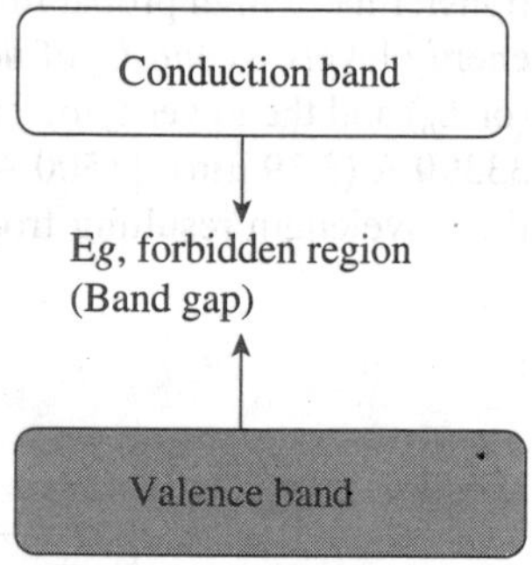

Fig. 4.21 Band structure in an intrinsic semiconductor

The characteristics of pure semiconductors can be drastically altered in very important ways by the introduction of small amounts of impurities. Pentavalent or trivalent atoms are used as dopants, which easily substitute themselves in place of some of the host atoms in the semiconductor crystal. If the crystal is doped with phosphorus atoms, it will donate its fifth electron to the lattice, whereas the remaining four participate in bonding with the neighbouring host atoms.

In the energy-level diagram of the host crystal, the dopant atoms (e.g. phosphorus, pentavalent) give rise to an energy level within the band gap as shown in Fig. 4.22. A semiconductor with such a donor impurity is called n-type semiconductor, as electrons are more in number compared to holes. Similarly, a dopant impurity such as boron (trivalent) gives rise to p-type semiconductor as it accepts one electron from those electrons of the host atoms thereby increasing the number of holes. The impurity is called *acceptor impurity,*

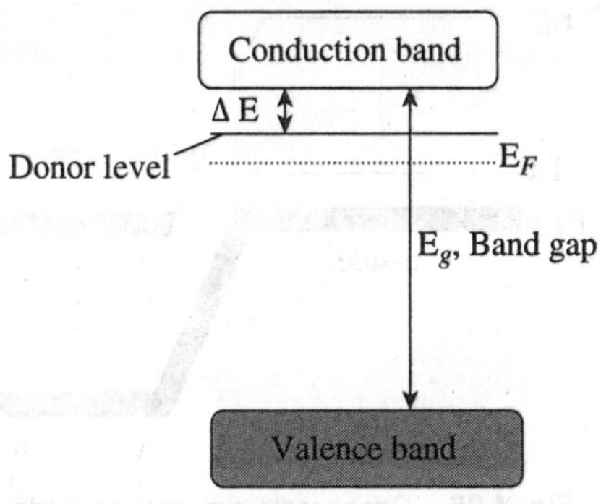

Fig. 4.22 Energy band diagram for n-type semiconductor

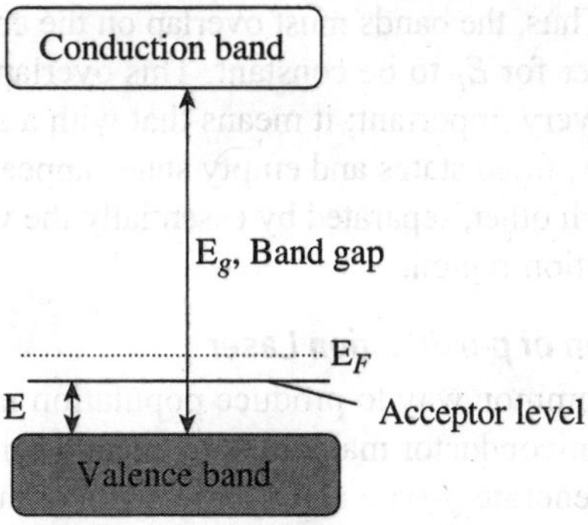

Fig. 4.23 Energy band diagram for p-type semiconductor

which also gives rise to an energy level within the band gap of the host crystal as shown in Fig. 4.23. The acceptor level lies very close to the valance band, whereas the donor level lies very close to the conduction band. The energy separation ΔE is about 0.01 eV, which is about one-hundredth of the band gap width E_g.

The energy level E_F is called the *Fermi energy level.* It is a reference energy level, which represents the average energy of the charge carriers participating in conduction.

Now, if we continue to dope a semiconductor with impurities, a point is reached at which the impurities become so closely packed within the lattice that interactions between them cannot be ignored. For example, if the concentration of donor atoms in the material is sufficiently high (e.g., 10^{20} donor/cm^3), then they are so close together that we can no longer consider the donor level as being composed of discrete, non-interacting energy states. Instead, the donor states form a band that may overlap the bottom of the conduction band and the Fermi level is no longer within the band gap, but lies within the conduction band. When this occurs, the semiconductor is said to be heavily doped or highly degenerate. The analogous case of degenerate p-type material occurs when the acceptor impurity concentration is very high and the Fermi level lies in the valence band. Thus, in degenerate n-type and p-type semiconductors the donor level and the acceptor levels become bands and encroach into conduction and valence band, respectively as shown in Figs 4.24 and 4.25.

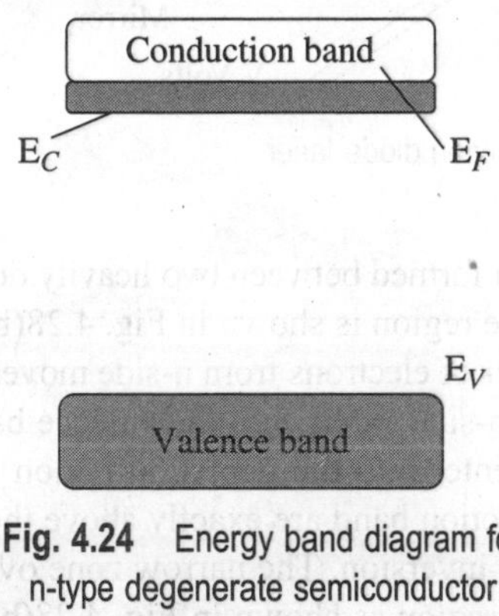

Fig. 4.24 Energy band diagram for n-type degenerate semiconductor

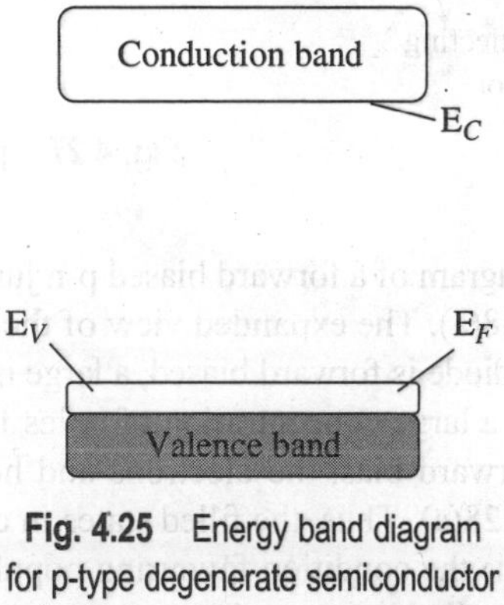

Fig. 4.25 Energy band diagram for p-type degenerate semiconductor

A p-n junction between two degenerate semiconductors is illustrated in terms of energy bands in Fig. 4.26.

This is the equilibrium condition for which the Fermi level is constant throughout the junction. The Fermi level E_{FP} lies below the valence band edge on the p-side and E_{FN} is above the conduction band edge on the n-side. Thus, the bands must overlap on the energy scale in order for E_F to be constant. This overlapping of bands is very important; it means that with a small forward bias, filled states and empty states appear opposite to each other, separated by essentially the width of the depletion region.

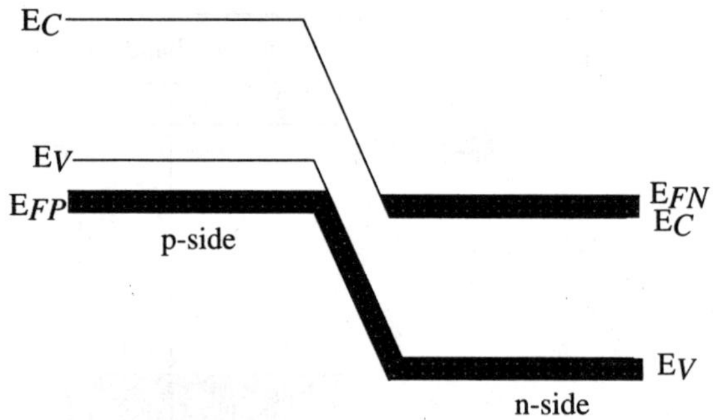

Fig. 4.26 Degenerate p-n junction diode in equilibrium

Construction of p-n Junction Laser

The most common way to produce population inversion in a semiconductor material is by metallurgically joining degenerate p-type and n-type semiconductor material, that is, through the use of p-n junction. GaAs p-n junction diode used is of cube shape with edge 1mm long and the junction lies in a horizontal plane through the centre. Current was passed through the junction by means of ohmic contacts attached to the top and bottom faces. The front and back faces are parallel to each other and perpendicular to the plane of the junction. The front and back faces were polished to form an optical cavity as shown in Fig. 4.27. The current was applied in the form of pulses of 5 to 20 μs duration. The active region consists of a layer of thickness ≈ 1 μm.

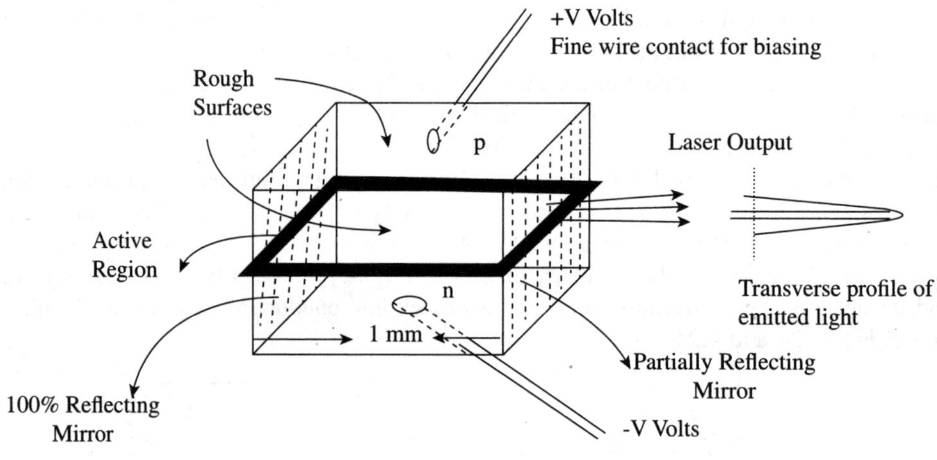

Fig. 4.27 p-n junction diode laser

Working

The energy band diagram of a forward biased p-n junction formed between two heavily doped semiconductors is shown in Fig. 4.28(a). The expanded view of the active region is shown in Fig. 4.28(b). When the heavily doped p-n junction diode is forward biased, a large number of electrons from n-side moves into the conduction band on p-side and a large concentration of holes from p-side enters into the valence band on n-side. Thus, with increase in forward bias, the electrons and holes enter into the depletion region in large numbers as shown in the Fig. 4.28(a). Thus, the filled states in conduction band are exactly above the empty states in the valence band. This is the condition favouring population inversion. The narrow zone over which population inversion occurs is called an *inversion region* or *active region* as shown in Fig. 4.28(b). The electrons and holes recombine in the active region emitting photons. The initial light output occurs due to spontaneous emission transition of carriers between the two energy bands.

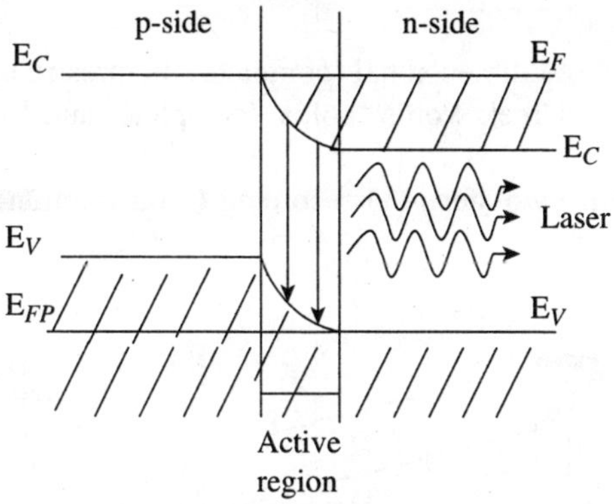

(a)

Fig. 4.28(a) Energy level diagram of degenerate p-n junction diode under forward bias

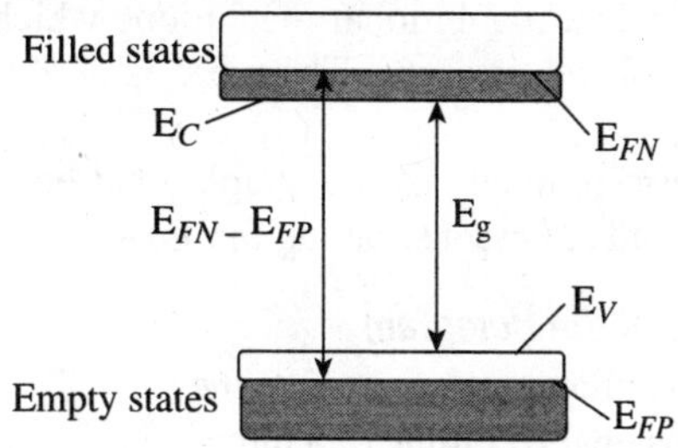

(b)

Fig. 4.28(b) Expanded view of active region showing a state of population inversion and laser transition

The spontaneous photons propagating in the junction plane induce the conduction electrons to jump into the holes of the valence band. The stimulated electron–hole recombination causes emission of coherent radiation. This stimulated radiation is selectively amplified as the light travels in the cavity between the end mirrors and is 'fed back'. During the round trip there is amplification as well as attenuation due to absorption in the light intensity. Lasing occurs when the gain provided by the medium overcomes the loss in the cavity for a round trip. The GaAs laser emits light at a wavelength of 9000 Å in the visible red region. Diode lasers are less monochromatic and highly temperature sensitive.

Beam Characteristics

The p-n junction lasers are remarkably small in size (0.1 mm long). They operate at low powers. However, the laser output is usually in the form of a wide beam 5 to 15° and as a consequence exhibits less monochromaticity as compared to other lasers. The advantage of semiconductor laser is that it is portable and easily controlled source of coherent radiation. The junction diode apart from being compact and simple has high efficiency because nearly every electron injected across the junction contributes a useful photon output. The power output of lasers, however is very small ~ 1 mW. It can be operated continuously or in pulsed mode.

4.5 Applications of Laser

The applications of lasers are widespread and range over various scientific disciplines. The numerous fields of laser applications can be broadly divided into two groups, one involving laser beams of high power and the other involving beams of low power. High-power gas and solid state lasers are used in material processing (say welding, heat treatment, cutting, etc.), nuclear fusion, medical field, defense, etc. The low power semiconductor lasers are used in CD players, laser printers, optical floppy discs, optical memory cards, data processing and information processing devices, hologram preparation and viewing, range finding, optical communications, etc. One of the most interesting and exciting applications of laser is *holography*. The principle, construction, and reconstruction of a hologram are discussed in the following section.

4.5.1 Principle of Holography

Holography is a revolutionary technique which allows three-dimensional (complete) pictures to be taken of a given object or scene. The word is derived from the Greek words 'holos' (complete) and 'graphos' (writing).

The basic principle of holography can be explained in two steps (1) recording (construction) of the hologram and (2) reconstructing the image.

Recording of the Hologram

An ordinary photograph represents the two-dimensional recording of a three-dimensional scene. The emulsion on the photographic plate is sensitive only to the intensity variations, (amplitude square) and hence while a photograph is recorded, the phase distribution that prevailed (exist) at the plane of the photograph is lost. Since only the intensity pattern has been recorded, the three-dimensional character of the object scene is lost. Dennis Gabor in 1948 conceived of an entirely new idea and proposed a method of recording not only the amplitude, but also the phase of the wave. Figure 4.29 illustrates the principle of holography. A beam splitter S divides the laser beam into two beams, A and B. Thus, in principle, holography is an interference-based technique and hence waves with a high degree of coherence are required for its realization. Laser beams with high spatial coherence ensures that the wave from every object point interferes with the reference wave. The interference pattern records on the photographic plate not only the amplitude distribution, but also the phase of the object wave. The developed photographic plate is known as the *hologram*. The hologram has little resemblance to the object. It has in it a coded form of the object wave, that is, interference pattern which is very fine, the spacing between the fringes being as small as 0.001 mm. The fine structure of the interference fringes requires photographic emulsion of high quality. The technique by which one reproduces the image is termed *reconstruction*.

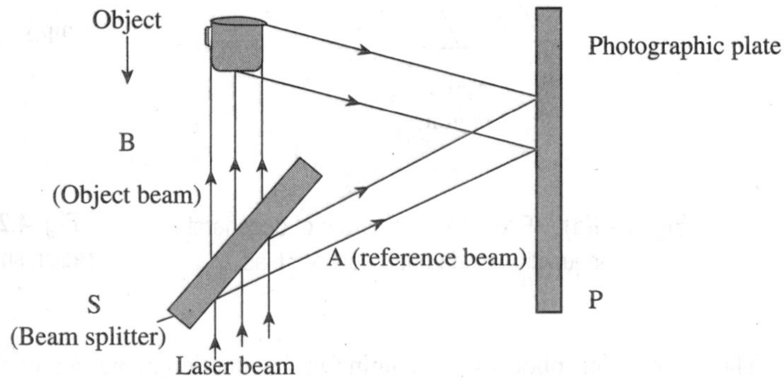

Fig. 4.29 Recording of a hologram

Reconstruction of Image

In the reconstruction process, the hologram 'P' is illuminated by a beam or wave called the reconstruction or readout beam or wave. The reconstruction beam in most cases is similar to the reference beam used for recording the hologram. If now the object is removed and the hologram is put in the place, the reconstruction beam interacts with the interference pattern on the plate and two images are produced by the diffracted waves as shown in Fig. 4.30. One image is the real image, which is formed in front of the hologram and the other, the virtual one, is

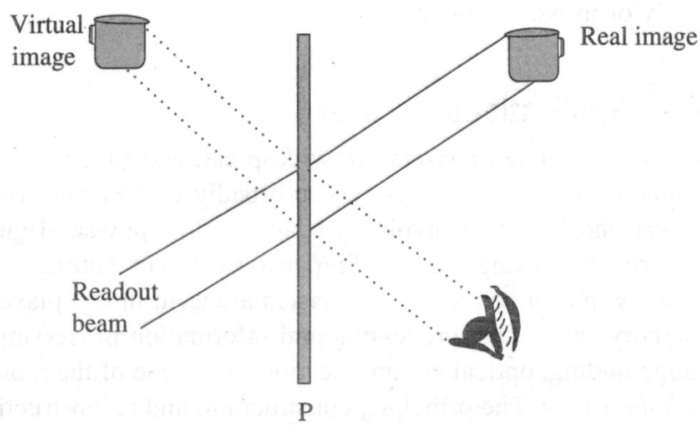

Fig. 4.30 Reconstruction of the image

formed behind the hologram at the original position occupied earlier by the object. The reconstruction of the object wave from a hologram is obviously independent of whether the recorded object has been a point or three-dimensional one. If a 3D object is recorded, then upon reconstruction the viewer will see not a plane, photograph-like image of the object, but rather a realistic 3D image indiscernible from the object. If the viewer shifts aside, other objects hidden behind the first one will become visible and new side features emerge. In other words, the viewer will see an absolutely real 3D scene.

Distinguishing Characteristics of Holographs

(a) When one records the hologram of a reflecting object, each point on the object scatters light on the complete surface of the hologram. Thus, every part of the hologram receives light from all parts of the object. Hence, even if one breaks the recorded hologram into various parts, each part is capable of reconstructing the entire object, but the resolution in the image decreases as the size of the hologram decreases. Thus, each part of the hologram contains information about the geometrical characteristics of the entire object and hence no matter how small the part is, it can reproduce the entire image. A hologram thus is a reliable method of data storage.

(b) A hologram may contain a number of consecutively recorded scenes that can be recorded independently. Therefore, the information holding capacity of a hologram is extremely high.

(c) The information about depth can also be stored in a hologram.

Checkpoint 2

1. In laser, stimulated emission is triggered by
   (a) stimulated photon
   (b) incident photon
   (c) spontaneously emitted photon
   (d) none of these
2. He–Ne laser employs which of the following pumping schemes?
   (a) Two-level
   (b) Three-level
   (c) Four-level
   (d) none of these
3. Ruby laser employs which of the following pumping schemes?
   (a) Two-level
   (b) Three-level
   (c) Four-level
   (d) none of these
4. In He–Ne laser, the role of He is
   (a) to activate Ne
   (b) to increase o/p power
   (c) to reduce temperature of medium
   (d) to help dissipate power
5. In ruby laser the pumping is
   (a) optical
   (b) chemical
   (c) thermal
   (d) electrical
6. In ruby laser, laser light is given by
   (a) aluminium ions
   (b) chromium ions
   (c) oxygen ions
   (d) Al and Cr ions
7. The relaxation of atoms to ground state by wall collision in He–Ne laser can be achieved by
   (a) reducing wall diameter
   (b) increasing wall diameter
   (c) maintaining He–Ne at different pressures
   (d) spontaneous emission
8. As compared to other lasers, the spectral purity and monochromaticity of the semiconductor laser is
   (a) extremely good
   (b) too poor
   (c) cannot be compared at all
   (d) none of these

Solved Numericals

1. Find the ratio of population of the two states in He–Ne laser that produces light of wavelength 6328 Å at 27°C.

Solution: Equation (4.14) gives the ratio of population as

$$\frac{N_2}{N_1} = \exp\ \exp\left[-\frac{E_2 - E_1}{kT}\right]\exp\left[\frac{\Delta E}{kT}\right]$$

Value of kT at room temperature (that is 27°C) is 0.026 eV. The difference in energy levels emitting light of wavelength λ is given by

$$\Delta E = h\nu = \frac{hc}{\lambda}$$

Substituting the values of $h = 6.63 \times 10^{-34}$ J-s, $c = 3 \times 10^8$ m/s and 1 eV = 1.6×10^{-19} J, we get

$$\Delta E = \frac{12400}{\lambda(A)} eV$$

Substituting wavelength of light in Armstrong unit, we get energy difference between the levels in eV unit as

$$\Delta E = \frac{12400}{6328} eV = 1.96 eV$$

Therefore, the ratio of population of the two states in He–Ne laser is given by

$$\frac{N_2}{N_1} = \exp\left[-\frac{1.96}{0.026}\right] = 1.82 \times 10^{-33}$$

2. With a He–Ne laser, fringes remain clearly visible when the path difference has increased upto 8 m. Deduce (a) coherence time, τ (b) spectral half width. Given wavelength $\lambda = 11.5 \times 10^{-7}$ m.

Solution: (a) Coherence time, $\tau = \frac{l_{coh}}{c}$

The coherence length is given as $l_{coh} = 8$ m

$$\therefore \tau = \frac{8}{3 \times 10^8}\ s = 2.7 \times 10^{-8} s$$

(b) Spectral half width, $\Delta\lambda = \frac{\lambda^2}{l_{coh}} = \frac{(11.5 \times 10^{-7})^2}{8}$

$$= 0.0016\ Å.$$

3. Calculate l_{coh} for Hg vapour lamp emitting in the green portion of spectrum at wavelength of 546.1 nm with emission bandwidth $\Delta\nu = 6 \times 10^{+8}$ Hz.

Solution: $\tau = \frac{1}{\Delta\nu} = 16.6 \times 10^{-10}$ s

The coherence length, l_{coh} is given by

$$l_{coh} = c\tau = 3 \times 10^8 \times 16.6 \times 10^{-10} m$$

$$= 0.49\ m$$

4. If the band width of the 10.6 μm transition of a CO_2 laser is 60 MHz; calculate the coherence length of the laser. If the cavity length is 2m, show that not more than one mode will oscillate.

Solution: The coherence time, $\tau = \frac{1}{\Delta\nu} = \frac{1}{60 \times 10^{-6}}$ s.

$$= 16.6 \times 10^{-9} s$$

The coherence length is $l_{coh} = \tau \times c$

$$= 16.6 \times 10^{-3} \times 3 \times 10^8\ m$$

$$= 49.8 \times 10^{-1} m = 5\ m$$

The length of the cavity L in terms of spatial width is given as

$$L = \frac{nc}{2\Delta\nu}$$

$$\therefore n = \frac{2L\Delta\nu}{c} = \frac{2 \times 2 \times 60 \times 10^6}{3 \times 10^8} \cong 1$$

Thus, in such a cavity of length 2m only single longitudinal mode will oscillate.

5. In a He–Ne laser the distance between two plane mirrors forming the resonant cavity is 0.5 m. What is the mode separation of longitudinal cavity in terms of frequency?

Solution: The mode separation $\Delta\nu$ is

$$\Delta\nu = \frac{c}{2L\mu} = \frac{3 \times 10^8}{2 \times 0.5 \times 1}$$

$$= 3 \times 10^8\ Hz$$

(Here is μ is assumed to be 1)

Summary

- Laser light is different from ordinary source of light in view of its exceptional directionality, high monochromaticity, intensity, and remarkable coherence.
- Laser light can be obtained by exceeding the rate of stimulated emission over stimulated absorption and spontaneous emission.
- The rate of stimulated emission can be maximized by creating population inversion in the medium and enclosing the medium between two mirrors.
- Out of three pumping schemes, the four-level scheme is more efficient than the three-level pumping scheme.
- Choice of pumping mechanism depends on the state of the material (solid, liquid, or gas) in which laser is to be obtained.
- Lasers are classified into solid, liquid, or gas laser based on the state of the material and pumping mechanism used.
- Ruby laser is a solid-state laser that uses the optical pumping mechanism and belongs to the three-level pumping scheme. It is a high wattage laser source that gives out highly directional and intense light.
- Helium–Neon laser is a gas laser, which uses electrical pumping mechanism and belongs to the four-level pumping scheme. It is a low-wattage laser source that gives out highly directional, monochromatic, and coherent light.
- Semiconductor laser is a low-wattage, highly compact laser source, which uses p-n junction diode, made of semiconducting material.
- Laser holograms contain three-dimensional information of the object. This is a reliable method of storing large information.

Self-study Questions

1. (a) What is laser? How does it differ from an ordinary source of light?
   (b) Discuss the properties of laser.
2. Explain spatial coherence and temporal coherence.
3. Explain the terms coherence length and coherence time. Derive the expression for the coherence length of a wave in terms of line width $\Delta\lambda$ corresponding to frequency band $\Delta\nu$.
4. (a) What is stimulated absorption?
   (b) Explain the three quantum processes. Which processes are maximized in laser operation and how?
5. Explain the terms: stimulated emission, population inversion, pumping and metastable states. What is their role in the working of the laser?
6. (a) What do you understand by negative temperature state? How can it be achieved?
   (b) What are the different mechanisms of pumping?
7. Explain in brief how the four-level laser is more efficient than the three-level laser.
8. (a) What is resonant cavity? Explain its working with neat sketches.
   (b) What is the role of resonant cavity length in supporting different frequency modes?
9. (a) What is the role of metastable state in laser media? Is it a necessary for occurrence of lasing?
   (b) In laser, active media should have preferably broad absorption band. Explain why?
10. (a) Describe the construction of a solid state ruby laser.
    (b) Describe the working of ruby laser using energy-level diagram.
    (c) Why are the end faces of the ruby rod silvered?
11. (a) Explain the construction and working of He–Ne laser. Explain the role of end mirrors in laser. What frequencies are amplified by this mirror system?
    (b) Increase in diameter of He–Ne laser may reduce lasing efficiency. Explain why.
    (c) What is the role of helium atoms in He–Ne laser?
12. Explain the construction and working of semiconductor laser. Draw the necessary energy-level diagrams.
13. Give two engineering applications of laser source and discuss one of them.

Numerical Exercises

1. White light has a frequency range from 0.4 x 10^{15} Hz to 0.7 x 10^{15} Hz. Find its coherence time and coherence length. (Ans: $\tau = 3.3 \times 10^{-15}$ s, $l_{coh} = 0.99\ \mu m$)
2. A sodium atom radiates for 4 X 10^{-9} s. What is the coherence length of light from a sodium lamp? (Ans: $l_{coh} = 1.2$ m)
3. Calculate l_{coh} for CO_2 laser whose line width is 1 x 10^{-5} nm at IR emission wavelength of 10.6 μm. (Ans: $l_{coh} = 11.24$ km)
4. If the half-width of the He–Ne laser operating at wavelength 6328 A° is 1500 MHz, what must be the length of the laser cavity to ensure that only one longitudinal mode oscillates? (Ans: L = 0.1m)
5. The ratio of population of two energy levels at 300 K is 10^{-30}, find the wavelength of the radiation emitted. (Ans: 6927.3 Å)
6. A typical helium neon laser emits radiation of λ = 6328 Å. How many photons per second would be emitted by a one milliwatt (mW) He-Ne Laser? (Ans: 3 – 2 x 10^{15} Photons/sec)
7. A certain ruby laser emits 1.00 J pulses of light whose wavelength is 6940 Å. What is the minimum number of Cr^{3+} ions in the ruby? (Ans: 3.49 x 10^{10} ions)
8. In a Ruby Laser, the total number of Chromium ions are 19·8 x 10^{16}. If the Laser emits radiation of wavelength 7000 Å, then calculate the (i) Energy of one emitted photon (ii) Total energy available per pulse. [(Ans: (i) E_{Photon} = 1.77eV (ii) E_{Photon}/Pulse = 56.07mJ]
9. The coherence length of CO_2 laser of central wavelength of 10.6 μm is 11 Km. Calculate the non-monochromaticity factor. (Ans: 10^{-9})
10. The light of wavelength 600 nm has wave train 13.2 × 10^{-6} m. Calculate the coherence time. (Ans: 4.4 x 10^{-14} sec)
11. Light of wavelength 4800 Å has a length of 25 waves. What is the coherence length and coherence time? (Ans: 12 μm)
12. The coherence length of sodium light is 2.945 x 10^{-2} m and its wavelength is 5890 Å. Calculate (i) the number of oscillations corresponding to coherence length and (ii) the coherence time. (Ans: (i) 5 x 10^4 (ii) 98 x 10^{-12} sec)
13. He-Ne laser, giving light at 6330 Å has a coherence length of 20 km. Determine its (i) coherence time and (ii) number of waves per wave train. (Ans: (i) 6.67 x 10^{-5} sec (ii) 3.16 x 10^{10}).
14. A quasi-monochromatic source emits radiations of mean wavelength λ = 5461 A˚ and has bandwidth Δv = 10^9 Hz. Calculate coherence time, coherence length and monochromaticity factor. [Ans: (i) 10^{-9} sec (ii) 0.3 m (iii) 1820 x 10^{-9}]
15. An optical filter has a line width of 1.5 nm and mean wavelength 550 nm. With white light incident on the filter, calculate (i) coherence length and (ii) the number of wavelengths in the wave train. [Ans: (i) 2.02 x 10^{-4} m (ii) 367]
16. The spectral purity of a source can be appreciated via the quantity Δv/v , the frequency stability. For example, a Hg^{198} low-pressure isotope lamp (λ_{air} = 546.078 nm) has a bandwidth of Δv = 1000 MHz. Compute the coherence length and coherence time of the light, as well as the frequency stability. [Ans: (i) 0.3 m (ii) 10^{-9} sec (iii) 182.03 x 10^{-9}]

Answers

Checkpoint 1

1 (a), 2 (b), 3 (d), 4 (a), 5 (a), 6 (c), 7 (b), 8 (c)

Checkpoint 2

1 (c), 2 (c), 3 (b), 4 (a), 5 (a), 6 (b), 7 (a), 8 (b)

Interference

5

Learning Objectives

After going through this chapter, you will be able to:

- Understand the basic concepts related to light
- Learn interference using thin film of uniform thickness
- Learn interference in non-uniform thin film
- Understand various applications of interference in non-uniform thin film
- Learn the application of interference for producing anti-reflection coatings

5.1 Introduction—Properties of Light

Light is produced by either incandescence or luminescence. Incandescence is a phenomenon where the emission of light is from the matter at temperatures above 800 K. In luminescence emission of light takes place when electrons jump from higher energy to lower energy state. In this section you will learn the fundamental properties of light.

5.1.1 Nature of Light

Light is a form of energy which according to wave theory propagates in the medium as a wave. Light is a transverse, electromagnetic wave that can be detected by the human eye. Wave theory of light successfully explains the commonplace phenomena exhibited by light such as interference, diffraction, and polarization, which we come across in our daily life.

5.1.2 Wave

Light is considered to have the wave-character. Whenever there is a disturbance from an equilibrium condition, the disturbance propagates with time from one point in space to another. The propagation of this disturbance is called *wave*. The wave motion is represented with the help of crests and troughs travelling away from the centre of disturbance. When every particle of the medium oscillates about its mean position at right angle to the direction of wave propagation, then such wave motion is known as *transverse wave motion*. Waves can be of any form. Harmonic waves are simple and ideal. They are described mathematically by a sine or cosine function. A wave is described by the displacement of a particle, y. The value of y is a function of space coordinates and a function of time t.

Thus

$$y = f(x, y, z, t)$$

A sinusoidal wave propagating in one direction, say x-direction may be represented by the wave function

$$y(x,t) = A\sin\left[\frac{2\pi}{\lambda}(x - vt)\right] \tag{5.1}$$

The constant A in Eq. (5.1) represents the maximum value of the displacement and is called the *amplitude* of the wave. The crest-to-crest distance or the distance between any two adjacent identical points is called the *wavelength* of the wave and is denoted by λ. It is the distance over which the waveform repeats itself. The time taken by the waveform to advance through a distance λ is called the *period, T.* The wave travels at a speed v and covers a distance vT in time T.

$$\lambda = \upsilon T \tag{5.2}$$

The reciprocal of period is called *frequency*. It may be defined as the number of crests per second.

$$\nu = \frac{1}{T} \tag{5.3}$$

Thus from Eqs (5.2) and (5.3) we get

$$\upsilon = \nu\lambda \tag{5. 4}$$

The wave equation represented by Eq. (5.1) may be rewritten as

$$y(x,t) = A\ \sin\left[2\pi\left(\frac{x}{\lambda} - \frac{t}{T}\right)\right] \tag{5.5}$$

As $\omega = 2\pi/\mathrm{T}$ and $x = 2\pi/\lambda$, Eq. (5.5) is expressed in a more convenient form as given below

$$y(x,t) = A\sin(kx - \omega t) = \mathrm{A}\ \sin\phi \tag{5.6}$$

ω is called the angular frequency and represents the number of cycles contained in the time interval equal to 2π seconds. k is called the wave number and represents the number of wavelengths contained in a distance equal to 2π metres. In Eq. (5.6) the argument of the sine function is $(kx - \omega t) = \phi$. Here ϕ is called the *phase angle* or simply the *phase*. It can be expressed in radians or degrees. The maximum possible value for ϕ is 2π radians or 360°, as shown in Fig. 5.1. Alternatively, it may be expressed as either the ratio t/T or x/λ. Amplitude, frequency, and phase are the three characteristics of a wave.

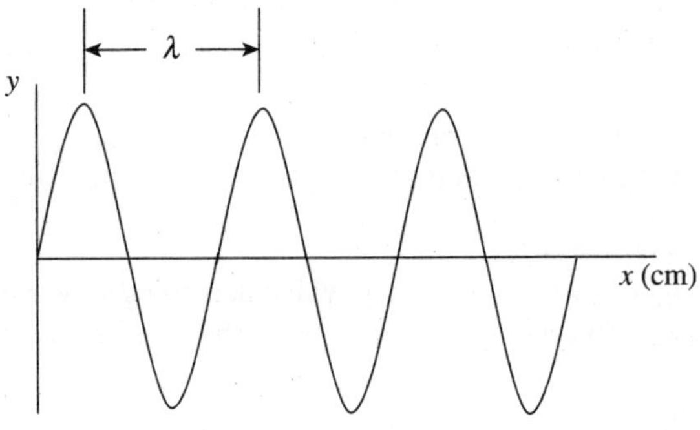

Fig. 5.1 Wave propagating in x-direction

The wave function defined by Eq. (5.6) assumes that displacement $y = 0$ at $x = 0$ and $t = 0$. Such a situation is a special case and in general 'y'may not be zero.

5.1.3 Light Wave

Light wave is an electromagnetic wave. It consists of periodically varying electric and magnetic fields, which oscillate at right angles to each other and also to the direction of propagation of the wave. A part of an electromagnetic wave is shown in Fig. 5.2.

In the wave the electric field strength vector E defines the vibrations of the electric field and magnetic field by the vector of magnetic induction *B*. The electric and magnetic field vectors, *E and B* are of equal importance to the wave. However, many of the effects produced by light are mainly due to its electric field. Hence, a light wave is often represented by E and electric field vector E is called light vector.

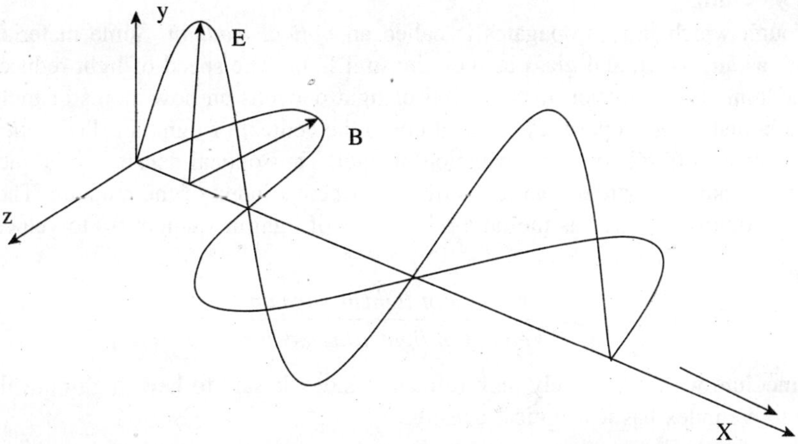

Fig. 5.2 Electromagnetic wave

Mathematically, the light wave depicted in Fig. 5.2 is represented by the expressions

$$E = E_0 \sin(kx - \omega t)$$

$$B = B_0 \sin(kx - \omega t)$$

We have to note at this stage that Fig. 5.2 and the above equations represent an ideal plane electromagnetic wave.

5.1.4 Visible Range

The electromagnetic spectrum consists of waves spanning a broad range of wavelength. It is bounded at one end by gigantic radio waves having wavelengths of a few hundred kilometres and on the other end, by γ-rays having a small wavelength of 10^{-12} m. The range of wavelength, which can be detected by the human eye is called *visible range*. It extends from the violet to the red colour. The limiting range of these waves varies approximately in the interval $\lambda = 4000\text{Å} - 8000\text{Å}$. The region on the longer wavelength side ($\lambda > 8000\text{Å}$) of the visible range is known as *infrared* and on the shorter wavelength side ($\lambda < 4000\text{Å}$) is *ultraviolet*. The wavelength of light waves, that is, length of a single wave is very small; hence smaller units are used to express them. They are usually expressed in angstroms, nanometres, or micrometres.

$$1 \text{ angstrom (Å)} = 10^{-10} \text{ m}$$
$$1 \text{ nanometre (nm)} = 10^{-9} \text{ m}$$
$$1 \text{ micrometre (}\mu\text{m)} = 10^{-6} \text{ m}$$

Thus, the wavelengths of visible light range can be expressed in different units as from 4000 Å to 8000Å or from 400 nm to 800 nm or from 0.4 μm to 0.8 μm.

Velocity of Light

In vacuum electromagnetic waves travel at a speed of 2.998×10^8 m/s. The velocity of light is denoted by the symbol 'c'. For all practical purposes, it is taken as

$$c = 3 \times 10^8 \text{ m/s}$$

From Eq. (5.4), we can write

$$c = \nu\lambda \tag{5.7}$$

5.1.5 Optical Medium

A material through which light propagates is called an *optical medium*. Some materials are said to be transparent such as air, water, and glass as they transmit light. The speed of light reduces when it travels through any medium. The reduction in the speed of light depends on how atoms or molecules or ions of the medium are bound. This property of the medium in the context of light is called *optical density*, which indicates the resistance offered to the propagation of light. Thus optical density of the medium determines the speed of light, which is expressed in terms of the refractive index of the medium. The refractive index μ of an optical medium is defined as the ratio of velocity of light in vacuum (c) to velocity of light in the medium (v).

$$\mu = \frac{\textit{Velocity of light in vacuum}}{\textit{Velocity of light in the medium}} = \frac{c}{v}$$

An optical medium with a relatively high refractive index is said to have high optical density and one with a low refractive index has low optical density.

Thus velocity of light decreases when it travels in an optically denser medium which attributes to the decrease in wavelength or frequency or both of the light wave. However, in view of the law of conservation of energy the frequency must remain constant. It implies that the decrease in velocity of light in a medium must lead to a decrease in the wavelength of light, as shown in Fig. 5.3.

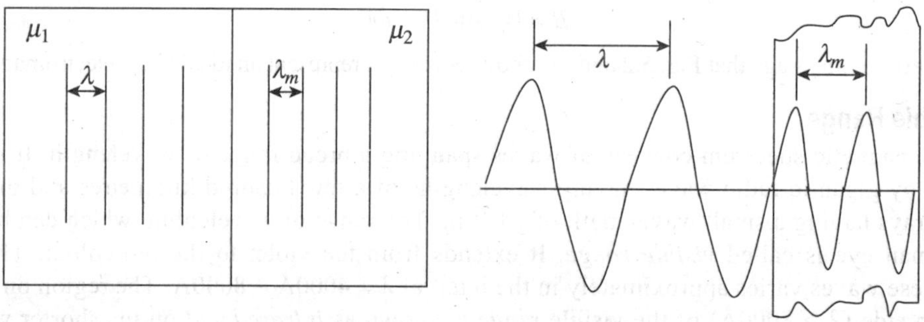

Fig. 5.3 Change in wavelength in different media

In tune with Eq. (5.7), the velocity of light in an optical medium may be expressed as

$$v = \nu\lambda_m \tag{5.8}$$

where λ_m is the wavelength of light in the medium. From Eqs (5.7) and (5.8), we obtain

$$\mu = \frac{\lambda}{\lambda_m} \tag{5.9}$$

5.1.6 Reflection and Refraction

When a light wave travels from one optical medium to another, at the boundary separating the two optical media there is splitting of the wave. Some part of the wave is reflected into the first medium and the remaining is transmitted into the second medium. Usually the waves are represented in terms of rays. If the surface irregularities of the boundary are small compared to the wavelength of the incident light, the boundary is said to be smooth as shown in Fig. 5.4.

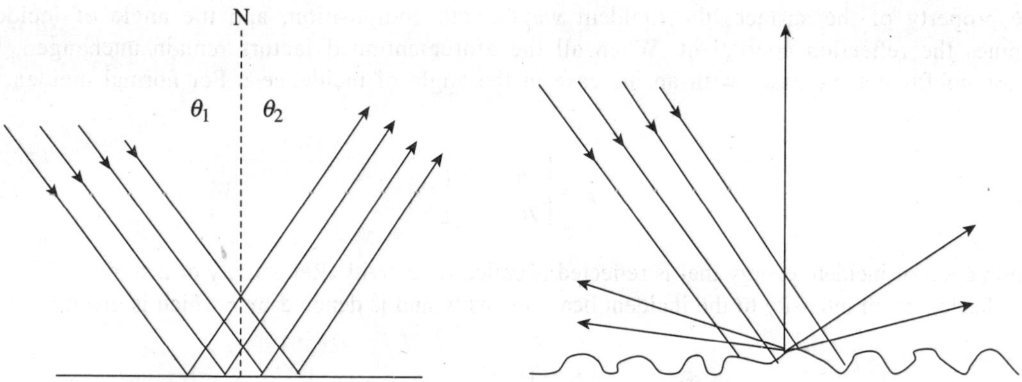

Fig. 5.4 Phenomenon of reflection (a) specular reflection occurs at smooth surface. the reflected rays are parallel. (b) diffuse reflection occurs at rough surface

In the case of rough surfaces rays are reflected in random directions, whereas a smooth surface reflects the light specularly. In *specular reflection* the angle of reflection of light is equal to the angle of incidence. When the light ray enters a transparent medium from another medium, it bends at the boundary and is said to be refracted. The boundary is known as *refracting surface*. The *angle of refraction r* depends on the properties of the two media and on the angle of incidence *i.* According to Snell's law

$$\frac{\sin i}{\sin r} = \frac{v_1}{v_2} = \frac{c/\mu_1}{c/\mu_2} = \frac{\mu_1}{\mu_2} = \mu_{21}$$

$$\frac{\sin i}{\sin r} = \mu_{21} \tag{5.10}$$

where v_1 is the velocity of light in medium 1 having a refractive index μ_1 and v_2 is the velocity of light in medium 2 having a refractive index μ_2. The ratio v_1/v_2 or μ_2/μ_1 is called the *relative refractive index*. It follows from Snell's law that an increase in the angle of incidence causes an increase in the angle of refraction, as refractive index is constant for the medium.

5.1.7 Reflectivity and Transmissivity

Whenever a light wave comes across a boundary, we use the laws of reflection and refraction to describe the direction of incident, reflected, and refracted rays. But these laws do not specify the intensities of these rays. If the incident energy on a transparent medium is E_i, then it is divided into two parts. The part of the incident energy, which stays in the same medium is a reflected part E_r and the leftover part of the incident energy is transmitted as E_t. The law of conservation of energy is stated as follows.

$$E_i = E_r + E_t$$

The reflection coefficient ρ characterizes the *reflectivity of a surface*. It is the ratio of the amplitudes of reflected electric energy to that of the incident energy.

$$\rho = \frac{E_r}{E_i}$$

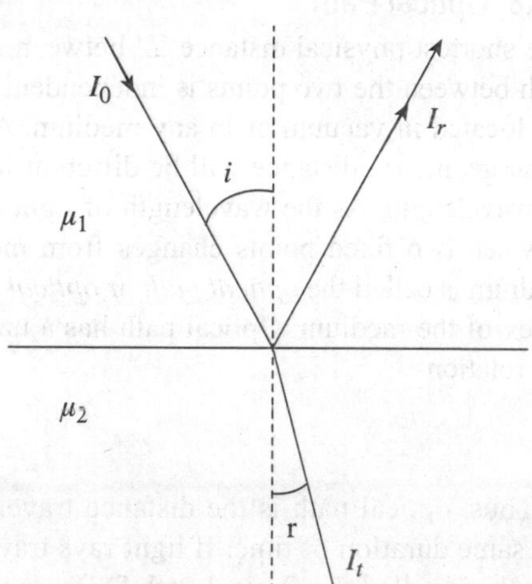

Fig. 5.5 Reflection of light rays at smooth surface

The property of the surface, the incident wavelength composition, and the angle of incidence determines the reflection coefficient. When all the aforementioned factors remain unchanged, then reflection coefficient increases with an increase in the angle of incidence i. For normal incidence, ρ can be written as

$$\rho = \left[\frac{\mu_1 - \mu_2}{\mu_1 + \mu_2}\right] \tag{5.11}$$

The fraction of incident energy that is reflected is called reflectivity. Reflectivity of a surface is the ratio of the reflected beam intensity to the incident beam intensity and is denoted by r which is given by

$$r = \frac{I_r}{I_i} = \left|\frac{E_r}{E_i}\right|^2 = |\rho|^2$$

$$r = \left[\frac{\mu_1 - \mu_2}{\mu_1 + \mu_2}\right]^2 \tag{5.12}$$

Thus from Eq. (5.11), when $\mu_2 < \mu_1$, the reflection coefficient ρ is positive. Thus the oscillations in the incident and reflected waves are in the same phase, implying that phase change is not introduced whenever there is a reflection from denser to rarer medium. For $\mu_2 > \mu_1$, ρ is negative which means that there is a phase difference of 180° between the incident and reflected waves. Hence, there is a phase change of π radians in the reflected wave.

The transmissivity t is given by $$t = \frac{4\mu_2\mu_1}{(\mu_1 + \mu_2)^2}$$

Because energy is conserved, $$r + t = 1$$

5.1.8 Optical Path

The shortest physical distance 'L' between any two points is known as the *geometric path*. The geometric path between the two points is independent of the medium. It will have the same value whether the points are located in vacuum or in any medium. As light travels with different velocities in different media, the same geometric distance will be different for light in different media. Light covers a distance in terms of its wavelength. As the wavelength of light changes from medium to medium, hence for light the distance between two fixed points changes from medium to medium. The distance spanned by light waves in a medium is called the *optical path* or *optical path length* and is denoted by 'Δ' that is related to the refractive index of the medium. Optical path has a unit of distance (metre) and is related to geometric path through the relation

$$\Delta = \mu L \tag{5.13}$$

Thus, optical path is the distance travelled by the light in a medium equivalent to the path in air in the same duration of time. If light rays travel along different paths, then a difference in their optical path lengths is called the Optical Path Difference (OPD). Optical path length is significant as it determines the phase of the light rays that controls the interference and diffraction of light when it propagates through a medium.

5.1.9 Phase Change and Path Difference

For a wave moving in space, its phase changes continuously. At a fixed time, phase difference between any two points at x_1 and x_2 on the wave differ by an amount

$$\delta = \phi_2 - \phi_1$$

Using Eq. (5. 6) we can write the above relation as

$$\delta = (kx_2 - \omega t) - (kx_1 - \omega t)$$

$$\delta = k\,(x_2 - x_1)$$

$$\delta = \frac{2\pi}{\lambda}(x_2 - x_1) \qquad \because k = \frac{2\pi}{\lambda}$$

$$\delta = \frac{2\pi}{\lambda}L \tag{5.14}$$

where geometric path, $(x_2 - x_1) = L$. Thus from Eq. (5.14) we can find the corresponding phase difference for the given path difference. For example, for $L = \lambda$ then $\delta = 2\pi$ and for $L = \lambda/2$ then $\delta = \pi$, and so on. After a distance of one λ the waveform repeats itself. Hence, the phase difference of 2π corresponds to zero phase difference. Thus, a displacement of wave by one complete wavelength does not change the waveform.

(i) When two waves moving along two different paths in a medium overlap at some point, then there exists a path difference between them. For two waves, the quantity $L = x_2 - x_1$ also represents the path difference. The corresponding phase difference can then be computed using Eq. (5.14) as

$$\delta = \frac{2\pi}{\lambda_m}L \text{ or } \delta = \frac{2\pi}{\lambda}\mu L \text{ since } \lambda_m = \frac{\lambda}{\mu}$$

Therefore,

$$\delta = \frac{2\pi}{\lambda}\Delta \text{ as } \Delta = \mu L \tag{5.15}$$

(ii) Whenever there is reflection, it also accounts for the path, or phase difference. A light wave travelling in a rarer medium suffers reflection at the boundary of a denser medium and undergoes a phase change of π radians as illustrated in Fig. 5.6. However, a light wave does not suffer phase change when it is reflected at the boundary of a denser to rarer medium.

Thus when a light wave is incident on a glass block, the wave reflected from the air-to-glass boundary experiences a phase change of π radians relative to the incident wave, because the reflection is from the rarer to denser medium ($\mu_g > \mu_a$), as shown in Fig. 5.6. The wave transmitted through the glass is again reflected at the bottom surface, that is, glass-to-air boundary. The reflection at glass–air boundary does not produce any phase change as the reflection is from denser to rarer medium. From Eq. (5.14) a phase change of π radians is equivalent to a path difference of $\lambda/2$. Thus a light wave gets inverted or there is a loss of half a wave when it is reflected from a rarer to denser medium.

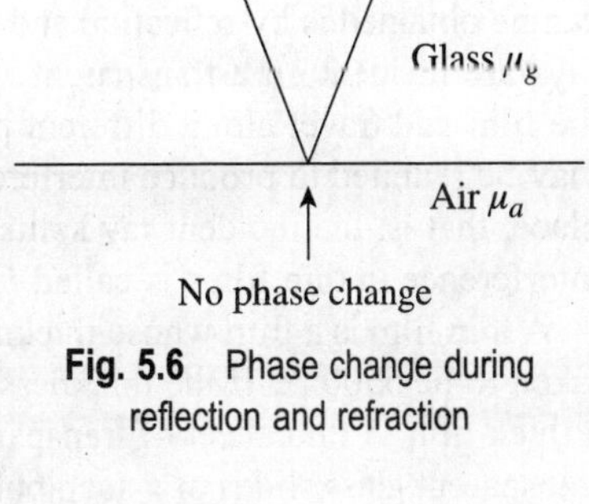

Fig. 5.6 Phase change during reflection and refraction

Checkpoint 1

1. On external reflection, i.e., from air to glass light waves suffer a phase change of
   (a) λ (c) π
   (b) zero (d) $\pi/2$
2. A light of wavelength 6200 Å travels through a film of water of 0.72 μm thick. If its refractive index is 1.33, the optical path will be
   (a) 0.96 μm (c) 0.48 m
   (b) 0.72 μm (d) 6200 Å
3. On internal reflection from glass to air or in refraction, the light wave will suffer phase change of
   (a) π (c) $\pi/2$
   (b) zero (d) none of these
4. The optical path difference represents
   (a) the physical path travelled in that medium
   (b) the number of wavelengths that can be accommodated in a given medium
   (c) the path travelled by light
   (d) none of above
5. A phase change of π introduces a path difference of
   (a) λ (c) $\lambda/2$
   (b) zero (d) none of above

5.2 Interference in Thin film

Interference can be observed in everyday phenomena such as iridescence and structural colouration. Certain surfaces which show iridescence phenomenon change colour when the angle of view or the angle of incidence of light beam is changed. Examples of iridescence include soap bubbles, butterfly wings, and seashells, as well as certain minerals. In a soap bubble, different colours are seen due to the interference of light reflecting off the front and back surfaces of the thin soap film. Different colours interfere constructively and destructively depending on the thickness of the film. Structural colouration is the production of colour by microscopically structured surfaces fine enough to interfere with visible light. For example, peacock tail feathers are pigmented brown, but their microscopic structure makes them also reflect blue, turquoise, and green light.

In interference there is superposition of waves that are coherent. The necessary condition for observing interference is that the waves should be coherent, that is, they should be derived from the same source or they must have the same or nearly the same frequency. All types of waves, for example, light, radio, acoustic, surface water waves, or matter waves show interference effect. In Young's double slit experiment, the coherent sources are obtained by physically dividing the primary source by the use of a pair of slits. The method of such splitting is known as *division of wavefront.* The other method by which coherent sources can be obtained is by reflection and this method is known as *division of amplitude*. For example, when light rays are incident on a transparent thin film, it undergoes reflections from the upper and lower surfaces of the film and travel along different paths. These rays reflected from the top and bottom portion of the film may be reunited to produce interference. Thus at each reflection from thin film, division of amplitude takes place, that is, the incident ray splits into two rays, one reflected and the other refracted ray. Therefore, the interference in thin films is called *interference by division of amplitude*.

A thin film is a film whose thickness is about the order of one wavelength of visible light, approximately taken to be 5500 Å. If the thickness of the film is more, that is, more than 100 μm, then it is considered as a thick film. A thin sheet of transparent material such as glass, mica, and an air film enclosed between two transparent glass slides or a soap bubble can be considered as thin film. When light falls on such a thin film, a small part of it gets reflected from the upper surface of the film and the remaining is refracted into the film. Again, a small part of the transmitted component is reflected back into the film by the lower surface and the rest of it emerges out of the film. A small portion of light thus gets partially reflected several times in succession within the film as shown in Fig 5.7. In a transparent thin film, the intensity of the transmitted light is more than that of the reflected light. In this case only the first reflection at the top surface and at the bottom surface will be considered as they are of appreciable strength. The amplitude of the reflected and refracted components depends on the refractive index of the reflecting medium. Normally only 4% gets reflected while the major portion (96%) gets transmitted. For example, if we consider a soap film with refractive index 1.3, the reflectivity of the top surface is given by

$$r = \left(\frac{1.3-1}{1.3+1}\right)^2 = \frac{0.09}{5.29} = 0.017$$

This means that approximately only 2% of the incident light is reflected by the surface of the soap film. After two reflections, the intensity will be 0.02 × 0.02 or 0.04% of the initial intensity. The successive reflected rays will be of very poor intensity and therefore from theoretical point of view they can be neglected. Hence, for deriving the interference conditions we consider only the first two reflected rays. At each reflection, the incident amplitude is divided into a reflected component and a transmitted component. The interference in thin films was first observed by Newton and Robert Hooke but Thomas Young gave a satisfactory explanation of the phenomenon.

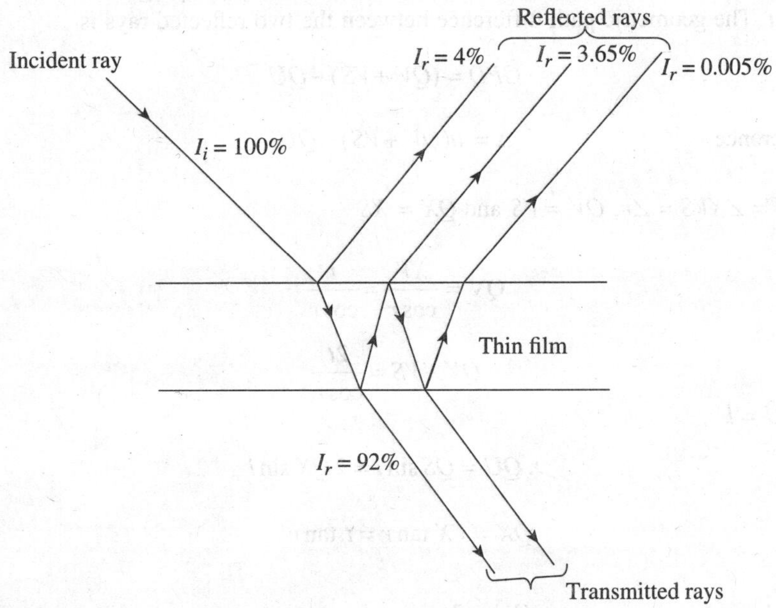

Fig. 5.7 Division of amplitude, multiple reflections in thin film

5.2.1 Plane Parallel Thin Film

Let us consider a transparent thin film of uniform thickness t and refractive index μ bounded by two parallel surfaces, as shown in Fig. 5.8. The film is surrounded by air on both the sides. Broad monochromatic sources illuminate the film obliquely. PQ represents one of the rays from a plane parallel monochromatic light beam incident on the upper surface of the film.

The ray PQ is partly reflected along QR and is partly transmitted into the film along QV. The transmitted ray QV makes an angle r with the normal and meets the lower bounding surface at V.

Again at V, it is partly reflected back into the film along VS, while a major portion refracts into the outer medium along VW. The reflected ray VS refracts at the upper surface and emerges out along ST, which is parallel to the ray QR. The waves travelling along the directions QR and $QVST$ are obtained from the same incident wave PQ. Therefore, they are coherent. They can produce interference if they are made to overlap by using a condensing lens. The optical path difference determines the condition of interference.

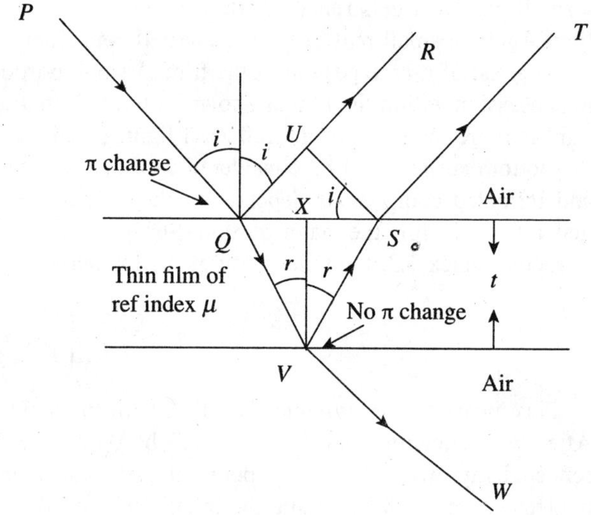

Fig. 5.8 Geometry of reflection of light from a parallel thin film

Calculation of Optical Path Difference

Let us now compute the optical path difference, Δ, between the reflected ray QR and the refracted ray $QVST$. A normal SU is drawn on line QR. From points U and S onwards the rays UR and ST travel equal distances. The reflected ray QR travels in air while the refracted ray ($QV + ST$) travels in the film of refractive index μ. The geometric path difference between the two reflected rays is

$$GPD = (QV + VS) - QU \tag{5.16}$$

Optical path difference

$$\Delta = \mu(QV + VS) - QU \tag{5.17}$$

In $\Delta QVS, \angle QVX = \angle XVS = \angle r$, $QV = VS$ and $QX = XS$

Therefore,

$$QV = \frac{XV}{\cos r} = \frac{t}{\cos r}$$

Thus,

$$QV + VS = \frac{2t}{\cos r} \tag{5.18}$$

In ΔQSU, $\angle QSU = i$

$$\therefore QU = QS \sin i = 2QX \sin i$$

Since,

$$QX = VX \tan r = t.\tan r$$

Therefore,

$$QU = 2t.\tan r.\sin i$$

From Snell's law,

$$\sin i = \mu \sin r$$

Therefore,

$$QU = 2t.\tan r\ (\mu \sin r) = \frac{2\mu t \sin^2 r}{\cos r} \tag{5.19}$$

Substituting Eqs (5.18) and (5.19) in Eq. (5.17), we get

$$\Delta=\mu\left[\frac{2t}{\cos r}\right]-2\mu t\frac{\sin^2 r}{\cos r}=\frac{2\mu t}{\cos r}\left(1-\sin^2 r\right)=\frac{2\mu t\cos^2 r}{\cos r}=2\mu t\cos r \tag{5.20}$$

The path difference given by Eq. (5.20) is not the actual optical path difference between the reflected rays as the phase change of π radians at point Q due to reflection at rarer to denser boundary has not been taken into account. Such a change does not take place at point V, as here the reflection is from a denser to a rarer boundary. Thus the correct path difference due to reflected rays is

$$\Delta = 2\mu t \cos r - \lambda/2 \tag{5.21}$$

Thus, the optical path difference between two reflected waves incident on a thin film is determined by Eq. (5.21).

Condition of constructive interference Constructive interference will take place between the reflected rays if the optical path difference between them is equal to an even multiple of half-waves. This means that

$$\Delta=2n\left(\frac{\lambda}{2}\right)=n\lambda \tag{5.22}$$

Thus, equating Eqs (5.21) and (5.22), we obtain the condition for constructive interference.

$$2\mu t\cos r=(2n+1)\frac{\lambda}{2} \tag{5.23}$$

where n is an integer; n = 0,1,2,3...

Condition of destructive interference Similarly, for the condition of destructive interference the optical path difference between the interfering waves should be an odd multiple of half-waves.

$$\Delta=(2n+1)\frac{\lambda}{2} \tag{5.24}$$

Equating Eqs (5.21) and (5.24), we get the condition of destructive interference.

$$2\mu t\cos r-\frac{\lambda}{2}=(2n+1)\frac{\lambda}{2}=(n+1)\lambda$$

Therefore,

$$2\mu t\cos r=n\lambda \tag{5.25}$$

As $(n + 1)$ is also an integer, we will denote $(n + 1)$ by n, where n = 1, 2, 3, 4...

The number n is called the *order of interference*. The rays after reflection from the surface of the uniform thin film become parallel. Parallel rays do not intersect at finite distances. The rays are to be focused with the help of a lens and then the interference pattern can be observed. Hence, the interference pattern formed due to a parallel thin film is located at infinity. The unaided eye accommodated to infinity can observe the interference pattern.

Salient Features of Interference Pattern in Uniform Thin Film

The following are the characteristic features of interference pattern formed in a uniform thin film.

Excessively thin film If the film is exceedingly thin, that is, $t \ll \lambda$, then from Eq. (5.21), $\Delta = \lambda/2$. Therefore, the condition of destructive interference is satisfied and the complete film appears dark in reflected light.

Normal incidence When monochromatic light is incident normal to the film, cos r = 0, therefore the condition of brightness and darkness becomes

$$2\mu t=(2n+1)\frac{\lambda}{2} \text{ and } 2\mu t=n\lambda$$

Thus the whole film will appear bright in reflected light when the thickness of the film is of $\lambda/4\mu$, $3\lambda/4\mu$, $5\lambda/4\mu$... and appears dark when its thickness is $\lambda/2\mu$, λ/μ, $3\lambda/2\mu$... etc. (Substitute $n = 0, 1, 2, 3....$)

Oblique incidence If the monochromatic parallel beam is incident at a fixed angle, then the whole film will appear uniformly dark or bright. The parameters μ, λ, t, and angle r decide which condition will be satisfied. For an incident parallel monochromatic beam on a uniform thin film, the film thickness t, the angle of refraction r, refractive index μ, and the wavelength λ have constant values. The condition of constructive interference or destructive interference will be satisfied depending on the constant values of these parameters. If the condition of constructive interference is satisfied, then it causes intensification of the incident colour. Thus if yellow light is incident on the film, intense yellow light will be observed in the reflected light.

Variation in angle of incidence A change in the angle of incidence of the rays causes a corresponding change in the angle of refraction and hence the optical path difference. The optical path difference decreases with the increase in the angle of incidence. Therefore, as the inclination of the film is changed, it will appear dark and bright (coloured) in succession.

White light If white light is incident on the film at a fixed angle, then the optical path difference will vary from one colour to another colour as wavelength is different for different colours. Accordingly, the film will appear to be coloured (mixture of a few colours) as those colours that satisfy the condition of constructive interference will be present in the reflected light. The colours that are not seen have interfered destructively. Further, the colours present in the reflected light will change if the inclination of the film with respect to the incident light changes.

If incident white light is not parallel, the optical path difference will change due to the change in the incidence angle. Hence, the film will show different colours when viewed from different directions.

5.2.2 Thin Film of Varying Thickness: Wedge-shaped Thin Film

In real life we come across films whose thickness is not constant. What kind of interference pattern is formed if we expose these films to light? We will study the interference of light in a film of varying thickness. One such film of varying thickness is called a *wedge film*. Its surfaces make an angle with each other as shown in Fig. 5.9. A wedge is thus a plate or a film of varying thickness. A wedge has zero thickness at one end and gradually increases to a particular thickness at the other end. An air wedge can be formed by inserting a thin spacer in between two microscope slides that are in contact with each other at one edge. For a film having thickness of the order of the wavelength of light used, the wedge angle θ will be very small. When a parallel beam of monochromatic light is incident on the wedge, the rays reflected from the top and bottom surfaces will be inclined with each other. The two reflected rays are not parallel because they are reflected from two surfaces that are not parallel but inclined. Thus, these two rays appear to diverge from a point near the film because of which the interference pattern is localized on the top surface of the thin film. As the path difference between these two interfering rays depends on the thickness of the non-uniform thin film, the interference pattern will show intensity variation. The interference pattern formed thus shows alternate dark and bright fringes.

From Fig. 5.10 a ray of light denoted by AB is incident on the wedge. The ray AB is partially reflected from the top surface as ray BC and a part of it is transmitted as ray BF. Ray BF is reflected back into the medium from the bottom surface as ray FD and a major portion of it, passes out. At point D, ray FD is refracted out as ray DE from the top surface. Both the reflected rays BC and ray DE are derived from the same incident ray AB by division of amplitude. Therefore they are coherent and on overlapping produce interference fringes. The fringes formed are equidistant, straight, and parallel to the contact edge of the wedge. The fringes are bright and dark when monochromatic light is incident.

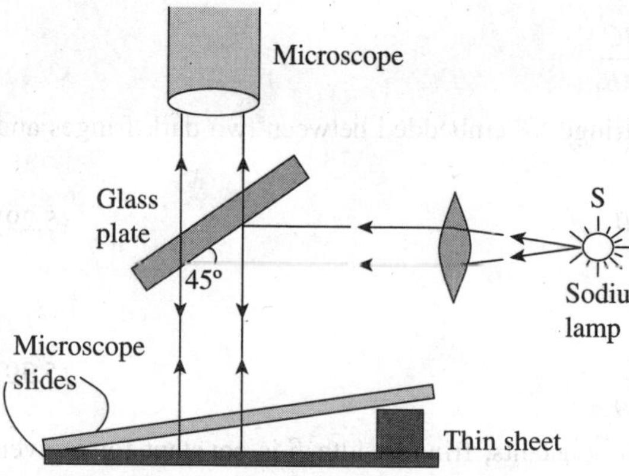

Fig. 5.9 Experimental arrangement to observe interference in wedge shaped thin films

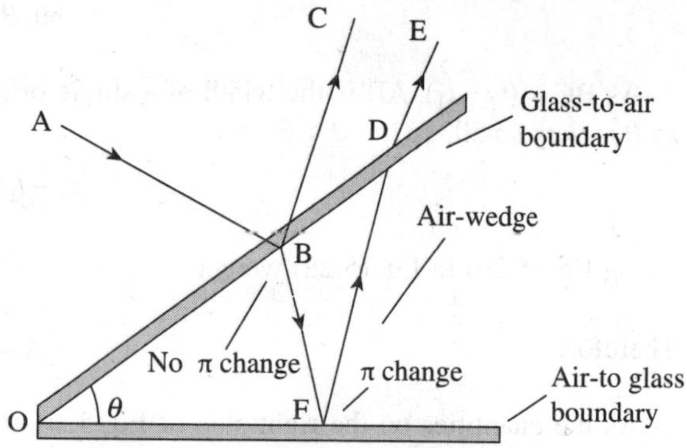

Fig. 5.10 Reflection of light incident on the wedge shaped air film

Salient Features of Interference Pattern in Wedge-shaped Film

The following are the characteristic features of interference patterns formed in a wedge-shaped film.

Fringes are equidistant The optical path difference is given in this case by the same equation (Eq. 5.21). The difference here is that the additional path difference of $\frac{\lambda}{2}$ (or phase difference of π) is introduced at the lower surface instead of the top surface as shown in Fig. 5.10.

Thus, $$\Delta = 2\mu t \cos r - \frac{\lambda}{2}$$

Let us consider that a dark fringe is formed at point A as shown in Fig. 5.11. The condition for dark fringe is

$$2\mu t \cos r = n\lambda$$

For normal incidence of light, cos $r = 1$ and if thickness of the wedge at A is considered to be, t_1 then the condition for dark fringe can be written as

$$2\mu t_1 = n\lambda \tag{5.26}$$

Now we consider that the next dark fringe will occur at point C, where the thickness of the wedge is t_2. The equation for dark fringe formed at point C can be written

$$2\mu t_2 = (n+1)\lambda \tag{5.27}$$

Subtracting Eq. (5.26) from Eq. (5.27), we get

$$2\mu(t_2 - t_1) = \lambda \tag{5.28}$$

Now by using the geometry of Fig. 5.11 we consider triangle CAB to find unknown $(t_2 - t_1)$ in terms of known or measurable parameters. From Fig. 5.11

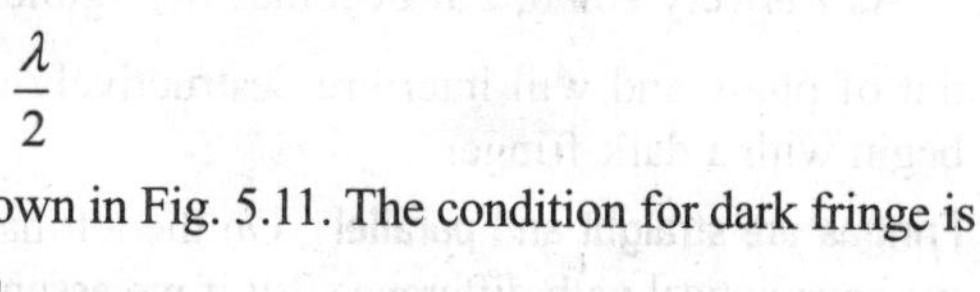

Fig. 5.11 Calculation of fringe width β

$$\tan\theta = \frac{BC}{AB}$$

As BC = $(t_2 - t_1)$, AB is the width of a single bright fringe 'β' embedded between two dark fringes and as θ is very small, so $\tan\theta \approx \theta$, then

$$(t_2 - t_1) = \beta \times \theta \tag{5.29}$$

Using Eq. (5.29) in Eq. (5.28) we get $$2\mu\beta\theta = \lambda$$

Therefore, $$\beta = \frac{\lambda}{2\mu\theta} \tag{5.30}$$

As the quantities on the right side of Eq. 5.30 are all constants, fringe width β is constant for a given wedge angle θ. It means that the interference fringes are equidistant from one another.

According to Eq. (5.30) an increase in the angle θ makes the fringes move closer. At an angle $\theta \approx 1°$ the interference pattern vanishes. On the other hand, if θ is gradually decreased, the fringe separation increases, and ultimately the fringes disappear as the faces of the film become parallel ($\theta = 0°$).

Fringe at the apex is dark The fringe at the apex of the wedge is always dark. At the apex, the thickness of the wedge is very small compared to λ, that is, $t \ll \lambda$. The optical path difference at the apex is given by

$$\Delta = 2\mu t - \frac{\lambda}{2}$$

As t is very small, $2\mu t$ becomes negligible. Hence, $\Delta = \frac{\lambda}{2}$. This implies that the interfering rays will be out of phase and will interfere destructively at the apex. Therefore, the wedge-fringe pattern will always begin with a dark fringe.

Fringes are straight and parallel On the left hand side of Eq. (5.23) and Eq. (5.25) of dark and bright fringes we have optical path difference $2\mu t$ if we assume normal incidence. As refractive index μ is constant, the path difference depends only on the thickness of the film. In wedge-shaped films, the thickness is constant along lines parallel to the apex line of the wedge. Hence, locus of all points having the same film thickness is a line parallel to apex line. Therefore the fringes are straight. Alternatively, we can say that as the path difference between the interfering rays is constant along a straight line parallel to the apex line, the fringes are straight. Since fringes are straight and are also equidistant, they are parallel.

Fringes are of equal path difference For each bright or dark fringe, the film thickness is constant. Hence, fringes are called *fringes of equal path difference.*

Fringes are localized Since fringes are formed or are located at the top surface of the wedge film, they are said to be localized.

Wedge angle θ The wedge angle θ can be experimentally determined with the help of a travelling microscope. The distance of the two dark fringes formed at two distinct points Q and R as shown in Fig. 5.12 can be measured by using a travelling microscope. Let the distances of the two fringes from apex O be x_1 and x_2, respectively.

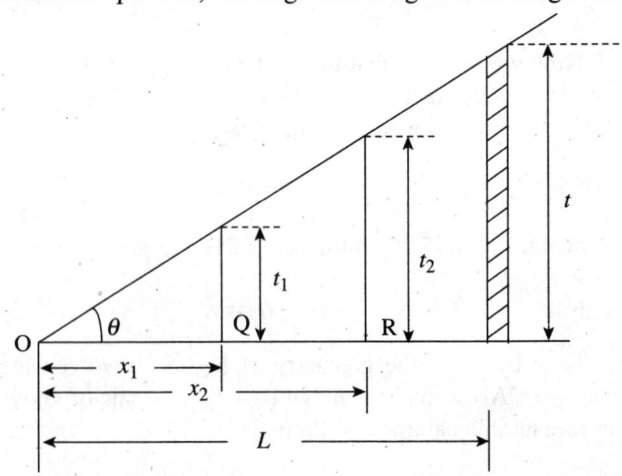

Fig. 5.12 Determination of the thickness of a thin object

Let t_1 and t_2 be the thickness of the wedge at points Q and R. Applying the condition of dark fringe at position Q, we get $2\mu t_1 = n\lambda$. But from Fig. 5.12, $t_1 = x_1 \tan\theta \approx x_1\theta$ as θ is very small.

Therefore,
$$2\mu\, x_1\theta = n\lambda \tag{5. 31}$$

Similarly for a dark fringe at R, $2\mu\, t_2 = (n+N)\lambda$, where N is the number of dark fringes between Q and R. Again we can express $t_2 = x_2 \tan\theta \approx x_2\theta$

Thus,
$$2\mu x_2\theta = (n+N)\lambda \tag{5.32}$$

Subtracting Eq. (5.31) from Eq. (5.32), we get

$$2\mu(x_2 - x_1)\theta = N\lambda$$

or
$$\theta = \frac{N\lambda}{2\mu(x_2 - x_1)}$$

If the film is of air, $\mu = 1$ then the above equation can be rewritten as

$$\theta = \frac{N\lambda}{2(x_2 - x_1)} \tag{5.33}$$

Spacer thickness The thickness 't' of the spacer used to form the air wedge can be determined from the above measurements. From Fig. 5.12 we can write

$$t = L\tan\theta \approx L\theta$$

where L is the length of the air wedge. Thus the spacer thickness can be written by using Eq. (5.33) as

$$t = \frac{LN\lambda}{2\,(x_2 - x_1)} \tag{5.34}$$

5.2.3 Newton's Rings

When a monochromatic light is reflected from a plano-convex lens of long focal length placed in contact with a plane glass plate, as shown in Fig. 5.13, concentric circular rings are formed which are called *Newton's Rings*. Newton discovered these circular rings hence they are called Newton's rings, but Thomas Young provided an accurate explanation for the formation of the rings. In the experimental arrangement, a plano-convex lens is kept on the glass plate so that a thin film of air is formed between them. The thickness of the air film varies from zero at the point of contact to some value 't' at the outer edge of the lens. If monochromatic light falls normally on the lens–plate system, then concentric bright and dark interference rings are observed in the reflected light.

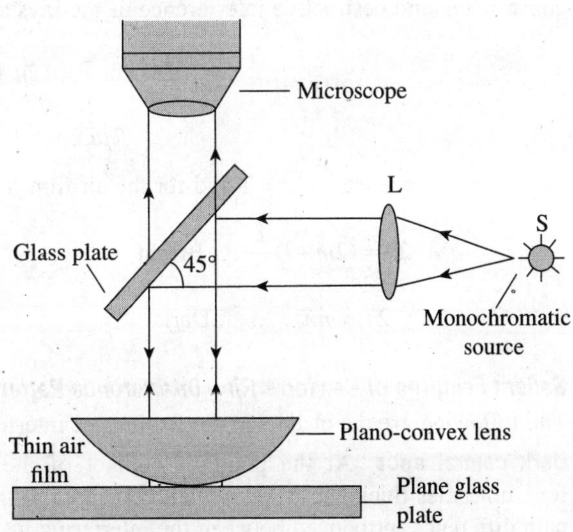

Fig. 5.13 Experimental arrangement for observing Newton's rings

In Fig. 5.13(a) when ray 1 is incident on the system, it gets partially reflected as ray 2 at the bottom curved surface of the lens (point A) and part of the transmitted ray is partially reflected as ray 3

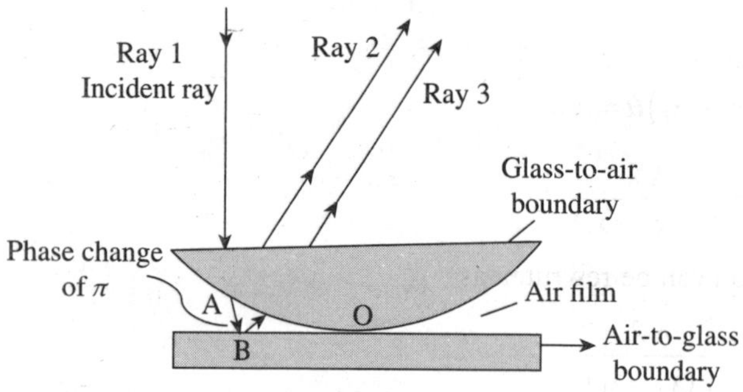

Fig. 5.13(a) Two reflected rays in Newton's ring set-up

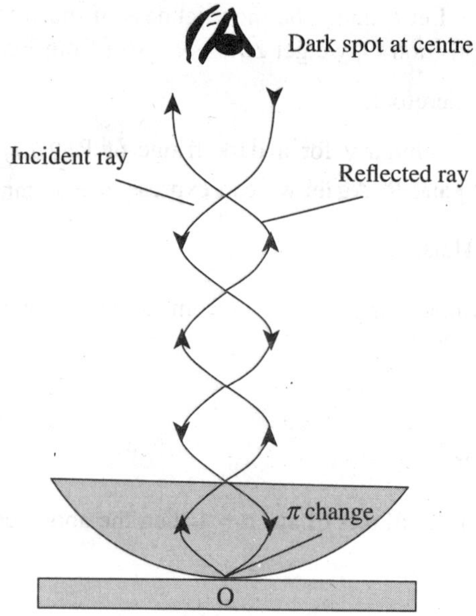

Fig. 5.13(b) Dark spot at centre

from the top surface of plane glass sheet (point B). The rays 2 and 3 are derived from the same incident ray 1 by division of amplitude and hence are coherent. Ray 2 undergoes no phase change, but ray 3 acquires a phase change of π upon reflection, since it is reflected from air-to-glass boundary. We do not consider the rays reflected from the surfaces of lens and glass plate owing to their thickness and further they do not show any interference. Thus, we consider only two reflected rays, which are reflected from top and bottom portions of the thin film formed between the plano-convex lens and glass plate. Hence, the conditions for constructive and destructive interference of the rays are same as given by Eqs (5.23) and (5.25).

$$2\mu t \cos r = (2n + 1)\frac{\lambda}{2} \text{ ... Bright}$$

$$2\mu t \cos r = n\lambda \text{ ... Dark}$$

For normal incidence cos $r = 1$ and for the air film $\mu = 1$. Therefore the conditions become

$$2t = (2m+1)\frac{\lambda}{2} \text{ .... Bright} \tag{5.35}$$

$$2t = m\lambda \text{ .... Dark} \tag{5.36}$$

Salient Features of Newton's Ring Interference Pattern

The following are the characteristic features of interference patterns formed by Newton's ring:

Dark central spot At the point of contact of the lens with the glass plate, the air film is only a few molecules thick and is very small compared to the wavelength of the visible light ($t << \lambda$). Thus, the path difference introduced between the interfering waves because of the thickness of thin film is zero, that is, 2 $\mu t = 0$. But the wave reflected from the glass plate suffers a phase change of π due to reflection, which is equivalent to a path difference of $\lambda/2$. Consequently the interfering waves at the centre are out of phase and thus interfere destructively as shown in Fig. 5.13(b). Therefore, at the point of contact between the lens and the plate, a dark spot is observed. The presence of dark spot at the centre is the direct evidence of the relative phase change of π between the two types of reflections at air-to-glass and glass-to-air boundaries.

Fringes of equal thickness Every alternate maxima and minima is due to the variation in the film thickness 't'. Each maximum or minimum is a locus of constant film thickness. Hence, the fringes are called fringes of equal thickness.

Circular fringes The air film extending from the point of contact to outward can be regarded as formed by an air wedge rotated around an axis passing through the point of contact. The locus of points having the same thickness then falls on a circle having its centre at the point of contact. Thus the thickness of the film is the same at all points on any circle having 'O' as the centre. The fringes are therefore circular. Alternatively as the thickness of the film is constant along the circle and the optical path difference depends on film thickness, the locus of all points having same path difference is a circle with the centre as a point of contact.

Fringes are localized If top and bottom surfaces of the thin film are not parallel, then the reflected rays from these surfaces are also not parallel. As the surface of plano-convex lens is inclined with the glass plate, the reflecting rays from these two surfaces are also inclined. When the system is illuminated with a parallel light beam, the inclined reflected rays interfere near the top surface of the air film. The fringes are seen near the upper surface of the film and hence they are said to be localized in the film.

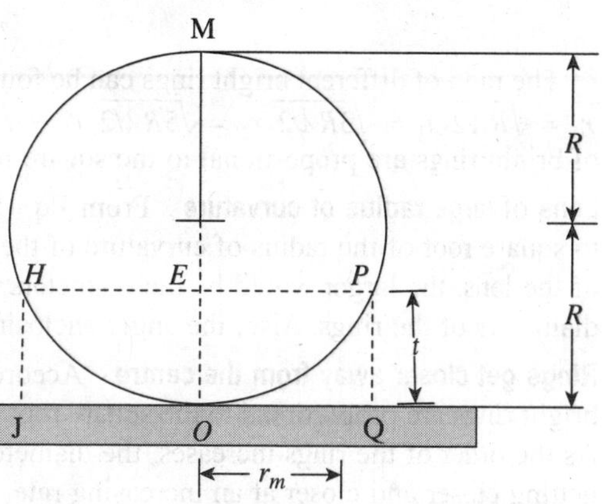

Fig. 5.14 Determination of the relationship between the radius of the dark ring and air film thickness

Radii of the dark rings Let R be the radius of curvature of the lens. Let a dark ring be located at point Q as shown in Fig. 5.14. The thickness of the air film at Q is $PQ = t$. The radius of mth ring at Q is $OQ = r_m$. By the theorem of intersecting chords,

$$EP \times HE = OE \times EM$$

$$EP = OQ = HE = r_m;\ OE = PQ = t$$

From the figure

$$EM = (OM - OE) = 2R - t$$

$$\therefore r_m^2 = t(2R - t) \text{ or } r_m^2 = 2Rt - t^2$$

As $2Rt \gg t^2$, t^2 can be neglected

$$\therefore r_m^2 = 2Rt \tag{5.37}$$

Substituting Eq. (5.36) in Eq. (5.37) we get

$$r_m^2 = m\lambda R$$

The radius of the dark ring is given by

$$r_m = \sqrt{m\lambda R} \tag{5.38}$$

The diameter of the dark ring is therefore given by

$$D_m = 2\sqrt{m\lambda R} \tag{5.39}$$

The radii of different dark rings can be found by substituting m = 1, 2, 3.... It can be seen that $r_0 = 0$, $r_1 = \sqrt{\lambda R}$, $r_2 = \sqrt{2\lambda R}$, $r_3 = \sqrt{3\lambda R}$, $r_4 = \sqrt{4\lambda R}$, and so on Thus, the radii (also diameters) of the dark rings are proportional to the square root of the natural numbers. Hence, the rings crowd away from the centre as the radius of dark rings does not increase in equal proportions as the ring number.

Radii of the bright rings Let us now suppose that a bright ring is located at point Q. Again by using the same geometry the radius of the m th bright ring is given by

$$r_m^2 = 2Rt$$

Substituting the condition of brightness represented by Eq. (5.35) in the above equation we get

$$r_m^2 = \frac{(2m+1)\lambda R}{2}$$

$$r_m = \sqrt{(2m+1)R\lambda/2} \qquad (5.40)$$

The radii of different bright rings can be found by substituting $m = 0, 1, 2, 3...$in Eq. (5.40). It is seen that $r_0 = \sqrt{R\lambda/2}, r_1 = \sqrt{3R\lambda/2}, r_2 = \sqrt{5R\lambda/2}, r_3 = \sqrt{7R\lambda/2}$, and so on. It is obvious that the radii (also diameters) of bright rings are proportional to the square root of the odd natural numbers.

Lens of large radius of curvature From Eq. (5.38) it is seen that the radius of a dark ring is proportional to square root of the radius of curvature of the lens $r_m \propto \sqrt{R}$. It implies that greater the radius of curvature of the lens, the larger would be the diameters of the rings. It leads to less error in the measurement of the diameters of the rings. Also, the angle enclosing the air film will be smaller as required in theory.

Rings get closer away from the centre According to Eqs (5.38) and (5.40), the diameters of the dark and bright rings are proportional to the square root of the natural numbers and odd natural numbers respectively. As the order of the rings increases, the diameter does not increase in the same proportion resulting in rings getting closer and closer at an increasing rate.

Alternatively, the wedge angle increases as one moves away from the centre as shown in Fig 5.15; hence the rings crowd as fringe width decreases with increase in wedge angle as per Eq. (5.30).

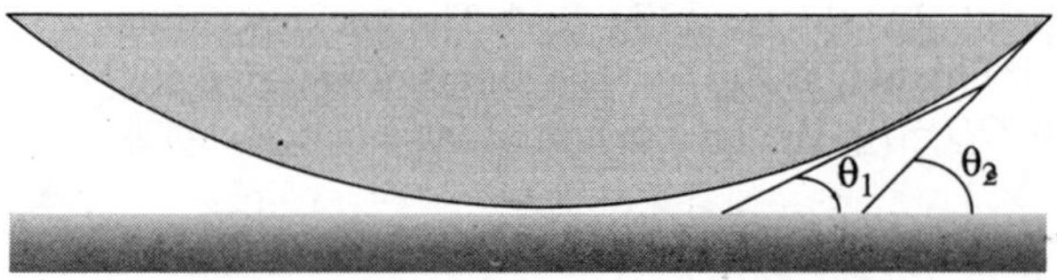

Fig. 5.15 Wedge angle 'θ' between the plano convex lens and glass plate.

Transmitted light If rings are seen from transmitted light then the condition of brightness and darkness is reversed. This is due to the change in the optical path difference by exactly $\lambda/2$ as the transmitted waves do not suffer any phase change at the boundary. Hence, the rings are exactly complementary to the reflected ring system, so that the centre spot is now bright. However, the contrast of the fringes is lesser in this case.

White light When Newton's ring apparatus is illuminated with white light, only a few coloured rings are observed near the point of contact. The thickness of the film is very small near the point of contact; hence only a few colours satisfy the condition of brightness. But as we go away from the centre, thickness of the film increases and hence more colours satisfy the condition of brightness. Due to overlap of colours we get white illumination away from the centre.

Determination of Wavelength

The wavelength λ of the incident monochromatic light can be determined with the help of Eq. (5.39). The square of the diameter of the m th dark ring can be written as

$$D_m^2 = 4m\lambda R \qquad (5.41)$$

For the (m + p)th dark ring, the diameter square becomes

$$D_{m+p}^2 = 4\,(m+p)\lambda R \tag{5.42}$$

Subtracting Eq. (5.41) from Eq. (5.42), we get

$$D_{m+p}^2 - D_m^2 = 4p\lambda R \tag{5.43}$$

Therefore,

$$\lambda = \frac{D_{m+p}^2 - D_m^2}{4pR} \tag{5.44}$$

In practice, the diameters of successive dark rings are measured using a low-power travelling microscope and a plot of the square of diameter versus m is drawn. The plot is a straight line, such as the one shown in Fig. 5.16. The slope of the line gives the value of $4\lambda R$.

$$\therefore \lambda = \frac{Slope}{4R}$$

The radius of curvature R of the lens may be determined using a spherometer and λ is computed with the help of Eq. (5.44).

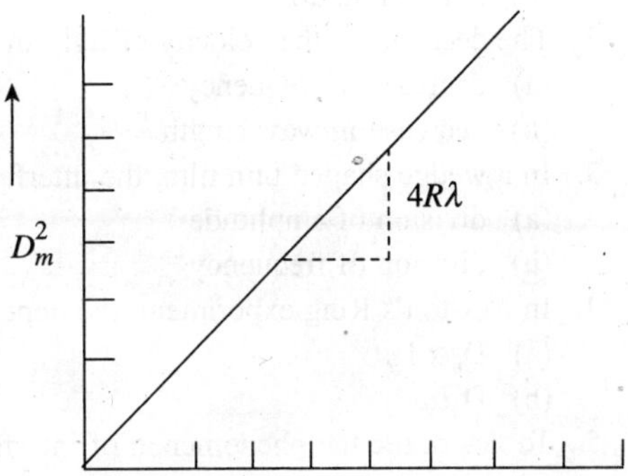

Fig. 5.16 Determination of wavelength of light from the plot of the square of the ring diameter versus the ring number

Determination of Refractive Index of a Liquid

Consider that the gap between the lens and the plane glass is filled with a transparent liquid, then the liquid film replaces the air film. For air film the diameter of the mth and $(m + p)$th dark ring is given by Eq. (5.38) as

$$(D_m^2)_{air} = 4\text{m}\,\lambda R \tag{5.45}$$

$$(D_{m+p}^2)_{air} = 4(\text{m}+\text{p})\lambda R \tag{5.46}$$

Subtracting Eq. (5.45) from Eq. (5.46), we get

$$(D_{m+p}^2)_{air} - (D_m^2)_{air} = 4\,p\lambda R \tag{5.47}$$

Now the air film is substituted by liquid film of which we want to find out refractive index. The diameter of the mth and (m + p)th dark ring is given by

$$\left(D_m^2\right)_L = \frac{4m\lambda R}{\mu} \tag{5.48}$$

$$(D_{m+p}^2)_L = \frac{4(m+p)\lambda R}{\mu} \tag{5.49}$$

Subtracting Eq. (5.48) from Eq. (5.49), we get

$$\left(D_{m+p}^2\right)_L - \left(D_m^2\right)_L = \frac{4p\lambda R}{\mu} \tag{5.50}$$

Dividing Eq. (5.47) by Eq. (5.50), we get

$$\mu = \frac{(D_{m+p}^2)_{air} - D_m^2)_{air}}{\left(D_{m+p}^2\right)_L - \left(D_m^2\right)_L} \tag{5.51}$$

Checkpoint 2

1. In Newton's Ring experiment when observed from transmitted light, contrast will be
   (a) very good
   (b) not seen at all
   (c) very poor
   (d) none of these
2. The decrease in the velocity of light in a medium may be attributed to
   (a) decrease in frequency
   (b) decrease in wavelength
   (c) decrease in its energy
   (d) none of the above
3. In a wedge shaped thin film, the interference is due to
   (a) division of amplitude
   (b) division of frequency
   (c) division of λ
   (d) division of energy
4. In Newton's Ring experiment, the dependance of the diameter of bright ring on refractive index μ is
   (a) $D_n \alpha\ 1/\mu$
   (b) $D_n \alpha \mu$
   (c) D_n does not depend on μ
   (d) $D_n \alpha\ 1/\sqrt{\mu}$
5. To determine the phenomenon of interference, we require two sources which emit radiation of
   (a) same frequency
   (b) nearly the same frequency
   (c) same frequency and should have a definite phase relationship
   (d) different frequency and should have a definite phase relationship
6. Two waves of the same frequency and of the same amplitude reach a point simultaneously. What should be the phase difference between the two waves so that the amplitude of the resultant wave is 2a?
   (a) 90°
   (b) 120°
   (c) 0°
   (d) 180°
7. A thin film having thickness $t \ll \lambda$ is seen in white light. It will appear
   (a) white
   (b) red
   (c) violet
   (d) black

5.3 Applications of Interference

An interference phenomenon has various engineering applications. These include testing of flatness of optical surfaces, testing chromatic aberrations of prism and lens, measurement of thickness of thin film, collimation testing, in interferometry, etc. Interferometry is an important investigation technique that is used in the fields of astronomy, engineering metrology, oceanography, seismology, spectroscopy, remote sensing, surface profiling, mechanical stress/strain measurement, and so on. We will discuss a few of these applications. The principle of interference is used for making precision measurements of the wavelength of light. The standard unit (SI) of length 'metre' was defined formerly as the distance between two marks on a platinum–iridium bar. The modern definition is that the standard metre is a length which contains exactly 1,650,763.73 wavelengths of orange-red light emitted by krypton-86. In astronomy, double-slit interference method is used to determine the angular separation of double stars and the diameter of fixed stars. In mechanical engineering, small displacements such as those produced by compression or elongation of a metal rod, crystals, etc. are measured with precision by using interference method. This measurement leads to accurate determination of Young's modulus and Poisson's ratio. The accurate determination of refractive index of liquids, gases, and thin transparent sheets is possible by using interference method.

Testing of Lens Surface

One of the important uses of Newton's rings is in the testing of the optical components manufactured for use in telescopes and other optical instruments. By keeping the test surface on a master, grinding (smoothing) of a lens surface can be verified. A master is an optically flat glass slide. The two faces of the optical flat glass slide are perfectly parallel to each other. The departure from flatness of each surface is less than a wavelength. If the lens is ground (smoothed/polished) perfectly, a circular fringe pattern is obtained as shown in Fig. 5.17(a). The formation of irregular ring pattern as shown in Fig. 5.17(b) indicates that the test surface is not ground properly. Thus the lens must be further ground and polished to remove the imperfections.

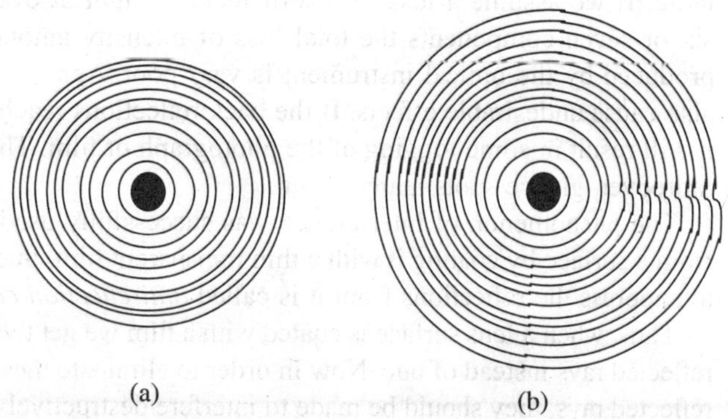

Fig. 5.17 Testing of grounding of plano-convex lens (a) perfect concentric rings (b) irregular ring pattern

Hence, by using interference phenomenon it is possible to obtain high-quality lenses ground with a precision of less than a wavelength.

Testing of Surface Finish

In modern technology interference is widely used for investigating the quality of a surface finish. Even after machining, components retain surface irregularities. The irregularities act as sources of stress leading to fatigue and cracks. The surfaces of components which are going to be subjected to high stresses and load reversals are therefore required to have a high surface finish. The smoothness of the test surface is examined by keeping it on an optical flat surface at an angle. If monochromatic light incident on the air wedge produces straight and equidistant fringes, then the surface of the component is smooth as shown in Fig. 5.18. If the fringes have a curvature then there are some irregularities on the surface and the surface needs further polishing for perfect optical flatness.

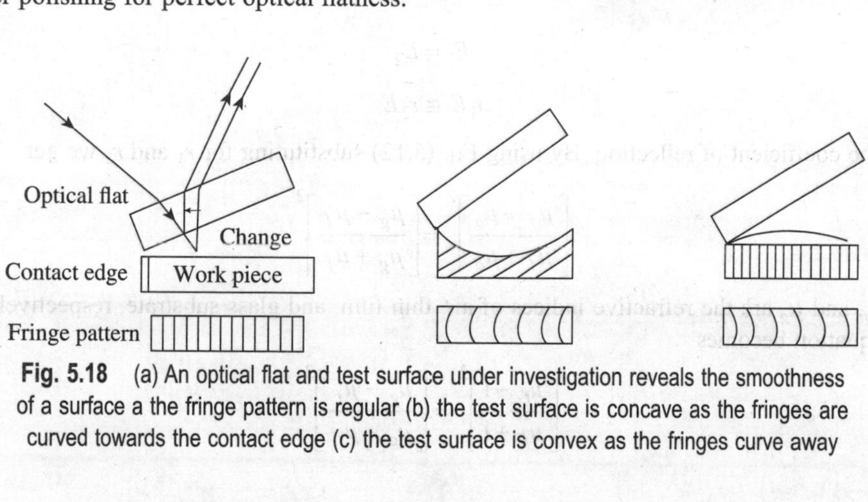

Fig. 5.18 (a) An optical flat and test surface under investigation reveals the smoothness of a surface a the fringe pattern is regular (b) the test surface is concave as the fringes are curved towards the contact edge (c) the test surface is convex as the fringes curve away

5.4 Anti-reflection Coatings

The compound lens system used in a camera or telescope contains six or seven optical components. When the light is incident on the glass component, it is reflected and that much amount of light is subtracted from the transmitted component. When the number of optical components is large, reflections are also large. If we assume a loss of 4% of incident light at every reflection from optical component, then for six or seven components the total loss of intensity amounts to almost 25%. Thus the quality of images produced by the optical instrument is very poor. Apart from reduction in intensity these back reflections also cause undesirable effects. If the back reflections reach the image formed by the lens in a camera, then it will result in some fogging of the photograph or film. There also will be a lack of contrast in the images; moreover, glance spots many occur.

The phenomenon of interference was successfully used by Alexander Smakula to reduce the reflection from a surface by coating it with a thin transparent film. Such a thin transparent film coated on a lens surface to suppress the reflections from it is called *antireflection coating* (AR coating).

Thus, when a lens surface is coated with a film we get two reflected rays instead of one. Now in order to eliminate these reflected rays, they should be made to interfere destructively. Hence, a thin film can act as an AR coating if (a) the waves reflected from its top and bottom surfaces are exactly 180° out of phase and (b) the waves have equal amplitudes. These two conditions are known as *phase condition* and *amplitude condition* respectively. Fulfilment of these two conditions ensures that the two reflected components nullify each other completely. These conditions help us in determining the required thickness of the film and the refractive index of the material to be used.

Let us discuss the two conditions by considering an example of a lens surface coated with a thin film of hard transparent material as shown in Fig. 5.19.

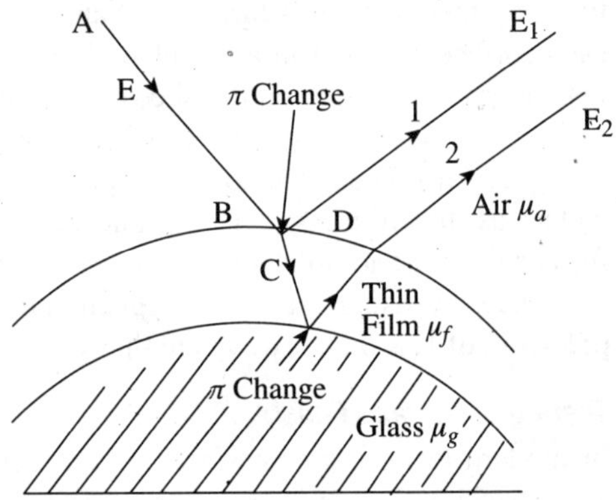

Fig. 5.19 Demonstration of principle involved in anti-reflection coating

Amplitude Condition

For eliminating the two reflected rays through destructive interference, the amplitudes of reflected rays, ray 1 and ray 2 must be equal, that is,

$$E_1 = E_2$$

or,

$$r_1 E \cong r_2 E$$

where r is the coefficient of reflection. By using Eq. (5.12) substituting for r_1 and r_2 we get

$$\left[\frac{\mu_f - \mu_a}{\mu_f + \mu_a}\right]^2 = \left[\frac{\mu_g - \mu_f}{\mu_g + \mu_f}\right]^2$$

where μ_a, μ_f, and μ_g are the refractive indices of air, thin film, and glass substrate, respectively. As $\mu_a = 1$, the above equation becomes

$$\left[\frac{\mu_f - 1}{\mu_f + 1}\right]^2 = \left[\frac{\mu_g - \mu_f}{\mu_g + \mu_f}\right]^2$$

Taking the square root of both sides of the above equation and then cross-multiplying, we get

$$\mu_f \mu_g - \mu_g + \mu_f{}^2 - \mu_f = \mu_f \mu_g - \mu_f{}^2 + \mu_g - \mu_f$$

By rearranging the terms, we get $$2\mu_f^2 = 2\mu_g$$

Therfore, $$\mu_f = \sqrt{\mu_g} \tag{5.52}$$

Thus, the amplitude condition gives the refractive index of the thin film relative to that of the substrate on which it is to be coated. Thus, the thin film should have less refractive index compared to that of the substrate and should be equal to its square root.

Phase Condition

The phase condition requires that the waves reflected from the top and bottom surfaces (ray1 and ray2) of the thin film should cancel each other. This is possible only when the path difference between them is 180°. Thus the optical path difference between them must be at least equal to half-wave or odd multiple of half-waves.

The optical path difference between the reflected waves 1 and 2 is

$$\Delta = 2\mu_f t \cos r - \lambda/2 - \lambda/2$$

The first $\lambda/2$ corresponds to the phase change of π due to reflection at the top surface and the second $\lambda/2$ is due to the phase change of π at the bottom surface. As the reflection from the bottom surface is from the rarer to the denser boundary (since μ_f is less than μ_g phase change of π radians is introduced). For normal incidence of light, cos $r = 1$ and the above equation simplifies to

$$\Delta = 2\mu_f t - \lambda = 2\mu_f t \tag{5.53}$$

Adding or subtracting λ does not change the optical path difference. From Eq. (5.53), the optical path difference depends only on the thickness of the thin film coated on the substrate. Hence, the two reflected waves 1 and 2 interfere destructively if the optical path difference is an odd multiple of $\frac{\lambda}{2}$.

Therefore, $$\Delta = (2n+1)\frac{\lambda}{2} \tag{5.54}$$

Using Eqs (5.53) and (5.54) we get $$2\mu_f t = (2n+1)\frac{\lambda}{2}$$

The thickness of the film will be minimum when $n = 0$. Hence, the minimum thickness of the thin film (AR coating) should be

$$2\mu_f t_{\min} = \lambda/2 \text{ or } t_{\min} = \frac{\lambda}{4\mu_f} \tag{5.55}$$

Thus the optical thickness $\mu_f t$ of the AR coating should be of a quarter of wavelength of the light used. Such quarter-wavelength coatings suppress the reflections and cause the light to pass into the transmitted component only.

Generally, for glass the refractive index is $\mu_g = 1.5$, then the thin film's refractive index will be $\mu_f = \sqrt{\mu_g} = 1.22$. The materials which are already available and have refractive index close to this value, are magnesium fluoride, MgF_2 ($\mu = 1.38$) and cryolite, $3NaF.AIF_3$ ($\mu = 1.36$). For selecting the material for coating in addition to the required value of refractive index, the material should also possess other properties. The requisite properties of the material are that the film formed of the material should adhere (stick) well. The film coated should be durable, scratch proof, and insoluble in ordinary solvents. Even though both MgF_2 and cryolite satisfy these requirements, magnesium fluoride is extensively used in AR coatings. Blooming technique is used for depositing such anti-reflection films on the glass by evaporating magnesium fluoride on the surface in vacuum.

It can be seen that the condition of minimum thickness as given by Eq. (5.55) is satisfied for only one particular wavelength. For visible spectrum, the wavelength normally considered is 5500 Å as the eye is most sensitive to this wavelength. This wavelength is located in the yellow–green portion of the spectrum. Consequently, when the white light is incident the reflection of red and violet light will be larger. Therefore, the optical component shows a purple hue by the reflected light.

Checkpoint 3

1. Which of the following is not suitable for non-reflecting coating
   (a) Magnesium fluoride
   (b) Sodium aluminum fluoride (cryolite)
   (c) Zinc sulphide
   (d) Titanium oxide
2. Which of the following is conserved when light waves interfere?
   (a) Intensity
   (b) Energy
   (c) Mass
   (d) Amplitude
3. Which of the following phenomena produces colours in a soap bubble?
   (a) Interference
   (b) Diffraction
   (c) Polarization
   (d) Dispersion
4. To observe interference in thin films with a light of wavelength λ, the thickness of the film should be
   (a) a few thousand times of λ
   (b) can be of any magnitude
   (c) should be of the order of wavelength of light used
   (d) should be much smaller than λ
5. Oil floating on water looks coloured due to interference of light. The approximate thickness of oil for such effect to be visible is
   (a) 100 Å
   (b) 10,000 Å
   (c) 1 mm
   (d) 1 cm
6. A thin film is observed in white light. The colour of the film seen at a particular point depends upon the
   (a) width of the source
   (b) distance of the source
   (c) location of the observer
   (d) none of the above
7. During the interference of light, energy is
   (a) created at the maxima
   (b) destroyed at the minima
   (c) not conserved
   (d) redistributed
8. Two sources of light are said to be coherent when both give out light waves of same
   (a) amplitude and phase
   (b) wavelength and constant phase difference
   (c) intensity and wavelength
   (d) phase and speed
9. Which of the following are coherent sources?
   (a) 60 W and 40 W bulbs
   (b) wo bulbs each of 40 W
   (c) Two halves of a 60 W bulb
   (d) Two virtual sources obtained by biprism
10. Two light waves of the same intensity are interfering. The intensity of light at the bright fringe compared to the intensity of one of the waves is
    (a) two times
    (b) four times
    (c) one-fourth
    (d) one-half
11. In a wedge shaped thin film, the bright fringe will change to dark for the thickness change of
    (a) double the wavelength of light
    (b) three-fourths of the wavelength of light

(c) one-fourth of the wavelength of light
(d) one-half of the wavelength of light

12. When a thin sheet of mica is introduced in the path of one of the interfering beams, then the fringe width
(a) increases
(b) decreases
(c) remains unchanged
(d) none of these

13. In Newton's ring arrangement with air film in reflected light, the diameter of *n*th dark ring is D_n. If air film is replaced by a liquid of refractive index, the diameter of *n*th dark ring will become
(a) $\sqrt{\mu}$ times
(b) $1/\sqrt{\mu}$ times
(c) $1/\mu$ times
(d) μ times

14. A wedge-shaped film is observed in reflected sunlight first through a red glass and then through a blue glass. The number of fringes in the latter case is
(a) less
(b) more
(c) equal in both cases
(d) none of these

15. With light of wavelength 490 nm in Newton's rings experiment, the radius of *n*th bright ring is R_n. With light of wavelength 640 nm in the same set-up, the corresponding radius will be
(a) 8 Rn/7
(b) 64 Rn/49
(c) Rn
(d) 7 Rn/8

16. In Newton's ring arrangement, bright and dark rings are obtained using sodium yellow light. If the entire arrangement is dipped in water, then the diameter of rings
(a) increases
(b) decreases
(c) fringe pattern disappears
(d) remains unchanged

17. Interference was observed in interference chamber when air was present. Now, the chamber is evacuated and if the same light is used, a careful observer will see
(a) no interference
(b) interference with brighter bands
(c) interference with darker bands
(d) interference in which width of fringes will be slightly increased

18. Interference occurs in
(a) longitudinal waves only
(b) transverse waves only
(c) electromagnetic waves only
(d) all the above

19. Newton's rings are
(a) loci of points of equal thickness
(b) loci of points of equal inclination
(c) loci of points of equal thickness and of equal inclination
(d) neither (a) nor (b)

20. To get the best result in anti-reflection coatings the refractive index of the coating should be equal to
(a) the refractive index of the glass on which it is to be coated
(b) the square root of the refractive index of the glass on which it is to be coated
(c) the square of the refractive index of the glass on which it is to be coated
(d) any value is suitable

21. If we wish to coat a flat slab of glass of refractive index 1.5 with the thinnest possible film of a transparent material so that light incident normally of wavelength 600 nm is transmitted, then the appropriate material should have refractive index equal to
(a) 1.6
(b) 1.5
(c) 1.25
(d) 2.25

22. In Newton's Ring experiment the central spot is bright when observed in
(a) white light
(b) absence of light
(c) transmitted light
(d) reflected light

Summary

- Light is a transverse, electromagnetic wave that is responsible for interference, diffraction, and polarization phenomena.
- The distance spanned by light waves in a medium is called the optical path or optical path length and is denoted by $'\Delta'$ that is related to the refractive index of the medium. Optical path has a unit of distance (metre).
- Interference by division of amplitude can be observed in a thin film is a film whose thickness is about the order of one wavelength of visible light, approximately taken to be 5500 Å.
- For a plane parallel thin film the conditions for constructive and destructive interference are

$$2\,\mu t\cos r = (2n+1)\frac{\lambda}{2} \text{ and } 2\,\mu t\cos r = (n+1)\lambda$$

where n is the order of interference which assumes the values 0,1, 2, 3, 4........

- The interference pattern formed in a wedge shaped thin film of varying thickness shows alternate dark and bright straight line equidistant fringes of width 'β' given by

$$\beta = \frac{\lambda}{2\mu\theta}$$

- In Newton's Ring, for a. non-uniform air film, concentric bright and dark interference rings are observed in the reflected light of diameter 'Dm' given by

$$D_m = 2\sqrt{m\lambda R}$$

- The applications of wedge shaped thin films and Newton's rings are to find unknown wavelength, refractive index of the thin film, thickness of the spacer, and radius of curvature of plano-convex lens respectively, as well as from testing the surface finish of the test surface.
- A thin transparent film coated on a lens surface to suppress the reflections from it is called antireflection coating (AR coating).
- The AR coating should satisfy the amplitude and phase conditions which result in finding the refractive index of the material to be used and the required thickness of the film as given by

$$\mu_f = \sqrt{\mu_g} \text{ and } t_{\min} = \frac{\lambda}{4\mu}$$

Solved Numericals

1. A parallel beam of light of wavelength 5893 Å is incident on a thin glass plate ($\mu = 1.5$) such that the angle of refraction into the plate is 50°. Calculate the smallest thickness of the glass plate, which will appear dark in reflected light.

Solution: The plate will appear dark when the condition of destructive interference is satisfied. The condition of darkness is

$$2\mu t\cos r = n\lambda$$

For calculating the smallest thickness of plate that causes destructive interference we will have to take $n = 1$.

$$t = \frac{\lambda}{2\mu\cos r}$$

$$t = \frac{5893\times 10^{-10}}{2\times 1.5\times\cos 50^\circ} = 3055.96 \text{ Å} = 0.31 \text{ µm}$$

2. Fringes of equal thickness are observed in a thin glass wedge of refractive index 1.55 with the fringes spacing of 0.2 mm when wavelength of light used is 5893 Å. Calculate the angle of wedge.

Solution: The angle of the wedge is given by

$$\theta = \frac{\lambda}{2\mu\beta}$$

where β is the fringe spacing

$$\theta = \frac{5893\times 10^{-10} m}{2\times 1.55\times 0.2\times 10^{-3} m} 9.5\times 10^{-4} rad = 0.05^\circ$$

3. In Newton's ring experiment the diameter of the 15 th dark ring changes from 1.45 cm to 1.34 cm when a liquid is introduced between the lens and the glass plate. Calculate the refractive index of the liquid.

Solution: The refractive index is given by

$$\mu = \frac{(D_m^2)Air}{(D_m^2)Liq.}$$

$$\mu = \frac{(D_{15}^2)_{Air}}{(D_{15}^2)_{Liq}} = \left[\frac{1.45}{1.34}\right]^2 = 1.17$$

4. Can a thin film of water ($\mu_f = 1.33$) formed on a glass window pane ($\mu = 1.62$) act as a non-reflecting film? If so, how thick should be the water film?

Solution: A film of refractive index μ_f can act as a non-reflecting film on a substrate having refracting index μ_g only if $\mu_f < \mu_g$. The refractive index of the coating is given by

$$\mu_f = \sqrt{\mu_g}$$

$$\therefore \mu_f = \sqrt{1.62} = 1.272$$

As the obtained value 1.272 is nearer to 1.33 hence, film of water can act as a non-reflecting film. The minimum thickness of the film is given by

$$t = \frac{\lambda}{4\mu_f}$$

It is assumed that for visible light λ = 5500 Å, as our eyes are more sensitive at this wavelength.

$$\therefore t = \frac{5500}{4 \times 1.33} = 1034 \text{ Å} = 0.31 \text{ μm}$$

Self-study Questions

1. What is the difference between optical and geometrical path difference?
2. What do you mean by thin film? Obtain the conditions for maxima and minima due to interference of reflected light in a thin film of uniform thickness.
3. In the phenomenon of interference in a thin film of uniform thickness due to reflected light, what happens when:
   (a) monochromatic light is incident normally on the uniform thin film?
   (b) white light is incident on the film?
4. The calculated path difference in interference due to reflected light is given by $2\mu\ t \cos r$, where symbols have their usual significance. What types of fringes do we get if:
   (a) t and μ are constant and
   (b) r and μ are constant?
   Explain how the former is different from the latter.
5. A wedge shaped film is illuminated by monochromatic light. Obtain an expression for the fringe width of interference pattern formed.
6. A wedge shaped air film is illuminated from top by a monochromatic light; discuss the salient features exhibited by the interference pattern formed due to the reflected light.
7. Explain how the phenomenon of interference is used in the determination of thickness of a thin paper. Derive the expression for the same.
8. Explain how the phenomenon of interference can be used for testing the optical flatness of a surface.
9. Describe Newton's rings experiment to determine the wavelength of incident monochromatic light.
10. In Newton's Ring experiment,
    (a) the plano-convex lens have larger radius of curvature.
    (b) the rings get closer away from the centre.
    (c) the central fringe is dark in reflected light.
    (d) the fringes are circular.
11. In a Newton's ring experiment, light of red colour is used first and then blue light. Which set of rings would have larger diameter and therefore greater spacing between them?
12. In Newton's rings experiment, if observed in reflected light, is it possible to obtain bright spot at the centre? Justify your answer.
13. How can Newton's rings experiment be used to determine the refractive index of a liquid?
14. Draw appropriate diagrams illustrating the interference in thin films in the following cases. Label the interfering rays as ray 1 and 2, and write down expressions for the total path difference between them:
    (a) constant thickness film
    (b) wedge shaped film
    (c) film enclosed between a plano-convex lens and plane glass plate in Newton's rings apparatus.
15. What are anti-reflection coatings? Explain the principle and applications.
16. Why do coated lenses look purple in reflected light?

Numerical Exercises

1. A monochromatic beam of light travels through a medium of refractive index 1.43 and thickness 0.78 μm. Calculate its optical path. (Ans. 1.12 μm)
2. A parallel beam of light of wavelength 5890 Å is incident on a thin glass plate (μ = 1.45) such that the angle of refraction into the plane is 40°. Calculate the smallest thickness of the glass plate that will appear dark. (Ans. 2651 Å)
3. A glass plate having parallel sides has thickness 4×10^{-4} mm and refractive index 1.5. If it is illuminated normally by white light, which wavelengths will be intensified in reflected beam in the visible spectrum? (Ans. 4800 Å)
4. A soap film of 8000 Å thickness is viewed at an angle of 25° to the normal. Find the wavelength in the visible spectrum which will be absent from the reflected light. The refractive index of the film is 1.433. (Ans. 5477 Å, 4382 Å)
5. A soap film of refractive index 1.33 and of thickness 1.6×10^{-6} m is illuminated by white light incident at an angle of 40°. The light reflected by it is examined by a spectroscope in which a dark band is found corresponding to a wavelength of 6×10^{-7} m. Calculate the order of interference of the dark band. (Ans. 6)
6. Find the thickness of a soap film (μ = 1.43) which gives constructive second order interference of reflected red light of λ = 500 mμ (millimicron). Assume normal incidence. (Ans. 2622.4 Å)
7. A thin film of soap solution is illuminated by white light at an angle of incidence, $i = \sin^{-1}(4/5)$. In reflected light, two dark consecutive overlapping fringes are observed corresponding to wavelengths 5.1×10^{-7} and 5.0×10^{-7}. Refractive index for the soap solution is 4/3. Calculate the thickness of the film. (Ans. 11.97 μm)
8. White light is reflected normally from a uniform oil film (μ = 1.33). Interference maximum for 6000 Å and a minimum for 4500 A, with no minimum in between, are observed. Calculate the film thickness. (Ans. 3383.45 Å)
9. A mixture of red light of wavelength 6600 Å and blue light of wavelength 4400 Å is incident normally on an air film formed by two glass plates. The thickness of the air film is 3300 Å. What is the colour of light reflected by the air film? (Ans. 4400 Å)
10. A parallel beam of sodium light strikes on a film of oil (μ = 1.46) floating on water (μ = 1.33). When viewed at an angle of 30° from the normal, eight dark bands are seen. Find the thickness of the film. (Ans. 1.72 μm)
11. White light is reflected from an oil film of thickness 0.01 mm and refractive index 1.4 at an angle of 45° to the vertical. If the reflected light falls on the slit of a spectrometer, calculate the number of dark bands seen between wavelengths 4000 Å and 5000 Å. (Ans. 12)
12. A parallel beam of white light is allowed to fall axially on the slit of a spectrometer and thin uniform air-film enclosed between two glass plates is placed in front of the slit with the surface of the air-film perpendicular to the path of light. On viewing through the telescope, 250 fringes are observed in the spectral region between 4000 Å and 6500 Å as measured in air. Calculate the thickness of the air-film. (Ans. 1.3×10^{-4} m)
13. Monochromatic light of wavelength 5500 Å falls on a wedge shaped air film enclosed by two glass plates which touch each other at one end. The wedge is formed by inserting a cylindrical wire 4.25 mm in diameter at a distance of 10 cm from this end. Find the fringe width of the pattern formed. (Ans. 6.47 μm)
14. Light (λ = 6300 Å) falls normally on a thin wedge-shaped film (μ = 1.6). There are ten bright and nine dark fringes over the length of film. By how much does the film thickness change over the length? (Ans. 1.7 μm)
15. In a certain region of a wedge shaped thin film, 17 fringes are obtained within region of the monochromatic light having 490 nm wavelengths. How many fringes will be observed in the same region of the film of 590 nm wavelength? (Ans. 14)
16. A cellophane film with index of refraction 1.5 has a wedge shape with the thickness on two opposite

sides being t_2 and t_1. If with normally incident light of wavelength 6000 Å, the number of fringes appearing on the film is 12, calculate the difference $t_2 - t_1$. (Ans. 2.4 μm)

17. Two plane glass surfaces in contact along one edge are separated at the opposite edge by a thin wire. If 20 interference fringes are observed between these edges in sodium light of normal incident, what is the thickness of the wire?

(Ans. 5.8 μm)

18. Newton's rings are formed with reflected light of wavelength 5893 Å with a liquid between the plane and the curved surface. The diameter of 15th dark ring is 0.5 cm and the radius of curvature of curved surface is 120 cm. Calculate the refractive index of the liquid.

(Ans. 1.7)

19. Newton's rings are produced by means of a plate and a lens of radius of curvature 4 meters and diameter 3.0 cm. How many rings are observed? How many rings would be observed if the arrangement were immersed in water (μ = 1.43)? Take λ = 5890 Å. (Ans. 96, 136)

20. In Newton rings experiment diameter of 18th ring was found to be 0.5 cm and that of 5th ring was 0.3 cm. If radius of Plano convex lens is 100 cm, calculate wavelength of light used. What happens to ring diameter if air film is replaced with liquid of refractive index 1.55.

(Ans. 3472 Å, 0.401 cm, 0.241 cm)

21. In Newton's rings experiment, what will be the order of the dark ring which will have double the diameter of that of 10th dark ring? The wavelength of incident light is 5890 Å. (Ans. 40)

22. A material having an index of refraction of 1.36 is used to coat a piece of glass. What should be the minimum thickness of this film in order to minimize reflected light at a wavelength of 550 nm? What should be the refractive index of the glass to get best effects? Why? (Ans. 101.1 cm, 1.85)

23. A glass microscope lens (μ = 1.5) is coated with magnesium fluoride (μ = 1.38) film to increase the transmission of normally incident light (λ = 6000 Å). What minimum film thickness should be deposited on the lens?

(Ans. 1086.95 Å)

24. A thin film of cryolite (μ = 1.35) is applied to a common lens. The coating is designed to reflect at blue end of the visible spectrum and transmit wavelength at the near infrared. What should be the minimum thickness of the film to be given to the lens to transmit highly monochromatic light at 8000 Å? (Ans. 1481.48 Å)

25. We wish to coat a flat slab of glass (μ = 1.55) with the thinnest possible film of a transparent material so that light of wavelength 550 nm incident normally is not reflected. We have two materials to choose from M_1 (μ = 1.45) and M_2 (μ = 1.6). Which one should be appropriate? What will be the smallest possible thickness of the coating?

(Ans. M_1, 94.82 ηm)

26. An engineer is interested in enhancing the transmission of the light incident on the glass lens. Which of the coatings, MgF_2 (μ = 1.38) or ZnS (μ = 2.37) will be appropriate? Calculate the minimum thickness of the coating. (Given: l = 5500 Å and μ for lens = 1.6)

(Ans. 996.38 Å)

Answers

Checkpoint 1

1(c), 2(a), 3(b), 4(b), 5(c)

Checkpoint 2

1(c), 2(b), 3(a), 4(d), 5(c), 6(c), 7(d)

Checkpoint 3

1(b), 2(b), 3(a), 4(c), 5(b), 6(c), 7(d), 8(b), 9(d), 10(b), 11(c), 12(c), 13(b), 14(a), 15(a), 16(b), 17(d), 18(d), 19(a), 20(b), 21(c), 22(c)

Polarization

6

Learning Objectives

After going through this chapter, you will be able to:

- Learn the different types of polarized lights
- Explain the different methods of production of plane-polarized light
- Learn the method of production of elliptically- and circularly-polarized lights
- Understand the methods to distinguish various types of polarized light
- Apply Malus law to calculate the intensity of transmitted polarized light
- Understand the applications of polarization in various fields

6.0 Introduction

Polarization is a fundamental property of light that is exclusively due to its transverse nature. The first evidence related to the existence of polarized light was given by Erasmus Bartholin. He observed two images when light was passed through a crystal of the mineral Iceland spar (a transparent, colourless variety of calcite). Natural light is unpolarized which can be converted into different types of polarized rays using simple optical components. Polarized light have many important practical applications in industry and engineering. One of the most familiar applications is in liquid crystal displays, which are widely used in wristwatches, calculators, TV screens, video displays, etc. In this chapter, we will learn the different types of polarization, the production and analysis of polarized light, and its applications.

6.1 Polarization of Light

Light is an electromagnetic wave with electric and magnetic vectors vibrating in mutually perpendicular planes, and also perpendicular to the direction of propagation of light. According to Maxwell's theory of electromagnetism, magnetic field in light is produced due to the changing electric field. However, all the optical effects shown by the light waves are only due to the electric field. Hence, electric field vector is also identified as the light vector. In ordinary or unpolarized light emanating from sources such as the sun, or a light bulb, the light vector vibrates in all possible directions in a plane perpendicular to the direction of propagation; as shown in Fig. 6.1. If, however, we confine the vibration of the light vector in a single plane or along a fixed straight line in the plane, then such a light is called *plane-polarized* or *linearly-polarized* light.

An ordinary light is incident normally on a pair of parallel plates P_1 and P_2 of tourmaline crystal (a naturally existing crystal which acts as a slit) cut parallel to the crystallographic axis (a direction inside the crystal only along which the light vector will propagate). The emergent light shows a variation in intensity as P_2 is rotated as shown in Fig. 6.2. The intensity is maximum when the axis of P_2 is parallel to that of P_1 (Fig. 6.2(a)), and zero when at right angles (Fig. 6.2(b)). This shows that the light emerging from P_1 is not symmetrical about the direction of propagation of light, but its vibrations are confined only to a single line in a plane perpendicular to the direction of propagation. Such light is called plane-polarized or linearly-polarized light as shown in (Fig. 6.2(c)).

Transforming unpolarized light into the polarized light is called the phenomenon of polarization.

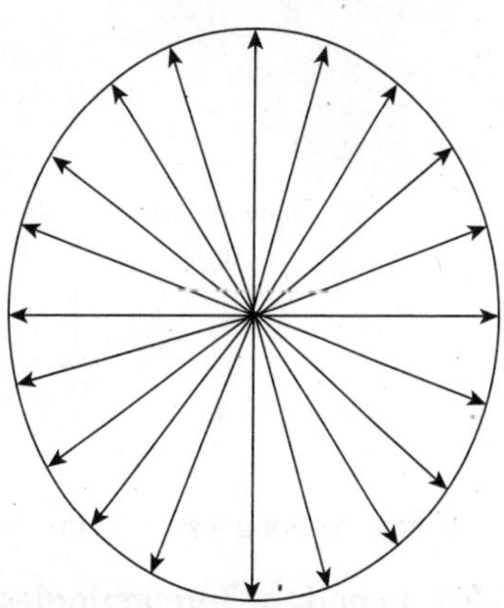

Fig. 6.1 Electric field vibrations in unpolarized light

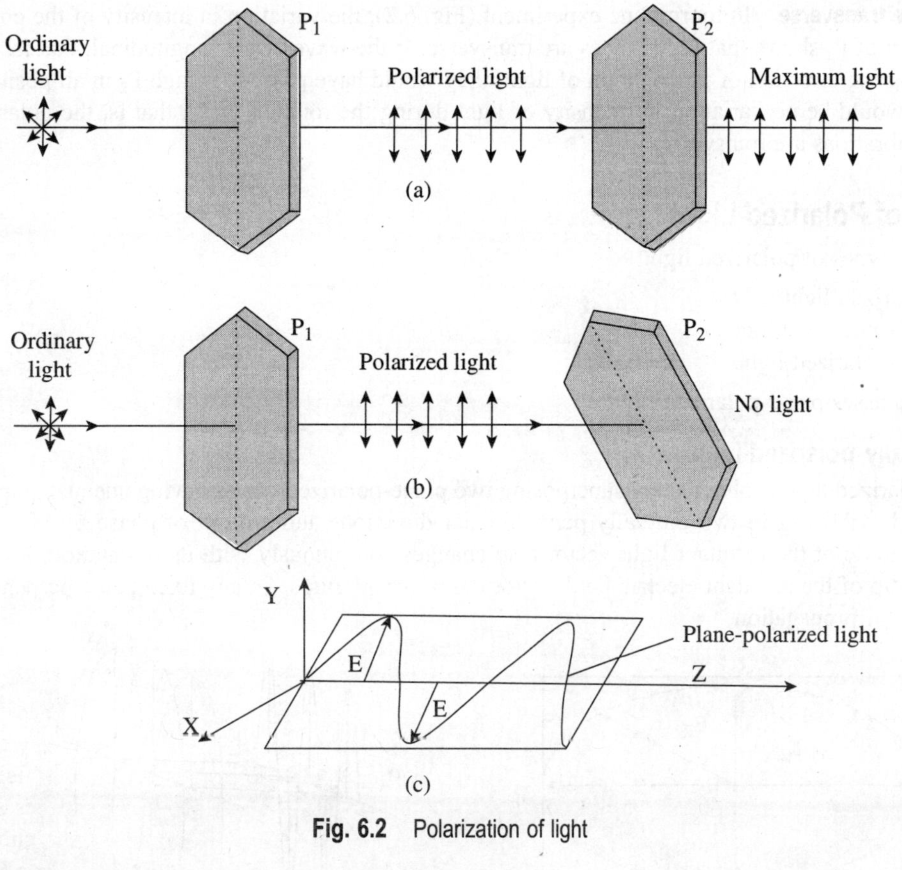

Fig. 6.2 Polarization of light

6.2 Representation of Plane-polarized Light

In a plane-polarized beam of light, the vibrations are along a single straight line. If the vibrations are parallel to the plane of the paper, they are represented by arrows (Fig. 6.3(a)).

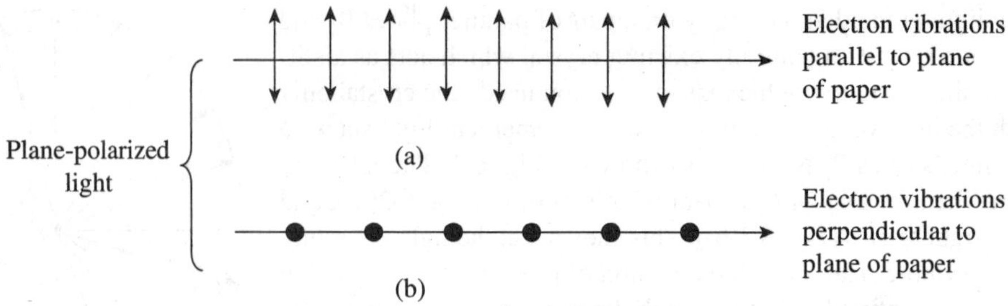

Fig. 6.3

If they are along a straight line perpendicular to the plane of the paper, they are represented by dots (Fig. 6.3 (b):

6.2.1 Important Terminologies

Plane of vibration This is the plane containing the direction of vibration and the direction of propagation of light.

Plane of polarization This is the plane passing through the direction of propagation and has no vibration.

Light waves are transverse In tourmaline experiment (Fig. 6.2), the variation in intensity of the emergent light on rotation of P_2 shows that light waves are transverse. If the waves were longitudinal, that is, having vibrations along the direction of propagation of light, they would have passed through P_2 in all positions of it. Then there would be no variation in intensity of light during the rotation of P_2, that is, the polarization requires that vibrations are transverse.

6.3 Types of Polarized Light

There are three types of polarized light.

1. Plane-polarized light
2. Elliptically-polarized light
3. Circularly-polarized light

Section 6.1 discusses plane-polarized light.

6.3.1 Elliptically-polarized Light

Elliptically-polarized light is obtained by superposing two plane-polarized waves having unequal amplitudes of electric fields vibrating in two mutually perpendicular directions and are out of phase by 90°. In this case, the magnitude of the resultant light vector also changes continuously with its orientation. As shown in Fig. 6.4 the tip of the resultant electric field vector describes an ellipse in any fixed plane perpendicular to the direction of propagation.

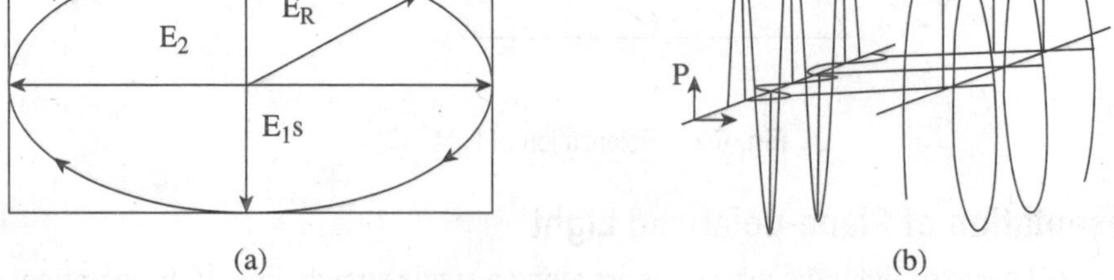

Fig. 6.4 Elliptically-polarized light

6.3.2 Circularly-polarized Light

Circularly-polarized light is obtained by superposing two plane-polarized waves having equal amplitude of electric fields vibrating in two mutually perpendicular directions and are out of phase by 90^0. The magnitude of the resultant light vector remains constant while its orientation changes continuously. The tip of the resultant electric field vector describes a circle in any fixed plane perpendicular to the direction of propagation as shown in Fig. 6.5.

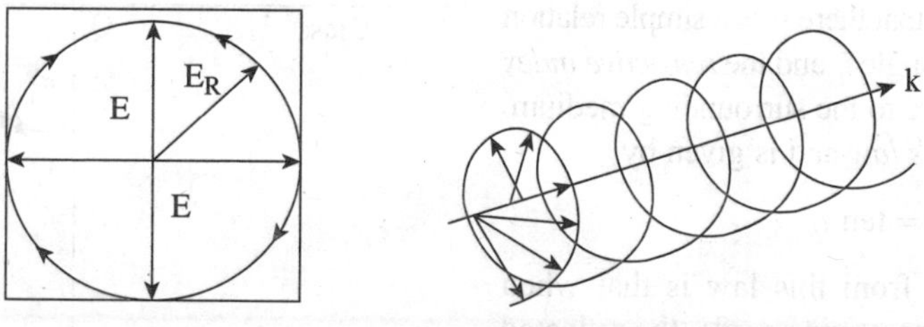

Fig. 6.5 Circularly-polarized light

Checkpoint 1

1. Natural light is
   (a) unpolarized
   (b) plane polarized
   (c) circularly polarized
   (d) partially polarized
2. Which of the following properties is not exhibited by sound waves?
   (a) Interference
   (b) Diffraction
   (c) Polarization
   (d) Scattering
3. Which of the following methods produces polarized light?
   (a) selective absorption
   (b) double refraction
   (c) reflection
   (d) all the above
4. An unpolarized light when passed through a Nicol prism becomes
   (a) elliptically polarized
   (b) circularly polarized
   (c) plane polarized
   (d) partially polarized
5. When the light is incident at the polarizing angle, which of the following is completely polarized?
   (a) Refracted light
   (b) Reflected light
   (c) Both reflected as well as refracted light
   (d) Neither reflected nor refracted light
6. Plane-polarized light can be produced by
   (a) simple reflection
   (b) Nicol prism
   (c) pile of plates
   (d) all of the above

6.4 Production of Plane-polarized Light

Plane-polarized light can be produced by different methods, namely reflection, refraction, double refraction, selective absorption (dichroism), and scattering. These methods are explained as follows.

6.4.1 Production by Reflection

Malus, in 1808, discovered that when unpolarized light is incident on the surface of any transparent material, the reflected and refracted beams are partially plane polarized. For a particular angle of incidence i_p, called the 'polarizing angle' the reflected light is completely plane-polarized with the plane of vibration perpendicular to the plane of incidence.

Brewster discovered that there was a simple relation between the polarizing angle i_p and the *refractive index* μ of the material relative to the surrounding medium. This is called *Brewster's law* and is given by

$$\mu = \tan i_p$$

A direct deduction from this law is that when light is incident at the polarizing angle, the reflected beam is at right angles to the refracted beam. This can be proved as below:

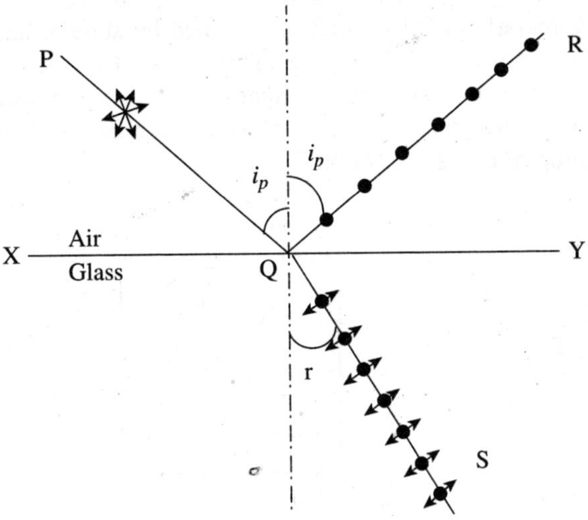

Fig. 6.6 Reflection of light for producing plane-polarized light

Let a beam *PQ* (Fig. 6.6) of un-polarized light be incident on the surface *XY* of a transparent substance at the polarizing angle i_p. Let *QR* be the reflected beam and *QS* the refracted beam. Let *r* be the angle of refraction. Then from Brewster's law, we have

$$\mu = \tan i_p = \frac{\sin i_p}{\cos i_p} \tag{6.1}$$

By Snell's law, we have

$$\mu = \frac{\sin i_p}{\sin r} \tag{6.2}$$

Comparing Eqs (6.1) and (6.2), we get

$$\frac{\sin i_p}{\cos i_p} = \frac{\sin i_p}{\sin r}$$

As $\sin (90° - i_p) = \cos ip$

$\therefore \sin (90° - i_p) = \sin r$, which implies $(90° - i_p) = r$

Hence, $(r = i_p) = 90°$

This is the case when reflected and refracted beams are at right angles to each other.

6.4.2 Polarization by Refraction (Pile of Plates)

When ordinary unpolarized light is incident on the upper surface of a glass slab at the polarizing angle i_p (Fig. 6.7), the reflected light is completely plane polarized with vibrations perpendicular to the plane of incidence, while the refracted light is only partially polarized having vibrations both in the plane of

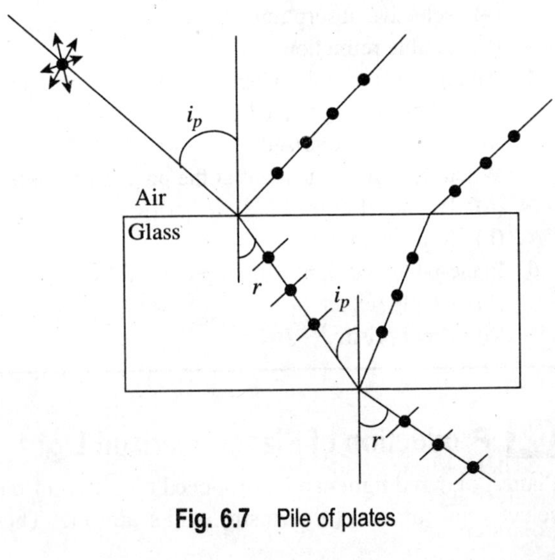

Fig. 6.7 Pile of plates

incidence as well as perpendicular to the plane of incidence. The refracted light is incident again at the lower face at an angle i_p and again the vibrations parallel to the plane of incidence get reflected. Thus in the refracted ray, the number of parallel vibrations decreases. If we keep a number of plates, then at each plate the light is reflected which is completely plane-polarized light and the refracted light also becomes completely plane polarized.

6.4.3 Polarization by Selective Absorption (Dichroism)

In 1815, Biot discovered that certain mineral crystals which are anisotropic, absorb light selectively depending on its direction of vibration. When natural light passes through such crystals, it splits into two components that are polarized in mutually perpendicular planes. The crystals strongly absorb light that is polarized in a direction parallel to a particular plane in the crystal, but freely transmit the light polarized in the perpendicular direction. This absorption of the rays along a particular direction is called *selective absorption* or *dichroism*. Thus for a crystal of particular thickness, one of the components is totally absorbed and the other component which emerges from the crystal is linearly polarized. An example of a naturally existing dichroic crystal is tourmaline crystal (an alumino-borosilicate containing Al_2O_3, B_2O_3, and SiO_2).

In 1852 W.B. Herapath discovered a synthetic dichroic crystal herapathite (quinine sulphate per iodide). However, these crystals did not find practical applications, as they cannot be grown to sufficiently bigger size. Later in 1928 E. H. Land invented a method of aligning the crystals to obtain an efficient and economic polarizing material called polaroid sheet. This method was refined further by Land and Rogers. Polaroid sheets are of two types, H-sheet and K-sheet. *H-sheet polaroid* sheets are fabricated by taking a plastic sheet of long-chain molecules of PVA (polyvinyl alcohol). It is then heated and stretched to many times its original length. During the stretching process the PVA molecules align themselves along the direction of stretching. The sheet is then laminated to a rigid sheet of plastic to stabilize its size. It is then impregnated with iodine, the molecules of which attach themselves to the PVA molecules and consequently form long parallel conducting chains. When natural light is incident on the sheet, the polarization component that is parallel to the iodine atom chains induces current and is therefore strongly absorbed as shown in the Fig. 6.8. Consequently, the transmitted beam is polarized in the direction perpendicular to the direction of the molecular chains. The direction of vibration of E vector in the transmitted beam corresponds to the transmission axis of the polaroid sheet.

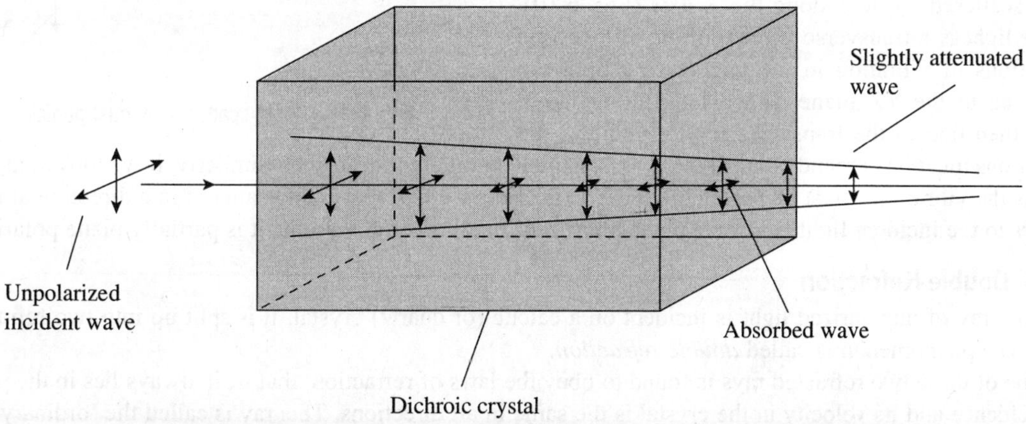

Fig. 6.8 Polarization of light by dichroic crystal (selective absorption process)

The K-sheet is fabricated by heating a transparent PVA sheet in the presence of an active dehydrating catalyst such as hydrogen chloride. The sheet darkens slightly and becomes strongly dichroic. Such sheet polarizers are cheap and can be fabricated in large sizes. Polaroid sheets are widely used in sunglasses and camera filters to eliminate unwanted glare from objects. They are also used as polarizer and analyser to produce polarized light and also to examine the nature of polarization of light.

6.4.4 Polarization by Scattering

When a beam of white light passes through a medium consisting of small particles of about the same size as the wavelength of light, the light seen in a direction at right angles to the incident beam appears bluish. This phenomenon is known as *scattering of light*. It can be demonstrated by the following experiment:

A few drops of dilute sulphuric acid is added to a dilute solution of hypo (sodium thiosulphate) prepared in a glass tank. A precipitate of fine sulphur particles is formed, the particles grows in size with time. Now a beam of light from a bright source is sent through the tank. The light scattered by the sulphur particles in a direction at right angles to the incident light appears blue.

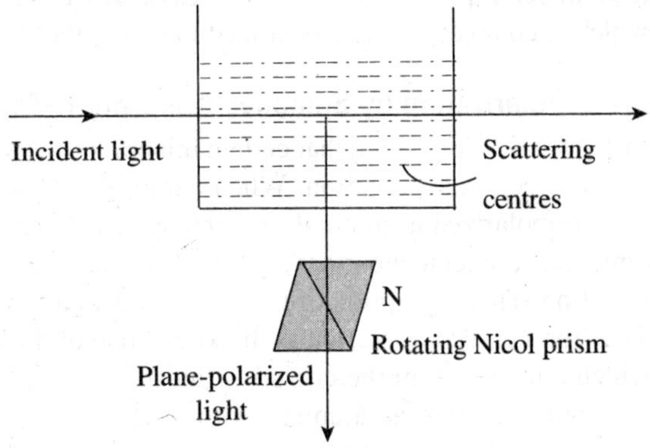

Fig. 6.9 Polarization by scattering of light

When the blue light is seen through a rotating Nicol (Fig. 6.9), the intensity of transmitted light varies from maximum to zero. This shows that the light scattered in this direction is plane polarized. The position of maximum intensity is obtained when the principal section of the Nicol is perpendicular to both the direction of the incident beam and the direction of observation. Polarization by scattering can be explained as below.

Light falling on objects such as particles of dust, smoke, and air molecules is scattered in all directions. Suppose that a beam of unpolarized light is incident on a scattered particle along the X axis (Fig. 6.10). Since light is a transverse wave motion, all possible directions of vibration in the unpolarized light are confined to the YZ plane. If we look along the Y axis, then due to the transverse nature of light, the vibrations must be perpendicular to the Y axis, namely parallel to the Z axis. Similarly, if we look along the Z axis the vibrations must be parallel to the Y axis. Thus we find that light scattered in a direction at right angles to the incident light is always plane polarized. In all other directions, it is partially plane polarized.

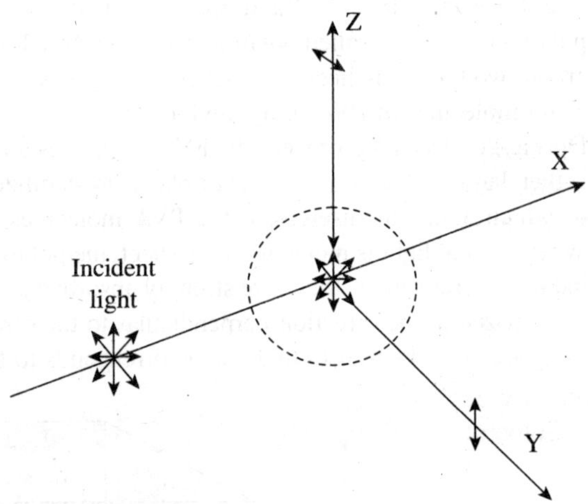

Fig. 6.10 Light scattered by dust particle

6.4.5 Double Refraction

When a ray of unpolarized light is incident on a calcite (or quartz) crystal, it is split up into two refracted rays. The phenomenon is called *double refraction.*

One of these two refracted rays is found to obey the laws of refraction, that is, it always lies in the plane of incidence and its velocity in the crystal is the same in all directions. This ray is called the 'ordinary ray' (O-ray). The other refracted ray does not obey the laws of refraction. It travels in the crystal with different speeds in different directions. Hence, it is called the 'extraordinary ray' (E-ray). Along the optic axis, however, the O-ray and the E-ray both have the same velocity and hence the same refractive index.

A ray of light is incident normally on a crystal, a principal section of which is as shown in Fig. 6.11. The ray is split up into two rays—O and E. O-ray passes through the crystal without deviation, whereas the E-ray is displaced by a distance proportional to the thickness of the crystal. Inside the crystal, the O-ray

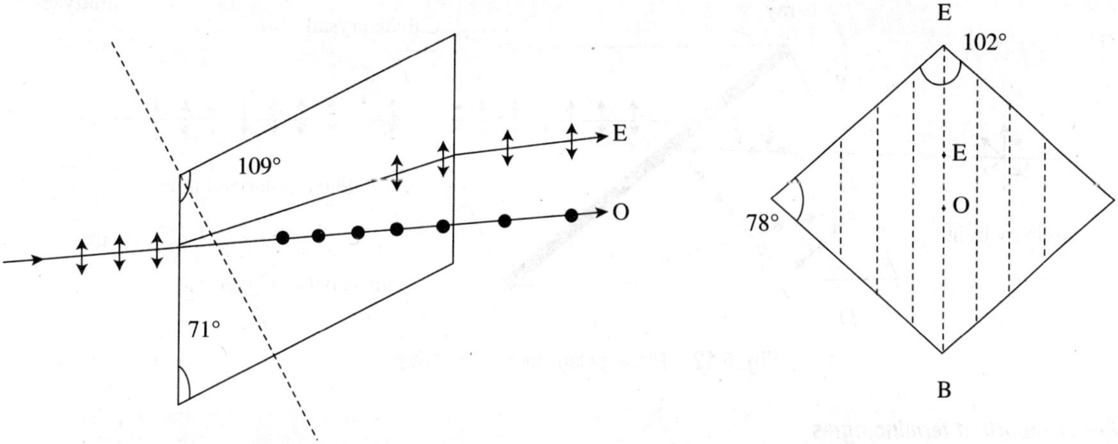

Fig.6.11 Representation of the phenomenon of double refraction

always lies in the plane of incidence. Only when the plane of incidence is a principal section, the E-ray also lies in the plane of incidence.

Polarization of rays The ordinary and extraordinary rays obtained by double refraction are plane-polarized. The O-ray is polarized in the principal section (i.e., it has vibrations perpendicular to the principal section), whereas the E-ray is polarized perpendicularly to the principal section (i.e., it has vibrations parallel to the principal section).

Nicol prism is made by using the doubly refracting material viz. calcite. To produce plane-polarized light, a beam of ordinary light is sent through a Nicol prism. Inside the prism it is broken up into two components—the O-component is internally reflected at the Canada Balsam layer and is absorbed as shown in Fig. 6.12; the E-component emerges out. It is plane polarized with its vibrations parallel to the shorter diagonal of the end face of the Nicol. Thus, we can convert ordinary light into plane-polarized light by using Nicol Prism.

Some of the properties and differences between O-ray and E-ray are highlighted in Table 6.1.

Table 6.1 Properties and differences between O-ray and E-ray

O-ray	E-ray
O-ray obeys Snell's law.	E--ray does not obey Snell's law.
It is plane polarized and has vibrations of electric field perpendicular to optic axis.	E-ray is plane polarized and has vibrations of electric field parallel to optic axis.
It does not deviate from its path when it comes out of the doubly refracting crystal.	E-ray deviates from its path when it comes out of the doubly refracting crystal.
Wavefront of O-ray is spherical as velocity of O-ray is constant in all directions.	Wavefront of E-ray is ellipsoidal as velocity of E-ray is different in different directions.
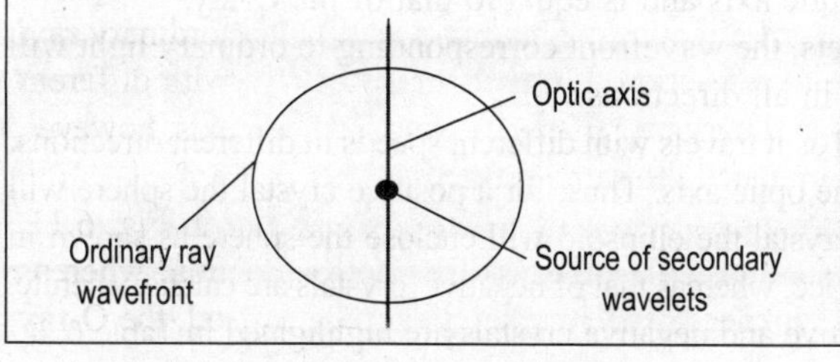	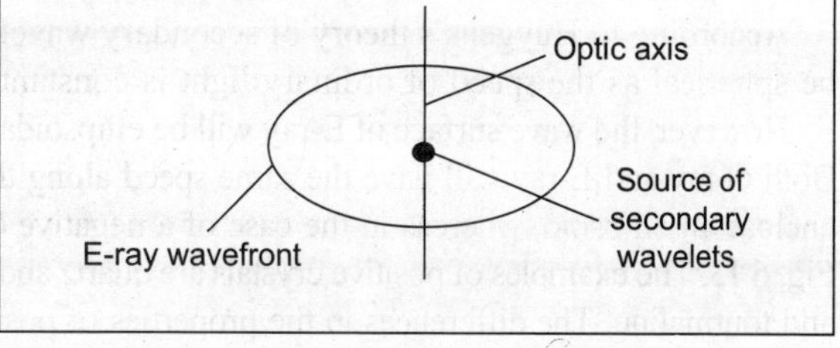

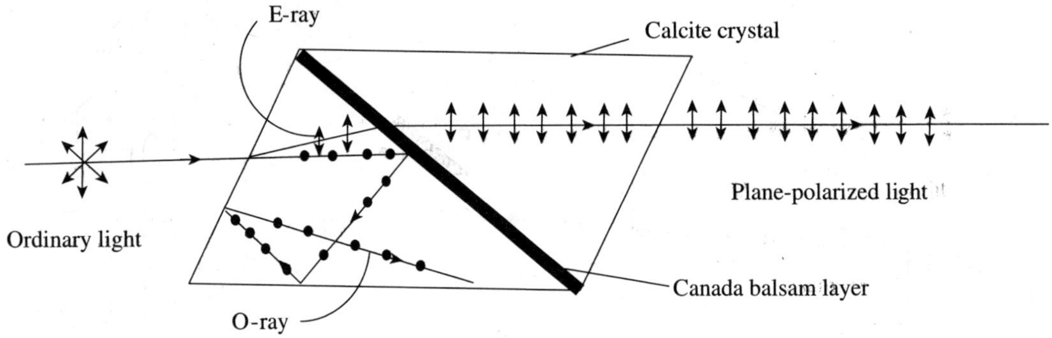

Fig. 6.12 Plane polarization using Nicol prism

Some important terminologies

Optic axis of the crystal It is the direction inside the doubly refracting crystal along which there is no double refraction, that is, ordinary and extraordinary rays will travel with the same speed.

Principal refractive indices of a crystal The refractive index of a medium is defined as the ratio of the velocity of light in vacuum to the velocity of light in the medium. In case of doubly-refracting crystals, the O-wave travels with the same velocity in all directions, but the E-wave has different velocities in different direction. The velocity of E-wave along the optic axis is same as that of O-wave, but in a direction perpendicular to the optic axis the two velocities have the maximum difference; hence there are two principal refractive indices, one corresponding to the O-wave, and the other corresponding to the E-wave travelling normal to the optic axis. The principal refractive index of the O-wave is given by

$$\mu_o = \frac{\text{Velocity of light in vacuum}}{\text{Velocity of O-wave in the crystal}} \tag{6.3}$$

The principal refractive index of the E-wave is given by

$$\mu_E = \frac{\text{Velocity of light in vacuum}}{\text{Velocity of E-wave in the crystal normal to the optic axis}} \tag{6.4}$$

Two types of uniaxial crystals: Negative and positive

Uniaxial crystals are the doubly refracting crystals having only one optic axis. This means that inside the crystal O-ray and E-ray will travel in only one direction with the same velocity. As doubly-refracting crystals have two refractive indices, the difference in them is defined as *birefringence,* given by $\Delta\mu = \mu_E - \mu_o$. For a doubly-refracting crystal when $\Delta\mu$ is positive it is called a *positive crystal* and when $\Delta\mu$ is negative it is called a *negative crystal*. In a positive crystal, such as quartz, $\mu_E > \mu_o$; hence, the velocity of the E-ray is least in the direction perpendicular to the optic axis and greatest along the direction of the optic axis and is equal to that of the O-ray. In a negative crystal, such as calcite, the refractive index of E-ray is less than that of O-ray ($\mu_E < \mu_o$). In such a crystal, the velocity of the E-ray is the maximum in the direction perpendicular to the optic axis and least along the direction of the optic axis and is equal to that of the O-ray.

According to Huygens's theory of secondary wavelets, the wavefront corresponding to ordinary light will be spherical as the speed of ordinary light is constant in all directions.

However, the wave surface of E-ray will be ellipsoidal as it travels with different speeds in different directions. Both O-ray and E-ray will have the same speed along the optic axis. Thus, for a positive crystal the sphere will enclose an ellipsoid, whereas in the case of a negative crystal the ellipsoid will enclose the sphere as shown in Fig. 6.13. The examples of positive crystals are quartz and ice, whereas that of negative crystals are calcite, selenite, and tourmaline. The differences in the properties of positive and negative crystals are highlighted in Table 6.2.

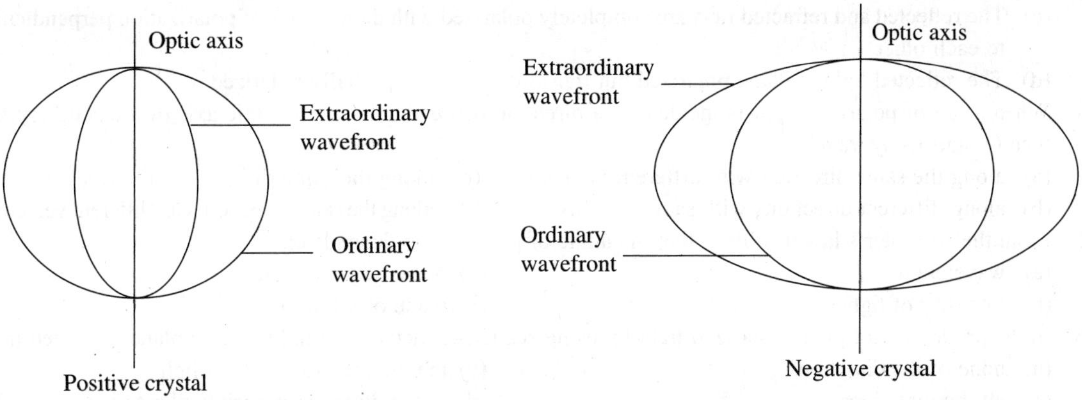

Fig. 6.13 Positive and negative crystals

Table 6.2 Differences between positive and negative crystals

Positive crystal	Negative crystal
In positive crystals birefringence $\Delta\mu$ is positive; that is, $\mu_E > \mu_o$.	In negative crystals, birefringence $\Delta\mu$ is negative; that is, $\mu_E < \mu_o$.
O-ray travels faster in positive crystals.	E-ray travels faster in negative crystals.
For a positive crystal, the sphere encloses an ellipsoid. 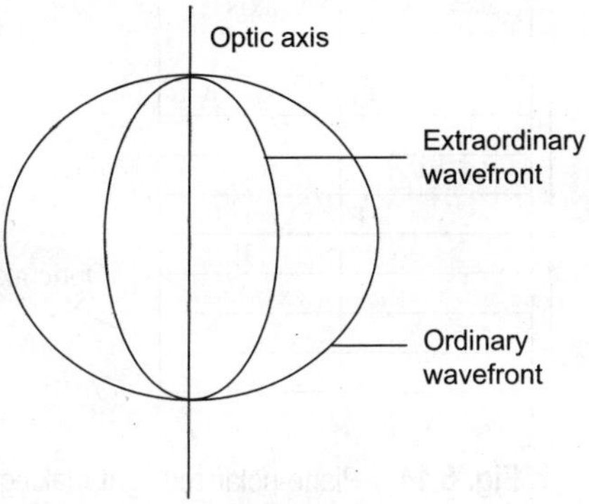	For a negative crystal, the ellipsoid encloses a sphere. 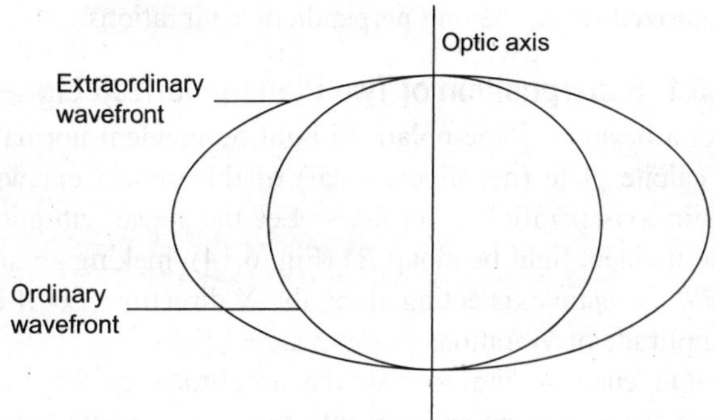

Checkpoint 2

1. For negative uni-axial crystals
   (a) sphere lies inside the ellipsoid.
   (b) ellipsoid lies inside the sphere.
   (c) sphere and ellipsoid do not contain each other
   (d) none of the above
2. For positive uni-axial crystals
   (a) sphere lies inside the ellipsoid.
   (b) sphere and ellipsoid do not touch each other.
   (c) ellipsoid lies inside the sphere.
   (d) none of the above
3. When unpolarized beam is incident on a parallel glass plate at Brewster's angle, which of the following statements is correct?
   (a) Both the reflected light and transmitted lights are partially polarized.
   (b) The reflected and refracted rays are completely polarized with their planes of polarization parallel to each other.

(c) The reflected and refracted rays are completely polarized with their planes of polarization perpendicular to each other.
(d) The reflected light is plane polarized but transmitted light is partially polarized.

4. When a ray of polarized light is incident in a direction perpendicular to the optic axis of a calcite crystal, then O- and E-ray travel
   (a) along the same direction with different velocities
   (b) along different directions with same velocity
   (c) along the same direction with same velocity
   (d) along the same direction with different velocities
5. From the Brewster's law it follows that the angle of polarization depends on
   (a) wavelength
   (b) intensity of light
   (c) plane of polarization
   (d) plane of vibration
6. In the production of plane-polarized light by using scattering method the light will be plane polarized in
   (a) none of the directions
   (b) all the directions
   (c) the direction of incident light
   (d) in the direction perpendicular to incident light

6.5 Production of Elliptically- and Circularly-polarized Light

When two plane-polarized waves are superposed, under certain conditions, the resultant light vector rotates with a constant magnitude in a plane perpendicular to the direction of propagation. As the resultant traces a circle, the light is said to be 'circularly polarized'. However, if the magnitude of the resultant light vector varies periodically during its rotation, then the tip of the vector traces an ellipse, and the light is said to be 'elliptically polarized'.

Let us study the result of superposition of two plane-polarized waves having perpendicular vibrations.

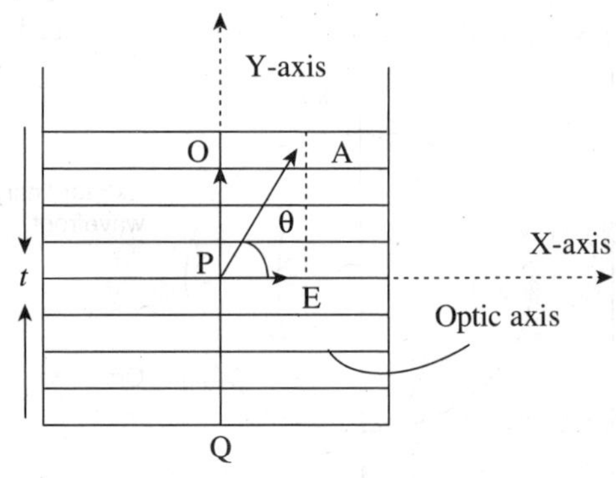

Fig. 6.14 Plane-polarized light making an angle with respect to optic axis

6.5.1 Superposition of Two Plane-polarized Lights

Let a beam of plane-polarized light be incident normally on a calcite plate (negative crystal) of thickness t cut with its optic axis parallel to its faces. Let the linear vibrations in the incident light be along PA (Fig. 6.14), making an angle θ *with the* optic axis acting along the X direction. Let A be the amplitude of vibrations in the incident light.

On entering the crystal, the amplitude of the incident light wave will be split up into two components—A cos θ along PE and A sin θ along PO. The component A cos θ having vibrations parallel to the optic axis forms the E-wave and the component A sin θ having vibrations perpendicular to the optic axis forms the O-wave. According to the Huygens' construction for double refraction, as soon as the incident wave strikes the doubly-refracting crystal, the two waves travel in the plate along the same path but with different velocities, the E-wave being faster as shown in Fig. 6.15.

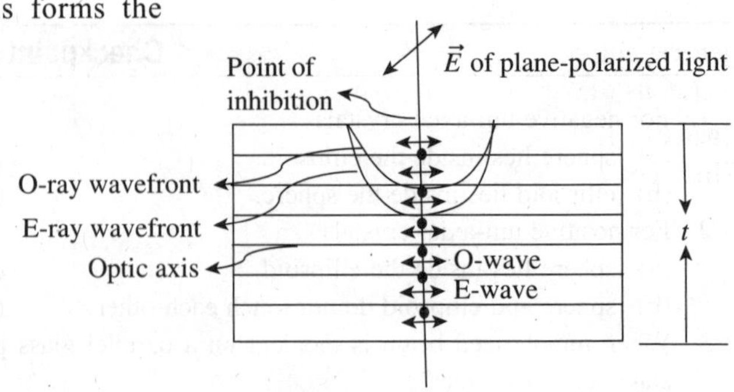

Fig. 6.15 Doubly refracting crystal of thickness t

Hence, the two waves emerging from the plate will have optical path difference Δ (or the phase different δ) between them. Thus, if $A \sin \omega t$ be the incident wave, the two emergent plane-polarized *E- and O*-waves would be represented respectively by

$$x = A \cos \theta \sin (\omega t + \delta) \tag{6.5}$$

$$y = A \sin \theta \sin \omega t \tag{6.6}$$

Let us substitute

$$a = A \cos \theta \tag{6.7}$$

$$b = A \sin \theta \tag{6.8}$$

Therefore, Eqs (6.5) and (6.6) can be written as

$$x = a \sin (\omega t + \delta) \tag{6.9}$$

$$y = b \sin \omega t \tag{6.10}$$

The nature of the resultant vibration can be obtained by eliminating ωt from Eqs (6.9) and (6.10). Expanding Eq. (6.9) we get

$$\frac{x}{a} = \sin \omega t \cos \delta + \cos \omega t \sin \delta$$

$$\frac{x}{a} = \sin \omega t \cos \delta + \sqrt{1 - \sin^2 \omega t}\, \sin \delta \tag{6.11}$$

From Eq. (6.10)

$$\sin \omega t = y/b \tag{6.12}$$

Substituting Eq. (6.12) in Eq. (6.11) we get

$$\frac{x}{a} = \frac{y}{b} \cos \delta + \sqrt{1 - \frac{y^2}{b^2}}\, \sin \delta$$

Rearranging the terms and squaring both sides, we get

$$\left(\frac{x}{a} - \frac{y}{b} \cos \delta\right)^2 = (1 - \frac{y^2}{b^2}) \sin^2 \delta$$

Simplifying further we get

$$\frac{x^2}{a^2} + \frac{y^2}{b^2} - \frac{2xy}{ab} \cos \delta = \sin^2 \delta \tag{6.13}$$

Equation (6.13), in general, represents an ellipse. Hence, light emerging from the crystal plate is, in general elliptically polarized.

Let us discuss the nature of the emergent light for different values of δ as special cases.

Case-(i) If the thickness of the plate is such that $\delta = 0, 2\pi, 4\pi, 6\pi$……
Then, $\cos \delta = 1$ and $\sin \delta = 0$ and Eq. (6.13) becomes

$$\frac{x^2}{a^2} + \frac{y^2}{b^2} - \frac{2xy}{ab} = 0$$

$$\left(\frac{x}{a} - \frac{y}{b}\right)^2 = 0$$

$$\pm\left(\frac{x}{a}-\frac{y}{b}\right)=0$$

$$\pm y=\pm\frac{b}{a}x \qquad (6.14)$$

This represents a pair of coincident straight lines through the origin having a positive slope b/a as shown in Fig. 6.16. This means that the emergent light is linearly polarized with the same direction of vibration as the incident light.

Case-(ii) If the thickness of the plate is such that $\delta = \pi, 3\pi, 5\pi\ldots$ … Then, $\cos\delta = -1$ and $\sin\delta = 0$ and Eq. (6.13) becomes

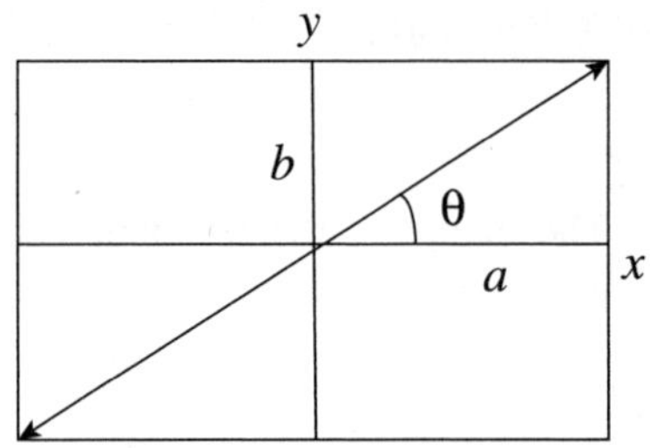

Fig. 6.16 Linearly-polarized light with E vector inclined at an angle θ

$$\frac{x^2}{a^2}+\frac{y^2}{b^2}+\frac{2xy}{ab}=0$$

$$\left(\frac{x}{a}+\frac{y}{b}\right)^2=0$$

$$\pm\left(\frac{x}{a}+\frac{y}{b}\right)=0$$

$$\pm y=\mp\frac{b}{a}x \qquad (6.15)$$

This again represents a pair of coincident straight lines but having a negative slope ($-b/a$) as shown in Fig. 6.17. Hence, the emergent light is again linearly polarized but with the vibration of electric field vector making an angle 2θ with that of the incident light. This case is the basis of a half-wave plate.

Case-(iii) If the thickness of the plate is such that $\delta = \frac{\pi}{2}, \frac{3\pi}{2}, \frac{5\pi}{2}, \frac{7\pi}{2}\ldots\ldots$

Then, $\cos\delta = 0$ and $\sin^2\delta = 1$ and Eq. (6.13) becomes

$$\frac{x^2}{a^2}+\frac{y^2}{b^2}=1 \qquad (6.16)$$

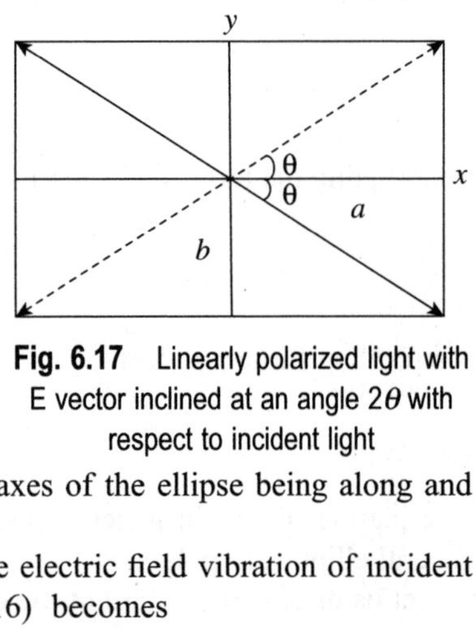

Fig. 6.17 Linearly polarized light with E vector inclined at an angle 2θ with respect to incident light

This represents an ellipse with its axes along X and Y directions. Hence, the light emerging from the plate is elliptically polarized, the axes of the ellipse being along and perpendicular to the optic axis.

In Eqs (6.7) and (6.8) if the angle θ, which is an angle made by the electric field vibration of incident plane-polarized light with the optic axis is 45°, then $a = b$ and Eq. (6.16) becomes

$$x^2+y^2=a^2 \qquad (6.17)$$

Hence, in this case the emergent light is circularly-polarized light. This case is the basis of a quarter wave plate.

Thus, we see that plane-polarized and circularly-polarized lights are special cases of elliptically-polarized light.

6.5.2 Quarter Wave Plate and Half Wave Plate

In the previous section we have learned that if we take a doubly refracting crystal of certain thickness so as to introduce the desired phase difference in the extraordinary and ordinary waves, then we can produce different types of polarized lights. In this section, we will learn to find the required thickness of the crystal plates to produce the desired polarization in the incident light.

Quarter Wave Plate

A doubly refracting crystal plate which produces a phase difference of $\pi/2$ or a path difference of $\lambda/4$ between the ordinary and the extraordinary waves is called a *quarter wave plate* or $\lambda/4$ plate.

Let us consider a plane parallel plate of thickness t is cut from a doubly-refracting crystal such that its reflecting faces are parallel to the optic axis. Let a beam of monochromatic light of wavelength λ be incident normally on the plate. The incident light breaks into ordinary and extraordinary waves which travel along the same path perpendicular to faces, but with different velocities as shown in Fig. 6.18.

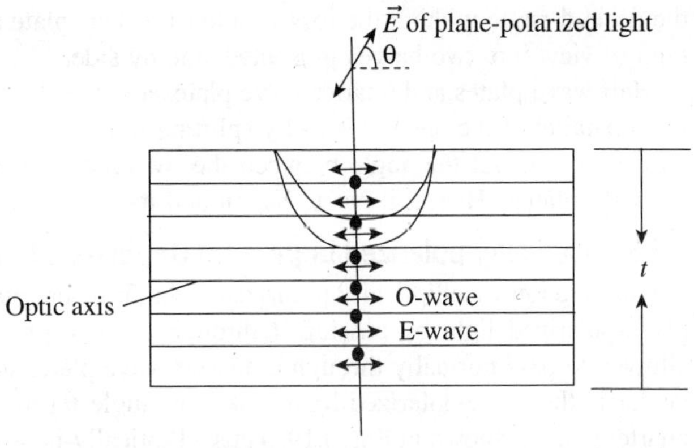

Fig. 6.18 Doubly-refracting crystal as a quarter wave plate

In case of a negative crystal, such as calcite, E-waves travel faster than O-waves, so that $\mu_O > \mu_E$ where μ_O and μ_E are the refractive indices of the crystal for the O- and E-waves. If t is the thickness of the plate, then the optical path of O- and E-waves in the plate are $\mu_O t$ and $\mu_E t$. Therefore the path difference between the two waves on emerging is

$$\Delta = \mu_O t - \mu_E t = t\,(\mu_O - \mu_E)$$

If the plate has to act as the quarter wave plate, then the path difference should be equal to $\lambda/4$.

$$(\mu_O - \mu_E)t = \frac{\lambda}{4}$$

$$t = \frac{\lambda}{4(\mu_O - \mu_E)} \tag{6.18}$$

For a positive crystal, such as quartz, $\mu_E > \mu_o$, so the thickness of quarter wave plate is

$$t = \frac{\lambda}{4(\mu_E - \mu_o)} \tag{6.19}$$

Hence, a plate of thickness given by Eqs (6.11) and (6.12) will serve as a quarter wave plate for a particular wavelength λ. The quarter wave plate is used for producing circularly- and elliptically-polarized lights. In conjunction with Nicol prism it is also used for analysing all kinds of polarized lights.

Half Wave Plate

A doubly-refracting crystal plate having thickness so as to produce a phase difference of π or a path difference of $\lambda/2$ is called a 'half wave plate'. If t is the thickness of such a plate, then for a negative crystal, for example, calcite, t is found as follows.

$$(\mu_O - \mu_E)t = \frac{\lambda}{2}$$

$$t = \frac{\lambda}{2(\mu_O - \mu_E)} \tag{6.20}$$

If a linearly-polarized light is passed through a half wave plate, then the emergent light is also linearly polarized, but its direction of electric vector will be inclined at an angle 2θ to that of the incident light. Here θ is the angle between the incident vibrations of electric field vector with the optic axis of the crystal plate. Hence, this plate is used to rotate the plane of vibration of the incident plane-polarized light through the desired angle without the loss of intensity. This plate also finds application in polarimeter to divide the field of view into two halves presented side by side.

Half wave plates and quarter wave plates are usually made either of quartz or calcite crystals by cutting them parallel to the optic axis or by splitting thin sheets of mica along cleavage planes. Mica is a negative biaxial crystal but the angle between the two axes is very small. Quartz is a positive crystal but has no cleavage planes. Hence, it has to be cut and its faces have to be polished to make them optically plane.

6.5.3 Elliptically-polarized Light from Unpolarized Light

For production of elliptically-polarized light from unpolarized light, we will have to first convert it into plane-polarized light by passing it through a Nicol prism or polaroid. Then the plane-polarized light is allowed to pass normally through a quarter wave plate, such that the direction of vibration of electric field vector in the plane-polarized light makes an angle θ (other than 0, 45°, and 90°) with the optic axis of the quarter plate as shown in Fig. 6.19. Thus, elliptically-polarized light can be produced from unpolarized light.

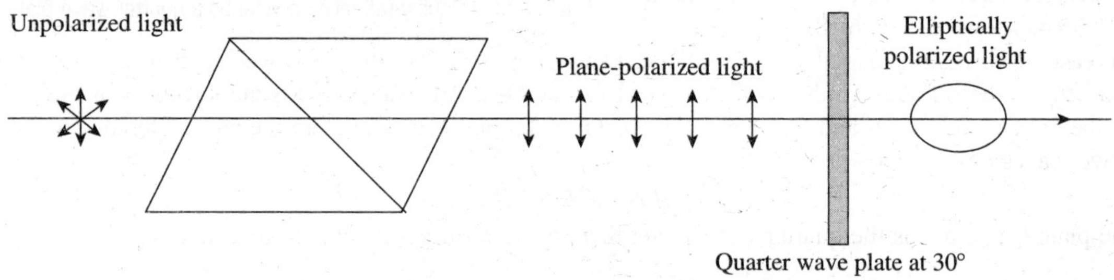

Fig. 6.19 Production of elliptically-polarized light

When $\theta = 0$, from Eqs (6.7) and (6.8), $a = 1$ and $b = 0$. For these values of a and b, Eqs (6.5) and (6.6) become $x = \sin(\omega t + \delta)$ and $y = 0$, respectivley. Therefore, the emergent light is plane-polarized with vibration of electric field parallel to the optic axis. Similarly, for $\theta = 90°$, from Eqs (6.7) and (6.8), $a = 0$ and $b = 1$. For these values of a and b, Eqs (6.5) and (6.6) become $x = 0$ and $y = \sin \omega t$. Therefore, again the emergent light is plane polarized with vibration of electric field now perpendicular to the optic axis. When $\theta = 45°$ then a = b and the resultant vibration is circular so the emergent light is circularly polarized. Hence, to produce elliptically-polarized light, θ must be different from 0, 45°, and 90°. An appropriate value for θ is about 30°. Thus for $\theta = 30^0$ the magnitudes of E- and O-waves will be unequal, that is, a ≠ b.

6.5.4 Circularly-polarized Light from Unpolarized Light

For production of circularly-polarized light from unpolarized light, we will have to first convert it into plane-polarized light by passing it through a Nicol prism or a polaroid. Plane-polarized light is made to fall normally on a quarter wave plate, such that the vibration of the electric field vector in plane-polarized light makes an angle of 45° with the optic axis of the plate. Hence, the light emerging from the quarter wave plate is circularly polarized as shown in Fig. 6.20.

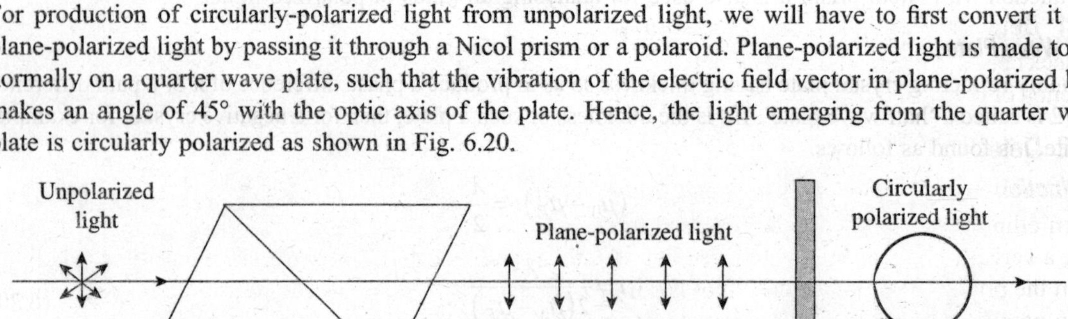

Fig. 6.20 Production of circularly polarized light

Checkpoint 3

1. In circularly-polarized light the phase difference between E-ray and O-ray is
   (a) 0 (c) 180°
   (b) 90° (d) 45°
2. In double refraction we get two refracted rays called O-ray and E-ray. Which of the following statements is true?
   (a) Only the O-ray is polarized. (c) Both O- and E-rays are polarized.
   (b) Only the E-ray is polarized. (d) Neither O-ray nor E-ray is polarized.
3. Along the optical axis in negative crystals
   (a) E-ray has maximum speed and O-ray has minimum speed.
   (b) E-ray has minimum speed and O-ray has maximum speed.
   (c) E-ray and O-ray have the same speed.
   (d) E-ray and O-ray have small differences in their speed
4. Circularly-polarized light is observed through a rotating Nicol prism. The observation indicates
   (a) variation in intensity with minimum zero (c) no variation in intensity
   (b) variation in intensity with minimum non-zero (d) irregular variation in intensity
5. In order to detect elliptically-polarized light, quarter wave plate should be kept
   (a) in any position (c) perpendicular to maximum intensity
   (b) parallel to maximum intensity (d) parallel to minimum intensity
6. If a half wave plate designed for 3600 Å is used for 7200 Å, then
   (a) it will work as quarter wave plate. (c) it will allow the entire light.
   (b) it will work as half wave plate. (d) it will block the entire light.
7. A light of wavelength 4200 A is incident on a calcite crystal. What will be the wavelength in Armstrong unit of an extraordinary ray (R.I = 1.4) moving perpendicular to optic axis inside the crystal?
   (a) 300 Å (c) 480 Å
   (b) 3000 Å (d) 240 Å

6.6 Detection of Different Types of Polarized Lights

The state of polarization of light cannot be perceived by our naked eyes. Hence, to detect and differentiate different types of polarized light we will have to use different optical components. In this section various methods of detection are discussed.

6.6.1 Detection of Plane-polarized Light

To detect plane-polarized light, it is examined through another Nicol prism or polaroid sheet rotating about the direction of propagation of light. If the intensity of light varies with zero minimum, the light is plane polarized.

6.6.2 Detection of Elliptically-polarized Light

Distinction Between Elliptically-polarized and Partially Polarized Light

When elliptically-polarized light is passed through a Nicol prism rotating about the direction of light as axis, a variation in intensity is observed but the minimum intensity is not zero. The intensity is maximum when the principal section of the Nicol is parallel to the major axis of the elliptic vibration and minimum when parallel to the minor axis. An exactly similar variation in intensity is observed when a mixture of unpolarized and plane-polarized light is passed through the Nicol prism. Intensity is maximum when the principal section of the Nicol prism is parallel to the vibrations in the plane-polarized light and minimum when it is at right angles. The minimum intensity is not zero because the unpolarized part always contributes to some intensity.

To distinguish between the two, light is first passed normally through a quarter wave plate and then examined for maximum intensity, so that the optic axis is also parallel to the major axis of the elliptic vibration. If the given light is elliptically polarized, then on entering the quarter wave plate it will be broken up into two unequal components vibrating perpendicular to each other and having a phase difference of $\pi/2$. The quarter wave plates introduce a further phase difference of $\pi/2$. Hence, the phase difference on emergence from the plate is either 0 or π. In either case the light emerging from the plate is plane polarized. Hence, on being examined through the rotating Nicol, the intensity will vary with zero minimum as shown in Fig. 6.21(a). If the given light is a mixture of unpolarized and partially plane-polarized, it would have remained unchanged by the quarter wave plate. Hence on passing thorough Nicol prism the light would have shown a variation in intensity but never completely cut off as shown in Fig. 6.21(b).

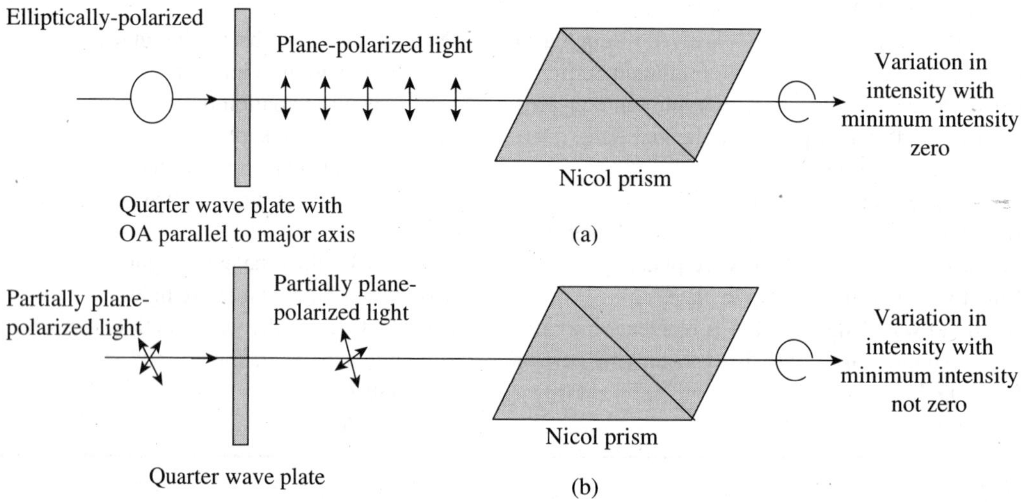

Fig. 6.21 Differentiation of elliptically polarized light from partially polarized light

6.6.3 Detection of Circularly-polarized Light

Distinction Between Circularly–polarized and Unpolarized Light

The circularly-polarized light, when seen through a rotating Nicol prism, shows no variation in intensity. It thus resembles unpolarized light. This is because inside the Nicol prism the circularly-polarized light is broken up into two plane-polarized components of amplitudes, having vibrations parallel and perpendicular to the principal section of the prism. The perpendicular component is lost by total reflection, whereas the parallel component is transmitted. Since the same happens in all positions of the Nicol prism, the intensity of the emergent light is always the same. For exactly the same reason an unpolarized light shows no variation in intensity on rotation of the Nicol prism.

To distinguish between the two, light is first made to pass normally through a quarter wave plate, and then examined through the rotating Nicol. If the given light is circularly polarized, then on entering the quarter wave plate it will be broken up into two components of equal amplitudes having a phase difference of $\pi/2$. The quarter wave plate will introduce a further phase change of $\pm \pi/2$ so that the phase difference between the two components on emergence will be either 0 or π. In either case the emergent light will now be plane-polarized. Hence if this is now examined through a rotating Nicol prism, the intensity will be maximum when the principal section of the Nicol prism is parallel to the vibrations in the emergent plane-polarized light. Light will be completely cut off when the principal section of the Nicol prism is perpendicular to the vibration in the emergent plane-polarized light. Thus variation in intensity with zero minimum shows that the given light is circularly-polarized as shown in Fig. 6.22 (a).

If the given light is unpolarized, then on passing through the quarter wave plate, it will remain unchanged. (This is because the unpolarized light has vibration in all directions, never making a constant angle with the optic axis of the plate). Therefore when examined through a rotating Nicol prism, it shows no variation in intensity as shown in Fig 6.22 (b).

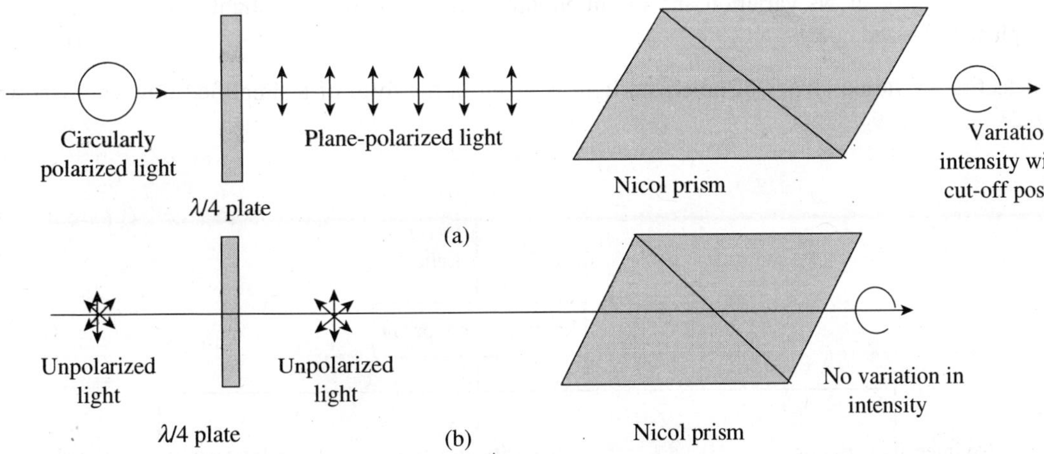

Fig. 6.22 Differentiation of circularly-polarized light from unpolarized light

Distinction Between Quarter Wave and Half Wave Plates

To distinguish between a quarter wave plate and half wave plate, plane-polarized light is made to fall normally on each of them. The emergent light is then examined with a Nicol prism rotating about the direction of light as axis for different orientations of the plates. In case of quarter wave plate, the light emerging from the plate may be elliptically-polarized, circularly-polarized, or plane-polarized depending upon the orientation of the plate. Hence, on examining through the rotating Nicol we shall observe (i) variation in intensity with non-zero minimum, when the emergent light is elliptically polarized (ii) variation in intensity with zero minimum when the emergent light is plane polarized. In case of half wave plate which creates a phase difference of π between the two components, the light emerging from the half wave plate will be plane polarized for all orientations of the plate; hence rotation of the Nicol prism will always give variation in intensity with zero minimum.

Differentiation Between Ordinary, Plane-polarized, Circularly-polarized or Elliptically-polarized Lights by using only Nicol Prism

The human eye cannot distinguish between ordinary and polarized light. A single Nicol prism rotating about the light as axis can distinguish between ordinary light and completely plane-polarized light as the latter is completely cut off in a particular position of the prism. It cannot however, distinguish between ordinary light and circularly-polarized light since in both cases there is no variation in intensity of light viewed through the rotating Nicol. Similarly, it cannot distinguish between elliptically-polarized light and partially plane-polarized light.

However, when the given light is passed through the Nicol prism and from the changes in intensity as the Nicol prism is rotated, there can be three possibilities:

(i) The intensity does not change at all. In such a case, the given light must be either un-polarized, or circularly polarized, or a mixture of the two.
(ii) The intensity shows variation, two maxima and two minima in one rotation, but minimum intensity is not zero. The light must be either elliptically polarized, or a mixture of plane-polarized light and unpolarized light (i.e., partially plane-polarized), or a mixture of elliptically-polarized light and unpolarized light.
(iii) The intensity shows variation and the minimum intensity is zero. The light must be completely plane polarized.

The detection of various types of polarized light is given in the form of a simplified flow chart shown in Fig. 6.23.

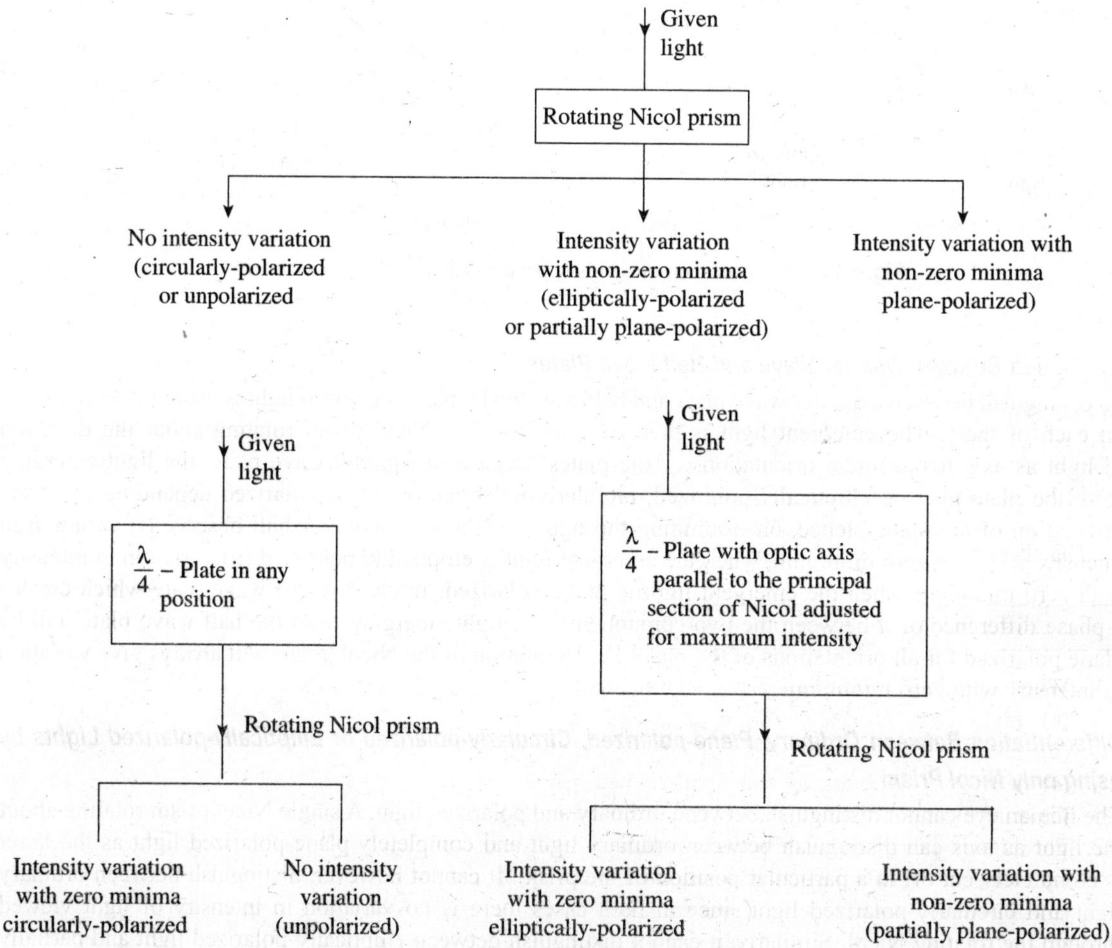

Fig 6.23 Flow chart for differentiation between various types of polarized lights

6.7 Law of Malus

When a completely plane-polarized light beam is incident on an analyser, the intensity of the emergent light varies as the square of the cosine of the angle between the planes of transmission of the analyser and the polarizer. This is known as 'Law of Malus' and holds for a combination of reflecting planes, Nicol prisms, polaroid, but not for the pile of plates (because in this case the light emerging from the polarizer is not completely plane polarized). It can be proved as follows.

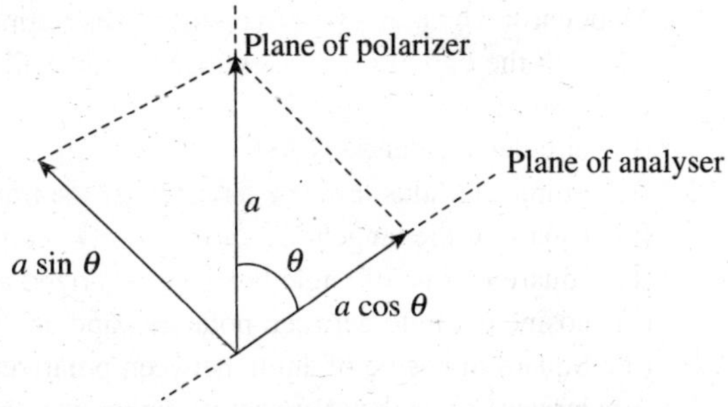

Fig. 6.24 Malus law

Let 'a' be the amplitude of the light transmitted by the polarizer (i.e., plane-polarized light) and θ the angle between the plane of transmission of the analyser and that of the polarizer. This light is incident on the plane of transmission of the analyser. Its amplitude can be resolved into a component $a \cos \theta$ parallel to the plane of transmission of the analyser and the other $a \sin \theta$ perpendicular to it as shown in Fig. 6.24. Only the former is transmitted by the analyser and hence, the intensity of light emerging transmitted from the analyser is given by

$$I_t = a^2 \cos^2 \theta = I_0 \cos^2 \theta \tag{6.21}$$

where $I_0 = a^2$, the intensity of the incident plane-polarized light. If the polarizer and the analyser are parallel ($\theta = 0°$ or $180°$), then $I_t = I_0$.

If they are perpendicular ($\theta = 90$)°, then $I_t = I_0$.

Let I_0 be the intensity of the unpolarized light incident on the polarizing sheet. Let 'θ' be the angle between the polarizing direction of the sheet and the direction of vibration of the electric vector of the incident light. By Malus law, the intensity of the plane-polarized light transmitted by the sheet would be given by $I_t = I_0 \cos^2 \theta$. As the incident light is unpolarized the electric vector in it vibrates randomly in all possible directions in a plane perpendicular to the direction of propagation. Hence, we shall put the 'average' value of $\cos^2 \theta$ in the above equation.

Thus $$I_t = I_0 \cos^2 \theta = \frac{1}{2} I_0 \tag{6.22}$$

The sheet thus reduces the intensity of unpolarized light by 50%.

Checkpoint 4

1. When a plane-polarized light is passed through a half wave plate, the emergent light is
   (a) plane polarized
   (b) elliptically polarized
   (c) a mixture of elliptically and circularly polarized
   (d) circularly polarized
2. For differentiating between circularly-polarized and unpolarized light
   (a) half wave plate and analyser are used.
   (b) quarter wave plate and analyser are used.
   (c) only quarter wave plate is used.
   (d) only analyser is used.

3. Monochromatic plane-polarized light falls normally on the surface of a quarter wave plate making an angle 90 degree with the optic axis. The emergent light will be
   (a) plane polarized
   (b) circularly polarized
   (c) partially polarized
   (d) elliptically polarized
4. Monochromatic plane-polarized light falls normally on the surface of a quarter wave plate making an angle 30° with the optic axis. The emergent light will be
   (a) plane-polarized light
   (b) circularly-polarized light
   (c) plane polarized light
   (d) elliptically-polarized light
5. According to Malus law, the intensity of the transmitted light is proportional to
   (a) square of the tangent of an angle between polarized and analyser
   (b) square of sine of angle between polarized and analyser
   (c) cosine of angle between polarized and analyser
   (d) Square of cosine of angle between polarized and analyser
6. A polarizer and analyser are set in such a way that the intensity of the emergent light is maximum. Now the analyser is rotated through 45° the percentage of maximum intensity of light transmitted is
   (a) 75
   (b) 100
   (c) 50
   (d) 25
7. For a wavelength of light of 3000 Å, a half wave plate is made of calcite. Neglecting variation in the indices of refraction with wavelength, how would this behave with respect to light of wavelength of 6000 Å?
   (a) half wave plate
   (b) quarter wave plate
   (c) as both quarter & half plate
   (d) none of these

6.8 Applications of Polarization

Devices based on the phenomenon of polarization have varied applications. The common uses of polariods are for reducing the glare viz. in sunglasses to reduce the intensity and the excess glare by cutting down the horizontally polarized light and in windshields of automobiles the polaroid sheets reduce headlight glare of approaching cars. Besides these uses, polarized light also finds many practical applications in industry and engineering science. In industry, polaroid filters are used to perform stress analysis tests on transparent plastics. When light is passed through a transparent plastic sheet, all colours of the visible light are polarized with their own specific orientation of electric field. The plastic sheet is then placed between two polarizing plates. A colourful pattern is formed when visible light falls on this assembly. This pattern of colour changes when the top plate is rotated as now different colours are blocked and the previously blocked ones are transmitted. The significance of this analysis is that where there is a large concentration of coloured bands, the structural stress in plastic is more and hence, the structural failure will most likely occur at these locations.

In the entertainment industry, polarization is used to produce and show 3D movies. Three-dimensional movies are produced by recording from two slightly different camera locations. The two movies are shown at the same time by using two projectors from different sides of the audience onto a metal screen. A polarizing filter is used for projecting movies. For the projector on the left side of the audience the polarizing filter used has its polarization axis aligned horizontally while the one used on the right would have its polarization axis aligned vertically. Thus, two slightly different movies casted by light that is polarized with an orientation perpendicular to the other movie are projected onto the screen. To watch the movie, audience will have to wear special type of glasses, which have polaroids with perpendicular polarization axes. The arrangement of projectors and filters leads to an effect where the left eye sees the movie that is projected from the right projector whereas the right eye sees the movie that is projected from the left projector, which gives the viewer a perception of depth.

Polarized light is used to enhance contrast in images observed through traditional microscopes or cameras. In order to reveal details that are poorly observed, the microscope or cameras must be equipped with a polarizer positioned in the light path before the specimen. An analyser should be placed in the optical pathway between the observation tubes or camera port and the objective rear aperture. Thus, the interaction of plane-polarized light with a birefringent specimen results in image contrast.

Polarization is significant for radio communications and radio frequency (RF) antennas. For most antennas, polarization is in the same plane as the elements of the antenna. Thus, a vertical antenna (i.e. one with vertical elements) will receive vertically-polarized signals and a horizontal antenna will receive horizontally-polarized signals. Vertical and horizontal antenna polarization belongs to a category of simple polarization known as *linear polarization*. To receive good strength of the signal it is necessary to match the polarization of the RF antenna to that of the incoming signal. If the RF antenna polarization does not match as that of the signal, then the strength of the signal reduces by a factor of cosine of the angle between the polarization of the RF antenna and the signal.

Hence, to provide and receive the optimum signal the polarization of the antennas located in free space is very important. In case polarization of the antennas were at right angles to one another (i.e., cross-polarized), then in theory no signal would be received. For terrestrial radio communications applications the polarization will broadly remain same once a signal has been transmitted, but the reflections from objects in the path can change the polarization slightly.

There is another category of antenna polarization viz. circular polarization, which is used for satellites radio communications.

Checkpoint 5

1. When a quartz half wave plate is placed between two crossed polaroids such that the half wave plate is oriented at 45° to the axes of polaroid, the combination will transmit
   (a) no light
   (b) one-fourth of the intensity of incident unpolarized light
   (c) half of the intensity of incident unpolarized light
   (d) one-eighth of the intensity of incident unpolarized light
2. Polaroid sunglasses decrease glare on a sunny day because they
   (a) block a portion of light
   (b) have a special colour
   (c) completely absorb the light
   (d) refract the light
3. We prefer polaroid sunglasses because they
   (a) reduce the intensity of light
   (b) have soothing colours
   (c) are cheaper
   (d) can change colours
4. Which of the following material can be used for manufacturing polaroid?
   (a) Calcite
   (b) Quartz
   (c) Tourmaline
   (d) Quinine iodosulphate
5. The intensity of light incident on a polarizer is I, and that of the light emerging from it is also I. What is the nature of light incident on the polarizer?
   (a) Polarized
   (b) Unpolarized
   (c) Partially polarized
   (d) Circularly polarized
6. When a polaroid is rotated, the intensity of light varies but never reduces to zero. It shows that the incident light is
   (a) completely plane-polarized
   (b) partially plane-polarized
   (c) unpolarized
   (d) none of the above

Summary

- In ordinary or unpolarized light emanating from sources such as the sun or a light bulb, the light vector vibrates in all possible directions in a plane perpendicular to the direction of propagation.
- If we confine the vibration of the light vector in a single plane or along a fixed straight line in the plane, then such a light is called plane-polarized or linearly-polarized light. The other two types of polarized lights are elliptically-polarized light and circularly-polarized light
- Plane-polarized light can be produced by different methods, namely reflection, refraction, double refraction, selective absorption (dichroism), and scattering.
- Elliptically-polarized light is obtained by superposing two plane-polarized waves having unequal amplitudes of electric fields vibrating in two mutually perpendicular directions and are out of phase by 90°.
- Circularly-polarized light is obtained by superposing two plane-polarized waves having equal amplitude of electric fields vibrating in two mutually perpendicular directions and are out of phase by 900.
- A doubly refracting crystal plate which produces a phase difference of $\pi/2$ or a path difference of $\lambda/4$ between the ordinary and the extraordinary waves is called a quarter wave plate or $\lambda/4$ plate.
- The quarter wave plate is used for producing circularly- and elliptically-polarized lights. In conjunction with Nicolprism it is also used for analysing all kinds of polarized lights.
- A doubly-refracting crystal plate having thickness so as to produce a phase difference of π or a path difference of $\lambda/2$ is called a 'half wave plate'. Itis used to rotate the plane of vibration of the incident plane-polarized light through the desired angle without the loss of intensity and also finds application in polarimeter to divide the field of view into two halves presented side by side.
- When a completely plane-polarized light beam is incident on an analyser, the intensity of the emergent light varies as the square of the cosine of the angle between the planes of transmission of the analyser and the polarizer. This is known as 'Law of Malus' and holds for a combination of reflecting planes, Nicol prisms, polaroid, but not for the pile of plates

Solved Numericals

1. A beam of light travelling in water strikes a glass plate, which is also immersed in water. When the angle of incidence is 51°, the reflected beam is found to be plane polarized. Calculate the refractive index of glass.

   Solution: By Brewster's law,

   $$\mu = \tan i_p = \tan 51 = 1.235$$

   This is the refractive index of glass with respect to water.

2. The critical angle of light in a certain substance is 45°. What is the polarizing angle?

   Solution: The incident light is moving in the denser medium. If c be the critical angle, then we have

   $$\sin c = \mu$$

   where μ is the refractive index of the second (rarer) medium.

   Here $c = 45°$

   $$\therefore \mu = \sin 45° = \frac{1}{\sqrt{2}} = 0.707$$

   If i_p be the polarizing angle, then by Brewster's law, we have

   $$\tan i_p = \mu$$

   $$i_p = \tan^{-1}(\mu)$$

   $$= \tan^{-1}(0.707) = 35.26°$$

3. A parallel pencil of light is incident on the plane surface of a glass of refractive index 1.5. Calculate the angle of incidence for which the reflected light is completely plane polarized. What is the plane of vibration of this light? What is the corresponding angle of refraction?

   Solution: By Brewster's law the reflected light is completely plane polarized when the angle of incidence is i_p,

   $$\tan i_p = \mu$$

where μ is the refractive index of glass. Thus

$$i_p = \tan^{-1}(\mu) = \tan^{-1}(1.5) = 56.3°$$

The plane of vibration of this polarized light is perpendicular to the plane of incidence.

If r be the angle of refraction, then we know that

$$r + i_p = 90°$$

$$r = 90° - i_p = 90° - 56.3° = 33.7°$$

4. Calculate the thickness of a plate which would convert plane-polarized light into circularly-polarized light. The principal refractive indices are μ_o = 1.658 and μ_E = 1.658 when the wavelength of light used is 5890 Å.

Solution: In order to convert plane-polarized light into circularly-polarized light, the quarter wave plate is used. The minimum thickness of quarter wave plate is

$$t_{\min} = \frac{\lambda}{4(\mu_o - \mu_e)}$$

$$t = \frac{5890 \times 10^{-10}}{4(1.658 - 1.486)} = \frac{5890 \times 10^{-10}}{4 \times 0.172}$$

$$= 8561 \times 10^{-10} \text{m} = 8561 \text{ Å}$$

This is the minimum thickness of the quarter wave plate or any odd multiple of this can be used to convert plane-polarized light into circularly-polarized light.

5. Find the thickness of a quarter wave plate when the wavelength of light is equal to 5890 Å and μ_o = 1.55, μ_e = 1.54.

Solution: The thickness of a quarter wave plate for which ($\mu_o > \mu_e$) such as calcite is given by

$$t = \frac{\lambda}{4(\mu_o - \mu_e)}$$

Substituting the given value,

$$t = \frac{5890 \times 10^{-10}}{4(1.55 - 1.54)} = \frac{5890 \times 10^{-10}}{4 \times 0.01}$$

$$= 1472 \times 10^{-8} \text{m} = 0.00147 \text{ cm}$$

6. Plane-polarized light is incident on a quartz plate cut parallel to the optic axis. Find the least thickness of the plate for polarized light on emergence.

$$(\mu_o = 1.5442,\ \mu_e = 1.5533, \lambda = 5893\text{Å})$$

Solution: In this case half wave plate is to be used. The thickness of half wave plate is

$$t = \frac{\lambda}{2(\mu_e - \mu_o)}$$

$$t = \frac{\lambda}{2(\mu_e - \mu_o)} = \frac{5893 \times 10^{-10}}{2(1.5533 - 1.5442)}$$

$$= \frac{5893 \times 10^{-10}}{2 \times 0.0091} = 323791.2 \times 10^{-10} \text{ m}$$

$$= 32.37 \ \mu\text{m}$$

7. For light of wavelength 4000 Å, half-wave plate is made of calcite. Neglecting variation in the indices of refraction with wavelength, how would this behave with respect to light of wavelength 8000 Å?

Solution: It will behave like a quarter wave plate for the wavelength 8000 Å.

$$\Delta T = 2000 \text{ Å} = \frac{\lambda}{4}$$

8. Plane-polarized light is incident normally on a plate of doubly-refracting uniaxial crystal with faces parallel to the optic axis. Calculate the ratio of intensities of extraordinary and ordinary rays, if the light is incident with vibrations making an angle of 30° with the optic axis,

(Given: $\lambda = 6000 \text{ Å}, \mu_e = 1.5532,\ \mu_o = 1.5442$)

Solution: According to the given data $\mu_o > \mu_e$, which means that the given crystal is positive (like quartz). Now, let A be the amplitude of the incident light wave and θ the angle, which the incident vibrations make with the optic axis. On entering the crystal, the light will be split upto into two components, an extraordinary component having amplitude $A \cos\theta$ having vibration parallel to the optic axis and an ordinary component $A \sin\theta$ having vibration perpendicular to the optic axis. Hence the ratio of intensities of the extraordinary and ordinary rays is

$$\frac{I_e}{I_0} = \frac{A^2\cos^2\theta}{A^2\sin^2\theta}$$

For $\theta = 30°$

$$\frac{I_e}{I_0} = \frac{\cos^2 30°}{\sin^2 30°} = \frac{(\sqrt{3}/2)^2}{(1/2)^2} = \frac{3}{1}$$

9. When two polarizing sheets, say A and B, are crossed i.e., have their polarizing directions perpendicular to each other, no light is transmitted through them. If, however, a third sheet C is introduced between them such that it is not parallel to either of the first two, then light begins to transmit through the assembly.

Solution: This can be explained by using Malus law: Let I_A be the intensity of the (plane-polarized) light transmitted by A. Then

$$I_A = \frac{1}{2} I_0$$

where I_0 is the intensity of unpolarized light falling on A. Now if be the angle between the polarizing directions of A and C, then the intensity of light transmitted by C would be given by

$$I_B = I_C \cos^2 (90 - 0)$$

$$I_B = I_C \sin^2\theta$$

$$I_B = \left(\frac{1}{2} I_0 \cos^2\theta\right)\sin^2\theta$$

Since the value of the above is not zero, light begins to transmit through the assembly.

10. Monochromatic plane-polarized light falls normally on the surface of a quarter wave plate with the plane of vibration making an angle of (a) 30°, (b) 45° and (C) 90° with the optic axis, Discuss the nature of emergent light in each case. How will you confirm your conclusion experimentally?

Solution: When plane-polarized light of amplitude A falls normally on a quarter wave plate with the plane of vibration making an angle θ with the optic axis of the plate, the resultant emergent vibration is given by

$$\frac{x^2}{a^2} + \frac{y^2}{b^2} = 1 \qquad \text{(i)}$$

where $a = A \cos \theta$ and $b = A \sin \theta$

(a) When $\theta = 30°$; $a = \text{A} \cos 30° = \frac{\sqrt{(3)}\,A}{2}$,

and $b = A \sin 30° = \frac{A}{2}$, Eq. (i) becomes

$$\frac{4x^2}{3A^2} + \frac{4y^2}{A^2} = 1$$

This show that the emergent light is elliptically polarized.

(b) When $\theta = 45°$: $a = \text{A} \cos 45° = \frac{A}{\sqrt{2}}$

and $b = \text{A} \sin 45° = \frac{A}{\sqrt{2}}$, then Eq. (i) becomes

$$x^2 + y^2 = \frac{A^2}{2}$$

This shows that the emergent light is circularly polarized.

(c) When $\theta = 90°$: $a = \text{A} \cos 90° = 0$ and $b = \text{A} \sin 90° = \text{A}$

This shows that the amplitude of vibration along the optic axis vanishes. The emergent light is therefore plane polarized with vibration perpendicular to the optic axis.

Self-study Questions

1. What is polarization of light? Distinguish between polarized and unpolar ized light.

2. Write a brief note on the different methods of production of plane-polarized light?

3. State Brewster's law. How can it be used to produce plane-polarized light?

4. Explain how circularly- and elliptically-polarized light can be produced from an unpolarized light.

5. How will you experimentally distinguish between plane-polarized, circularly-polarized, and elliptically-polarized light with the help of only a polarizer (or analyser)?

6. Differentiate between

a. E-ray and O-ray

b. Positive and negative crystal

7. What will be the nature of the emergent beam of light when a plane-polarized light from a monochromatic source falls on a quarter wave plate? Justify your answer.

8. Explain the terms:

a. Double refraction

b. Optic axis

c. Positive crystals

d. Negative crystals

9. How can you analyse different types of polarization by using a polarizer and a quarter wave plate?

10. Explain how can you identify plane-polarized and elliptically-polarized light?

Numerical Exercises

1. A ray of light is incident on the surface of water of refractive index 1.33. If the reflected light is completely plane polarized, calculate the angle of polarization and hence the angles of incidence and refraction. (Ans: 53.06°, 36.94°)
2. An unpolarized beam of light is incident on a group of four polarizing sheets, which are lined up so that the polarizing direction of each is rotated by 45° clockwise with respect to the preceding sheet. What fraction of incident light is transmitted? (Ans: 1/16)
3. Two Nicol prisms have parallel polarizing direction so that the intensity of transmitted light is maximum. Through what angle must either prism be turned if intensity is to drop by 1/5th of its maximum value? (Ans: 63.43°)
4. The electric vector of a beam of polarized light makes an angle of 55° with the axis of a polarizer sheet. What is the fraction of the beam intensity transmitted through the sheet? (Ans: 0.33)
5. A polarizer and an analyser are oriented so that the maximum amount of light is transmitted. To what fraction of its maximum value is the intensity of the transmitted light reduced when the analyser is rotated through (a) 30° (b) 45° (c) 60°? (Ans: 3/4, 1/2, 1/4)
6. It is desired to rotate the plane of vibration of plane-polarized light through 90°. How can we do so by using only polarizing sheets? Calculate the intensity of the light after rotation of vibration plane through 90°. (Ans: 3 polarizing sheets oriented at 30° with each other; 42.18)
7. We wish to use a plate of glass as polarizer. What is the polarizing angle? (Ans: $\mu = 1.5$ for glass).
8. A parallel beam of yellow light (λ = 5800 Å) enters the 60° triangular prism of quartz (refractive indices 1.54 and 1.55) through a rectangular surface at an incidence angle, which leads to minimum deviation of rays. Calculate the angular separation between the two emergent beams. (Ans: 0.902°)
9. A light of wavelength 5000 Å is incident on a calcite crystal. What will be the wavelength of ordinary and extraordinary rays along and perpendicular to the optic axis inside the crystal? ($\mu_o = 1.658$, $\mu_e = 1.486$). (Ans: 3015.68 Å, 3364.74 Å)
10. A beam of plane-polarized light is changed into circularly-polarized light by passing it through a slice of crystal 0.003 cm thick. Calculate the difference in the refractive indices of the two rays in the crystal, assuming this to be the minimum thickness that will produce the effect. (Wavelength of light is 6×10^{-7} cm). (Ans: 5×10^{-5})
11. Ice is a positive uniaxial crystal with indices of refraction of 1.309 and 1.310. What should be the minimum thickness of ice crystal so that it can act as a quarter wave plate for light of wavelength 6000 Å? (Ans: 150 μm)
12. The faces of a quartz plate are parallel to the optic axis of the crystal. (a) What is the thinnest possible plate that would serve to put the ordinary and extraordinary rays of λ = 5890 Å, a half-wave apart on their exit? (b) What multiples of this thickness would give the same result? The indices of refraction of quartz are μ_o =1.553, μ_e = 1.544. (Ans: (a)32.72 μm (b) 98.16 μm, 163.6 μm)

Answers

Checkpoint 1
1 (a), 2 (c), 3 (d), 4 (c), 5(b), 6 (d)

Checkpoint 2
1 (a), 2 (c), 3(d), 4 (a), 5 (a), 6 (d)

Checkpoint 3
1 (b), 2(c), 3 (c), 4(c), 5 (b), 6 (a), 7(b)

Checkpoint 4
1(a), 2(b), 3(d), 4 (a), 5 (d), 6 (c), 7 (b)

Checkpoint 5
1 (c), 2(a), 3 (a), 4. (d), 5 (a), 6 (b)

7 Semiconductor Physics

Learning Objectives

After going through this chapter, you will be able to:

- Demonstrate the knowledge of band theory of solids to distinguish various solids with respect to their electrical behaviour
- Understand and use basic terminologies related to semiconductor physics, namely intrinsic, and extrinsic semiconductor, and the current conduction in them
- Analyse the impact of temperature and doping concentration on the conductivity of extrinsic semiconductor
- Demonstrate the knowledge of semiconductor to use related component namely pn junction diode in various applications
- Explain Hall effect and its practical applications

7.0 Introduction

Semiconductor physics forms the core of modern solid-state device technology. Devices made from semiconductor materials are the foundation of modern electronics. Semiconductor devices are the electronic components that exploit the electronic properties of semiconductor materials. They are manufactured both as single discrete devices and as integrated circuits (ICs). These include many kinds of diodes including the light emitting diode, transistor, the silicon controlled rectifier, and digital and analog integrated circuits. In the field of alternate energy sources, solar photovoltaic panels, which are large semiconductor devices that directly convert light energy into electrical energy, are extensively used.

Semiconductor devices on account of their compactness, reliability, and low cost have found a wide range of applications. As discrete components, they are used as power devices, optical sensors, and light emitters, including solid-state lasers. These devices have a wide range of current- and voltage-handling capabilities. The current ratings range from a few nanoamperes to more than 5,000 amperes and voltage ratings are of approximately 100,000 volts. Thus, semiconductor devices are the key elements for the majority of electronic systems, including communications, sensor technology, data-processing, and industrial-control equipment.

Pure semiconductor materials have limited applications. However, doped semiconductors have a large scope of applications as we can manipulate their conductivities. In the following sections, we shall discuss the properties of semiconducting crystals, which give rise to their distinguishing band composition. Semiconductor fundamentals, their properties, and the effect of external parameters on it are also discussed along with a few of applications of semiconductor components.

7.1 States of Matter

Matter exists in three distinct states—solid, liquid, and gas. The smallest unit of matter is atoms or molecule (when atoms combine). In the gaseous state the distance between the adjacent atoms (inter-atomic distance) is large, approximately 30 A°. In liquid form due to more electrostatic interactions between neighbouring atoms or molecules the inter-atomic (or inter-molecular) distance decreases. In solid state, the inter-atomic forces are very strong; hence the inter-atomic distance further decreases and is approximately a few Armstrongs (3–6 A°) in length. Thus in a solid, atoms or molecules cannot move freely and are closely packed. If the atomic arrangement in solid is periodic (regular and repetitive), then the solid is called *crystalline solid*; else for random (haphazard) arrangement the resultant solid is called *amorphous*. Further, crystals depending on the atomic arrangement and its symmetry are classified into seven groups, namely cubic, tetragonal, hexagonal, trigonal, orthorhombic, monoclinic, and triclinic.

At room temperature in a crystalline solid like a metal that has cubic structure, the valence electrons are delinked from parent atoms and are free. Hence, these free electrons are shared by all atoms in the metal. However, in semiconductors, a solid with different crystal structure, valence electrons are located much nearer to the parent atom and cannot be associated with the entire collection of atoms. The free electron theory (a simple model which illustrates the behaviour of valence electron cannot be applied for such materials because the potential seen by valance electrons is not constant but varies rapidly, especially near the ion cores in the lattice.

The band theory of solids successfully explains the differing electrical properties of conductors, semiconductors, and insulators. Band theory explains conduction in semiconductors, based on whether the electronic states within an allowed energy band are empty or full.

7.2 Band Theory of Solids

7.2.1 Formation of Bands in a Solid

For an isolated atom, the energy levels of electron are discrete as shown in the Fig. 7.1. In solids energy bands exist instead of levels. The energy band formation can be understood by considering the atomistic (tight binding) approach in which, electrons are assumed to be tightly bound to individual atoms. As atoms are brought together to form a crystal, the interaction of the neighbouring atoms causes the electron energy levels of individual atoms to spread into bands of energies which is explained below in detail.

Fig. 7.1 Discrete energy levels of an isolated atom

According to Pauli's exclusion principle, not more than two electrons, each having opposite spin occupies an energy level. However, in any solid, atoms are packed very close to each other. Thus, they do not behave as isolated atoms, but interact with each other. As a result of this interaction between electronic orbitals of closely packed atoms, the energy levels of outermost electrons overlap, which leads to violation of Pauli's exclusion principle as more number of electrons share the same energy level. Therefore, the energy level which is overlapping splits to form a region of closely spaced energy levels to accommodate more electrons. According to band theory, the number of energy levels formed by splitting is given by $(2l + 1)$ N, where l is the orbital quantum number and N is the number of atoms forming the solid. For example, let us consider that two atoms have come close together such that their *s* orbitals overlap. Then the energy level corresponding to the *s* orbital will split into two energy levels (for *s* orbital, $l = 0$) as shown in Fig. 7.2 where r_o is the inter-atomic distance when the stable atomic arrangement is formed in a solid. Similarly, if four atoms come close then the overlapping of *s* orbital results in the formation of four energy levels. If there are *N* atoms in a solid, then each energy level splits into *N* allowed energy levels or states or quantum states. The energy levels formed by splitting are so close that they cannot be considered as discrete energy levels, but can be considered almost continuous and hence are known as energy band having width ΔE. Thus, when the atoms are brought close together to form a solid, the outermost energy level splits to form a band of energy levels, which are very closely spaced. Each energy band formed by overlapping of the *s* orbital has maximum capacity to accommodate $2N$ electrons as each quantum state can occupy a maximum of two electrons.

Similarly, if 'p' ($l = 1$), 'd' ($l = 2$) or 'f' ($l = 3$) orbitals are overlapping then they will form $3N$, $5N$, or $14N$ energy levels with double the capacity of accommodating electrons. Thus, various permissible energy levels of one electron represent allowed energy levels of all the electrons. Not all energy levels will split

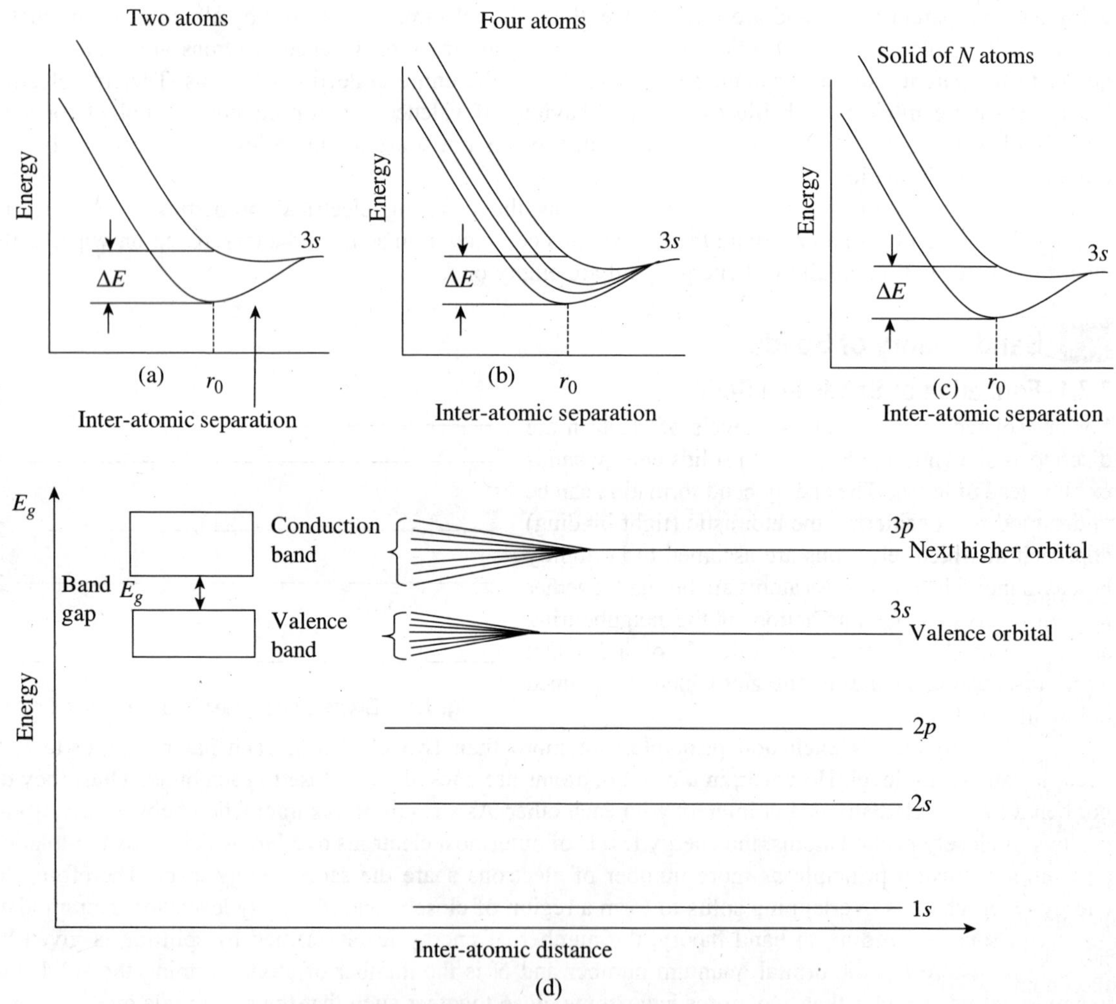

Fig. 7.2 Splitting of energy levels of electron in a solid containing (a) 2 atoms; (b) 4 atoms; (c) N atoms; (d) energy band diagram of a solid

to the same extent. As splitting is a function of the distance between atoms, the splitting of outer orbital starts at larger inter-atomic distances and splits more. The energy levels corresponding to only the outermost electrons split as energy levels corresponding to the inner shell electrons are shielded and hence are not affected by the presence of neighbouring atoms.

The electrons in the outermost orbits of atoms are known as *valence electrons*. The band formed by the splitting of orbitals or energy levels containing valence electrons is known as *valence band.* Thus, the valence band is the highest filled energy band in the solid. For example in Fig. 7.2 (d) the band corresponding to the valence electrons in 3*s* orbital is called the valence band which may be completely or half/partially filled. The empty band above the valence band corresponding to 3*p* electrons is known as *conduction band.* Thus, the conduction band is the lowest unfilled energy band. The forbidden (not allowed) energy band separates the lower valence band and the upper conduction band. The energy interval between the top edge of valence band and the bottom edge of conduction band is called the *band gap* or *energy gap* and is denoted by E_g.

7.2.2 Classification of Solids on the Basis of Band Structure

Solids can be grouped into conductors, semiconductors, or insulators based on their ability to conduct electricity. Based on the band theory of solids one can explain qualitatively the different types of electrical behaviour of these materials.

From the band structure of a given material we can find out which energy levels are occupied and which are empty, partially filled, or completely filled. Empty bands do not contain electrons and therefore, they are not expected to contribute to the electrical conductivity of the material.

In partially filled bands, lower energy levels are occupied by electrons, whereas slightly higher energy levels within the same band are empty. These unoccupied energy levels are available for electrons if they gain energy while moving in an applied electric field. Therefore, electrons in a partially filled band contribute to the electrical conductivity of the material.

Completely filled bands contain plenty of electrons but they do not contribute to the conductivity of the material, as all energy levels are filled and there are no empty levels in the same band. Thus, for good electrical conduction, a large number of electrons should be available (more filled states), but empty energy levels should exist immediately above the filled levels. Thus, we can classify the solids into three categories taking account of the arrangement of electrons within the bands and the gap between the filled and empty states, that is, width of forbidden energy gap.

Conductors

These are materials in which there is no energy gap between valence band and conduction band. The part of the valance band itself is the conduction band or both overlap on each other as shown in Fig. 7.3.

Figure 7.3 shows the energy bands of solid sodium. A sodium atom, Na has atomic number 11 and hence the electronic configuration $1s^2$, $2s^2\ 2p^6$, $3s^1$. Sodium has a single 3*s* electron in its valence shell [M-shell]. The corresponding band has 2N energy states and can occupy a maximum of 2N electrons. Since the 3*s* orbital itself is half-filled (3s orbital can occupy maximum 2 electrons of opposite spin), hence the corresponding band is also half-filled, that

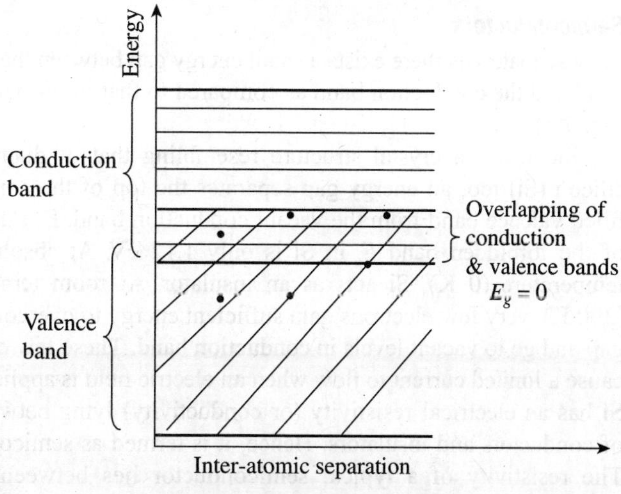

Fig. 7.3 Energy band diagram of a typical conductor

is, only N out of 2N energy states are filled while remaining N states are empty in the same band. Thus, the valence band is half-filled, that is, contains half-vacant levels. Now suppose an electric field is applied across a piece of solid sodium, then the electrons in the valence band easily acquire additional energy to move to the higher unoccupied energy levels within the same band as there is no energy gap. The remaining energy becomes the kinetic energy of the electron and thus these moving electrons constitute electric current. Sodium metal is therefore a good conductor of electricity. The resistivity of metals is less than 10^{-8} Ω m. Thus the conductor is characterized by partially filled valence band with zero energy gap.

Insulators

In these materials a large energy gap exists between valence band and conduction band.

Figure 7.4 shows the energy bands of diamond. At zero degree absolute temperature the valence band is completely filled and the conduction band is completely empty. There exists a large energy gap of 6 eV between valence band and conduction band. This means that at least 6 eV of additional energy must be supplied to an electron in the valence band so that it can go to conduction band. If an electric field is applied across diamond, the electrons in the valence band cannot move within the same band since there are no vacant levels in the valence band. However, they can move to the higher empty band provided they get energy at least equal to E_g. In order to give energy more than 3 eV, a large electric field must be applied. After application of electric field, still there are only a few electrons which move to the conduction band. Thus, there is almost no electric current and hence, the solid will be an insulator. The resistivity of an insulator is generally more than 10^{10} Ω m. Insulators are thus characterized by a completely filled valence band separated by a large band gap from a completely empty conduction band.

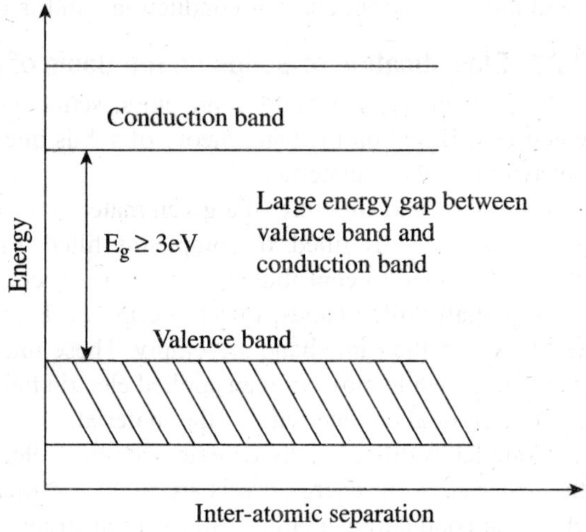

Fig. 7.4 Energy band diagram of an insulator

Semiconductors

In these materials there exists a small energy gap between the valence band and the conduction band as compared to that in insulators (see Fig. 7.5).

Silicon has a crystal structure resembling that of diamond. In silicon (Si) too, an energy gap separates the top of the completely filled valence band from the vacant conduction band. But the width of the forbidden band E_g in Si is only 1.12 eV. At absolute zero temperature (0 K), Si acts as an insulator. At room temperature (300 K), very few electrons gain sufficient energy to overcome band gap and go to vacant levels in conduction band. These few electrons cause a limited current to flow when an electric field is applied. Thus Si has an electrical resistivity (or conductivity) lying between that of conductors and insulators. Hence, it is termed as semiconductor. The resistivity of a typical semiconductor lies between 10^{-5} to 10^{10} Ω m. Thus, semiconductors are characterized by completely

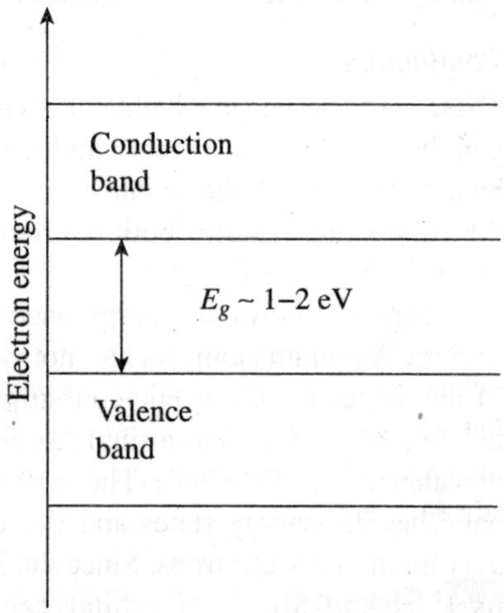

Fig. 7.5 Energy band diagram of a semiconductor

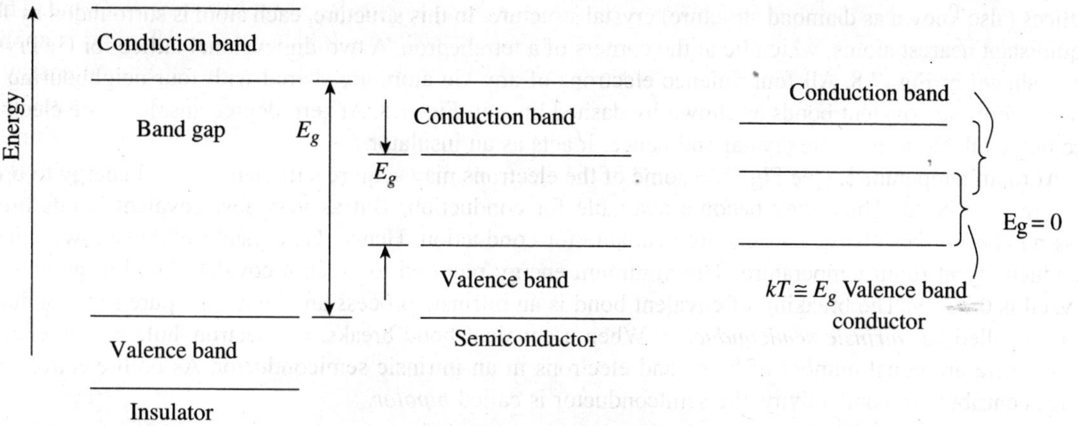

Fig. 7.6 Comparative band diagrams of three classes of materials

filled valence band and completely vacant conduction band at 0 K, with a small band gap. The conductivity is zero at 0 K and it increases with increase in temperature. Figure 7.6 shows the comparative band structures for the three classes of materials, namely conductor, semiconductor, and insulator.

7.3 Electrons and Holes in Semiconductor

Semiconductors have a band structure as shown in Fig. 7.7, which is basically similar to that of insulators, except that the gap energies are much smaller, typically of the order 1 eV. Hence, at ordinary temperatures, the electrons in the filled valence band acquire sufficient thermal energy to move to the empty conduction band creating vacancies in the valence band. These vacancies are known as *holes*. Thus absence of electron is called hole. They behave as particles with the same properties as that of electrons except that they carry a positive charge. Therefore, electrons are available for conduction in conduction band and holes in the valence band. In semiconductors, there is relatively smaller number of charge carriers viz. electrons and holes which account for their low conductivities as compared with metals. The fundamental difference between metals (conductors) and semiconductors is that metals are unipolar, that is, they have one type of charge carrier namely electrons and semiconductor is bipolar and thus have two types of charge carriers—electrons and holes.

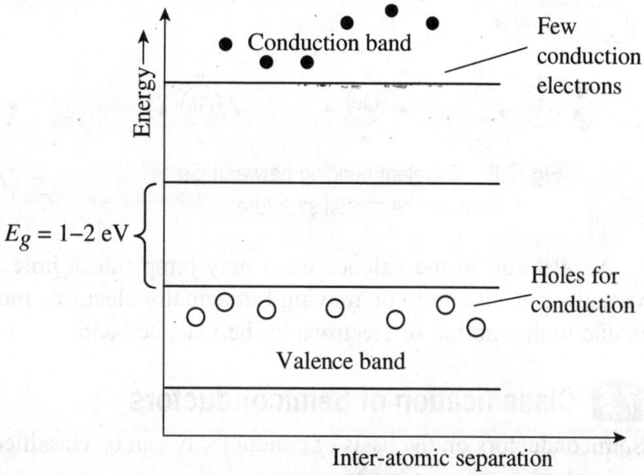

Fig. 7.7 Band diagram of a semiconductor at $T > 0$ K showing a few electrons and holes

7.4 Electrical Conduction in Semiconductor

The most commonly used semiconductor for fabricating solid state devices are germanium (Ge) and silicon (Si). Ge and Si both are tetravalent, that is, there are four valence electrons and has face-centred cubic lattices (also known as diamond structure) crystal structure. In this structure, each atom is surrounded by four equidistant nearest atoms, which lie at the corners of a tetrahedron. A two-dimensional lattice for Ge crystal is as shown in Fig. 7.8. All four valence electrons of any Ge atom are shared with four neighbouring Ge atoms, forming covalent bonds as shown by dashed lines in Fig. 7.8. At zero degree absolute free electrons are not available in pure Ge crystal and hence, it acts as an insulator.

At room temperature, (see Fig. 7.9) some of the electrons may acquire sufficient thermal energy to break the covalent bond. Thus, they become available for conduction. But as very few covalent bonds break, less number of free charge carriers are available for conduction. Hence, Ge crystal will have low value of conductivity at room temperature. The minimum energy required to break a covalent bond in germanium crystal is 0.72 eV. The breaking of covalent bond is an intrinsic process and hence the pure semiconductor is also called an *intrinsic semiconductor*. When a covalent bond breaks, an electron–hole pair is created. Thus, there are equal number of holes and electrons in an intrinsic semiconductor. As both electrons and holes contribute to conductivity, the semiconductor is called *bipolar*.

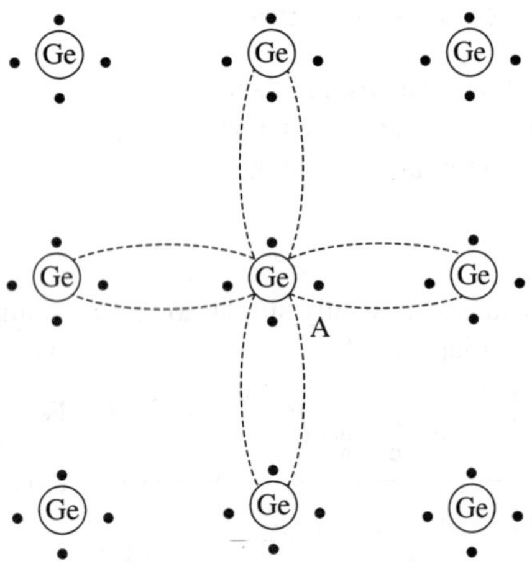

Fig. 7.8 Covalent bonding between Ge atoms in the crystal structure

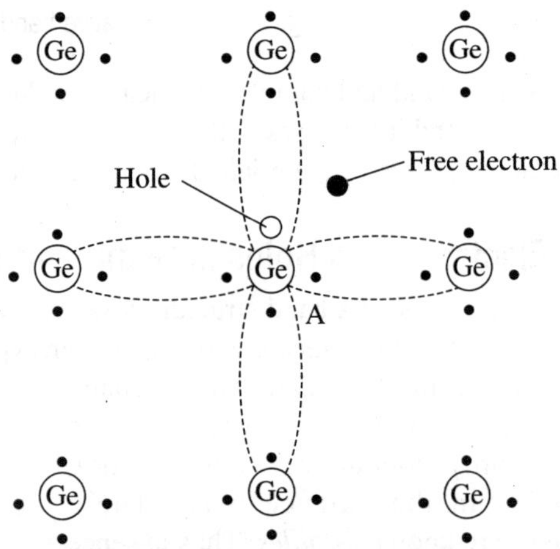

Fig. 7.9 A free electron and a hole in Ge crystal structure at room temperature

An electron in the valence band may jump into a hole and thus create a hole in its own place. Thus, virtually holes seem to be moving but actually electrons move to give conduction. Hence, the hole current is due to the motion of electrons in the valence band.

7.5 Classification of Semiconductors

Semiconductors on the basis of conductivity can be classified into two categories— intrinsic semiconductor and extrinsic semiconductor.

7.5.1 Intrinsic Semiconductors

A semiconductor in its pure form is known as intrinsic semiconductor. Silicon and germanium are the most commonly used intrinsic semiconductors.

In an intrinsic semiconductor, each broken covalent bond in the valence band generates an electron–hole pair. Thus

$$n = p = n_i \tag{7.1}$$

where n and p are the density of electrons and holes, and n_i, is called the intrinsic density. In intrinsic semiconductors very small current flows which increases with the increase in temperature as more number of covalent bonds breaks. Thus, conductivity of the intrinsic semiconductors can be increased by increasing the temperature. But this is not the prudent way of increasing the conductivity. Hence, there are limited applications of intrinsic semiconductors.

As temperature increases, free charge carriers become available. These free charge carriers cannot crowd a particular energy level. How the free charge carriers occupy available energy levels is explained by Fermi–Dirac distribution function which is discussed in detail in the next section.

Doping of Semiconductors

The addition of a small percentage of foreign atoms in the regular crystal lattice of Si or Ge produces dramatic changes in their electrical properties. This phenomenon is known as *doping* and the doped semiconductor is known as *extrinsic semiconductor*.

Insulators can be doped to change their optical properties. But we cannot make them good conductors of electricity as it is very difficult to overcome the large band gap. However, electrical conductivities of semiconductors can be changed remarkably by doping them.

7.5.2 Fermi–Dirac Distribution Function

The Fermi distribution function $f(E)$ is a probability function which determines the occupancy of energy states by electrons. This function governs the distribution of electrons among the energy levels as a function of temperature. It is given as

$$f(E)=\frac{1}{1+\exp[(E-E_F)/kT]} \qquad (7.2)$$

where E_F is a reference energy level called *Fermi energy level, K* is the Boltzman's constant, and T is the temperature. The function $f(E)$ is the probability of a state at energy E being occupied by an electron.

Analytical Treatment of Fermi Function

Let us study the behaviour of the Fermi function, that is, we will find the values of Fermi function for different energy levels at zero and at room temperature.

Case-I: For $E < E_F$ at T = 0 K

$$f(E)=\frac{1}{1+\exp[(E-E_F)/0]}=\frac{1}{1+\exp-[(E_F-E)/0]}=\frac{1}{1+e^{-\infty}}=\frac{1}{1+0}=1$$

The value one of fermi function indicates that all energy levels below E_F are occupied by electrons. Thus, in a semiconductor, valence band is completely filled. Hence, in semiconductor Fermi energy is defined as the highest energy which the electrons can have at zero temperature.

Case-II: For $E > E_F$ at T = 0 K

$$f(E)=\frac{1}{1+\exp[(E-E_F)/0]}=\frac{1}{1+e^{\infty}}=\frac{1}{1+\infty}=0$$

The zero value of Fermi function indicates that all energy levels above E_F are not occupied by electrons. Thus, in a semiconductor conduction band is completely empty.

Case-III: For $E = E_F$ at T = 0 K

$$f(E)=\frac{1}{1+\exp[(E-E_F)/0]}=\frac{1}{1+e^{\frac{0}{0}}}$$

This is an indeterminate function. Thus, we cannot find value of Fermi function for this case.

Case-IV: For $E < E_F$ at $T \neq 0$ K

$$f(E) = \frac{1}{1+\exp[(E-E_F)/KT]} = \frac{1}{1+\exp-[(E_F-E)/KT]}$$

$$= \frac{1}{1+\exp[-ve\ value/+ve\ value]} = \frac{1}{1+(+ve\ value)} < 1$$

The value of Fermi function is less than one which indicates that all energy levels below E_F are not occupied by electrons. Thus, in a semiconductor at temperature other than zero, the valence band is not completely filled.

Case-V: For $E > E_F$ at $T \neq 0$ K

$$f(E) = \frac{1}{1+\exp[(E-E_F)/KT]} = \frac{1}{1+\exp[+ve\ value/+ve\ value]} = \frac{1}{1+(+ve\ value)} > 0$$

The value of Fermi function is now greater than zero which indicates that few energy levels above E_F are occupied by electrons. Thus, in a semiconductor the conduction band is not completely empty but a few energy levels are filled. This is what exactly happens in a semiconductor, that is, when temperature increases covalent bonds are broken, hence valence electrons (bound electrons) become conducting electrons (free electrons).

Case-VI: For $E = E_F$ at $T \neq 0$ K

$$f(E) = \frac{1}{1+\exp[(E-E_F)/KT]} = \frac{1}{1+\exp[0/+ve\ value]} = \frac{1}{1+e^0} = \frac{1}{1+1} = \frac{1}{2}$$

This value of Fermi function indicates that there are 50% chance that the electron will have this energy. Thus if there is an electron in the conduction band then there is a hole in the valence band. Thus, Fermi energy E_F is also defined as the average energy of charge carriers at room temperature.

Graphical Analysis of Fermi Function

The Fermi function $f(E)$ is plotted for different temperatures as shown in Fig. 7.10.

For temperature T = 0 K the Fermi function is a step function. For energy levels below Fermi energy level the value of Fermi function decreases gradually from its maximum value 1 as temperature increases. However, for energy levels above Fermi energy level, the value of Fermi function increases gradually as temperature increases. The distribution curve will always pass through point 'C' as it has value 0.5 for $E = E_F$ at all temperatures. From the figure it is clear that the probability function is symmetrical about E_F.

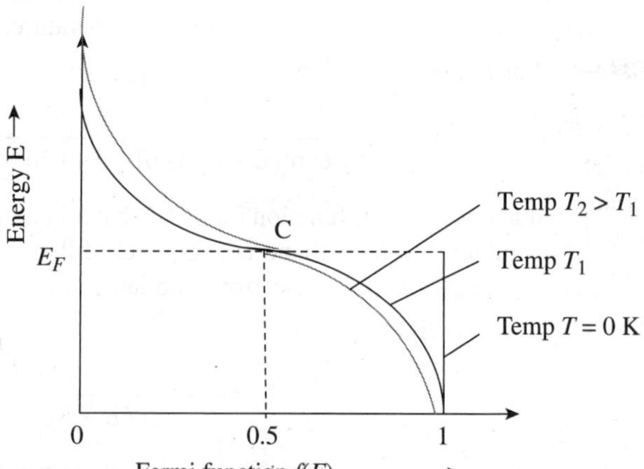

Fig. 7.10 Dependence of Fermi-Dirac distribution function on temperature

7.5.3 Electron and Hole Concentrations in a Semiconductor

Using Eq. (7.2), it is possible to find the electron concentration in conduction band and the hole concentration in valence band of a semiconductor.

The number of electrons in the conduction band is given by

$$n_C = N_c \cdot f(E_c)$$

where $$f(E_c)=\frac{1}{1+\exp[(E_c-E_F)/kT]}$$

$$\cong \exp\left[\frac{-(E_c-E_F)}{kT}\right]$$

and N_C is a temperature dependent material constant known *as effective density of states* in the conduction band. It represents the number of states required to be placed at energy E_C. After multiplying N_C with the probability of occupation of this level $f(E_C)$ we get the number of electrons in the conduction band.

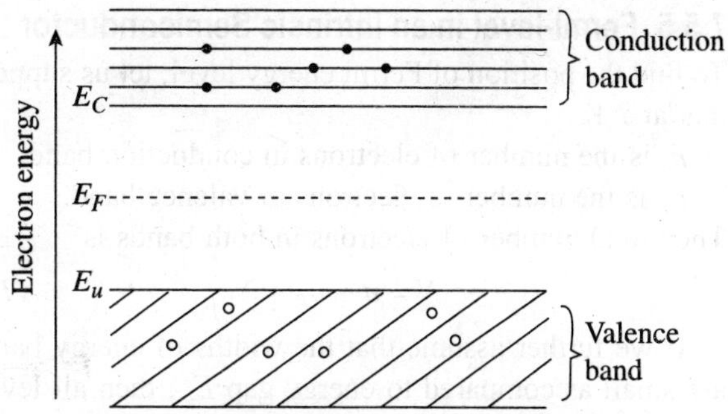

Fig. 7.11 A few electrons and an equal number of holes for conduction in intrinsic semiconductor at T > 0 K

We assume that $(E_C - E_F) \gg kT$, so we ignore one in the denominator. As a consequence of this Fermi–Dirac distribution can be approximated by the Boltzmann distribution.

Therefore, $$n_c = N_C\, e^{-(E_C-E_F)/kT} \tag{7.3}$$

Equation (7.3) gives the electron concentration n_c in the conduction band.

Similarly the number of holes in the valance band is given by

$$p_V = N_V \cdot f(E_V)$$

where $$f(E_V)=\frac{1}{1+\exp[(E_F-E_V)/kT]}\cong \exp\left[\frac{-(E_F-E_V)}{kT}\right]$$

Therefore, $$p_V = N_V\, e^{-(E_F-E_V)/kT} \tag{7.4}$$

where N_V is the effective density of states in the valence band.

7.5.4 Intrinsic Density, n_i

As the expressions for N_C and N_V are independent of doping of semiconductor, Eqs (7.3) and (7.4) are valid for any semiconductor, provided that E_F lies in the forbidden gap and is at least $3kT$ away from band edges. In an intrinsic semiconductor, at $T = 0$ K, the electron concentration in conduction band is equal to the hole concentration in valence band.

$$n = p = n_i \tag{7.5}$$

$$n\ \ p = n_i^2$$

$$n_i^2 = [N_C e^{-(E_C-E_F)/kT}][N_V e^{-(E_F-E_V)/kT}]$$

Thus, $$\therefore n_i = (N_C N_V)^{1/2} e^{-E_g/2kT} \tag{7.6}$$

Here E_g is the band gap energy, K is the Boltzmann's constant, and T is the temperature. From Eq. (7.6), it is clear that the intrinsic density n_i:

(a) is independent of Fermi level position
(b) is a function of E_g.
(c) depends on temperature T strongly.

7.5.5 Fermi-level in an Intrinsic Semiconductor

To find the position of Fermi energy level, let us suppose that at T K

n_c is the number of electrons in conduction band.

n_V is the number of electrons in valence band.

Then total number of electrons in both bands is

$$N = n_c + n_V \tag{7.7}$$

If we further assume that the widths of energy bands are small as compared to energy gap E_g, then all levels in a band can be considered to have same energy.

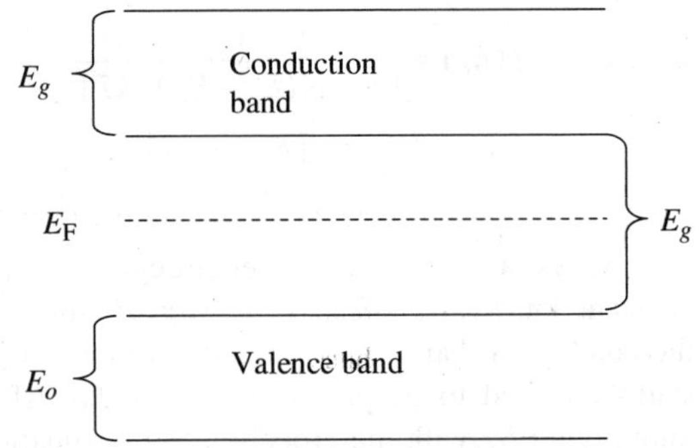

Fig. 7.12 Fermi level in intrinsic semiconductor

Let us assume that energy of all levels in valence band is zero, and energy of all levels in conduction band is E_g. Thus the number of electrons in the conduction band will be given by

$$n_c = N \cdot f(E_g)$$

where $f(E_g)$ is the probability of finding electron with energy E_g. Its value is determined by Fermi Dirac distribution function as

$$f(E_g) = \frac{1}{1 + e^{(E_g - E_F)/KT}}$$

Therefore,

$$n_c = \frac{N}{1 + e^{(E_g - E_F)/KT}} \tag{7.8}$$

and

$$n_V = \frac{N}{1 + e^{(0 - E_F)/KT}} = \frac{N}{1 + e^{-E_F/KT}} \tag{7.9}$$

Substituting Eqs (7.8) and (7.9) in Eq (7.7), we have

$$N = \frac{N}{1 + e^{(E_g - E_F)/KT}} + \frac{N}{1 + e^{-E_F/KT}}$$

or

$$1 = \frac{1}{1 + e^{(E_g - E_F)/KT}} + \frac{1}{1 + e^{-E_F/KT}}$$

where K is the Boltzman's constant, E_g is the band gap, E_F is the Fermi energy, and T is the temperature.

For temperatures above 0 K and as $E_g > E_F$, the exponential term in the denominator becomes very large. Hence, neglecting '1' in both the terms in the denominator, we can write the above equation as

$$e^0 = e^{-(E_g - E_F)/KT} + e^{E_F/KT}$$

Taking the log of both sides,

$$0 = -\left(\frac{E_g - E_F}{KT}\right) + \frac{E_F}{KT} \quad \text{or} \quad 0 = -E_g + E_F + E_F$$

Therefore,

$$E_F = \frac{E_g}{2} \tag{7.10}$$

Thus, in an intrinsic semiconductor, the Fermi level lies midway between the conduction band and the valence band as shown in Fig. 7.12.

At 0 K, all the energy levels are filled up to E_F. But since the energies in the forbidden gap are excluded, this indicates that the valence band is completely filled and conduction band is completely empty at 0 K.

Fermi level is the uppermost occupied energy level in a conductor at 0 K. Since in a semiconductor Fermi level is in the forbidden gap, it is not occupied by electrons and only serves as a reference level. In semiconductors Fermi energy can be defined as the average energy possessed by the charge carriers participating in electrical conduction at temperatures above 0 K. In intrinsic semiconductor there is a negligible effect of temperature on the position of Fermi energy level.

Substituting Eq. (7.10) in Eq. (7.8) and ignoring 1 in the denominator, we get

$$n_c = \frac{N}{1+e^{E_g/2KT}} = N e^{-E_g/2KT} \tag{7.11}$$

Equation (7.11) correctly represents the dependence of concentration of electrons on E_g as it gives lesser and lesser values of n_c for materials with higher band gap. Similarly, with increase in temperature, n_c increases which is the observed behaviour of the semiconductor. Thus, by using Eq. (7.11) we can correctly estimate the fraction of electrons in the conduction band for the given material (E_g) and at a given temperature.

7.5.6 Extrinsic Semiconductors

To enhance the conductivity of intrinsic semiconductor, it is doped with impurities from adjacent groups namely third (boron, gallium, indium, aluminium) and fifth group (bismuth, phosphorous, arsenic, antimony). These impurity atoms replace host atoms and thus act as substitution impurities. Depending on the type of impurity added to an intrinsic or pure semiconductor, extrinsic semiconductors can be of two types — n-type semiconductor and p-type semiconductor.

n-type Semiconductor

The addition of pentavalent impurities such as antimony, arsenic, or phosphorous to an intrinsic semiconductor such as Si or Ge gives n-type extrinsic semiconductor. The impurity atoms are added in a very small amount, that is in ppm (parts per million or 1 impurity atom per 10^6 host atoms).

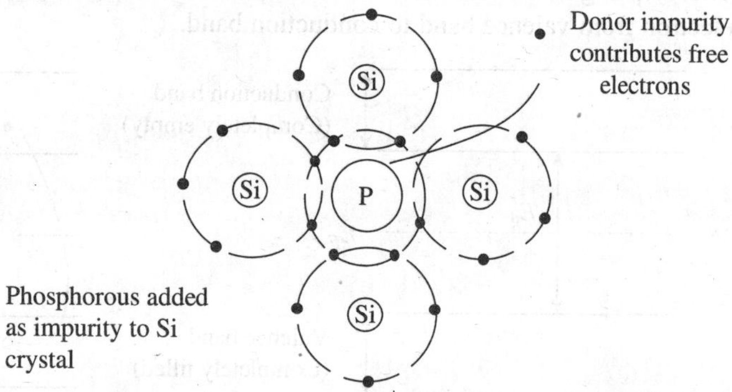

Fig. 7.13 Structure of portion of crystal lattice of Si doped with phosphorous atoms

Suppose that phosphorous impurity atoms are added to the Si melt from which the crystal is grown. At room temperature four out of the five outer electrons of phosphorus atom are used to form four covalent bonds with silicon (valency of Si is 4) as shown in Fig. 7.13. The fifth electron from each impurity atom is extra and at room temperature gets detached from the parent atom due to available thermal energy. In the band diagram these free electrons are shown to be promoted to conduction band. Thus each impurity atom contributes one extra free electron to the lattice and hence is called *donor atom*. Since the current in such crystals will be mostly due to electrons, which are negatively charged, this semiconductor is called n-type.

p-type Semiconductor

The addition of trivalent impurities such as boron, aluminium, or gallium to an intrinsic semiconductor such as Si or Ge gives p-type extrinsic semiconductor.

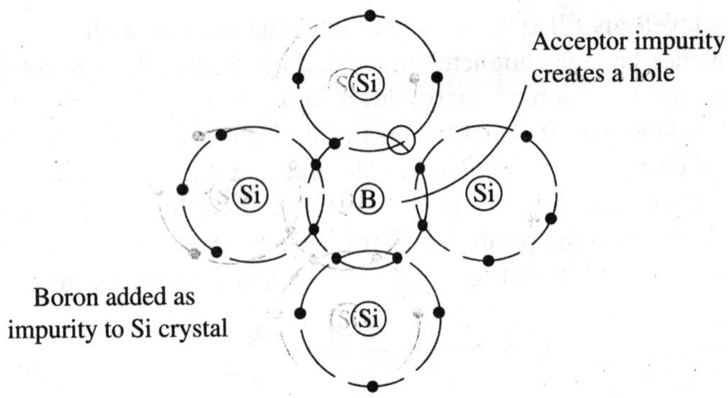

Fig. 7.14 Structure of a portion of crystal structure of Si doped with boron

From Fig. 7.14, a tetravalent Si is doped with trivalent boron. Since four electrons are needed by Si to fill the valence shell completely, this doping creates electron deficiency. Thus, each impurity atom added to Si creates one hole. Since the holes can accept electrons, the impurity atoms are called *acceptors*. Current in such semiconductors is due to large number of holes, and as holes are positive the semiconductor is called p-type.

7.5.7 Energy Band Structures

Energy Band Structure of Intrinsic Semiconductor

In intrinsic semiconductor at 0 K, valence band is completely filled and conduction band is completely empty as shown in Fig. 7.15 (a). At room temperature due to available thermal energy few covalent bonds break creating holes in valence band and free electrons in the conduction band as shown in Fig. 7.15 (b). The breaking of covalent band is conceptualized using energy band diagram by showing transition of an electron from valence band to conduction band.

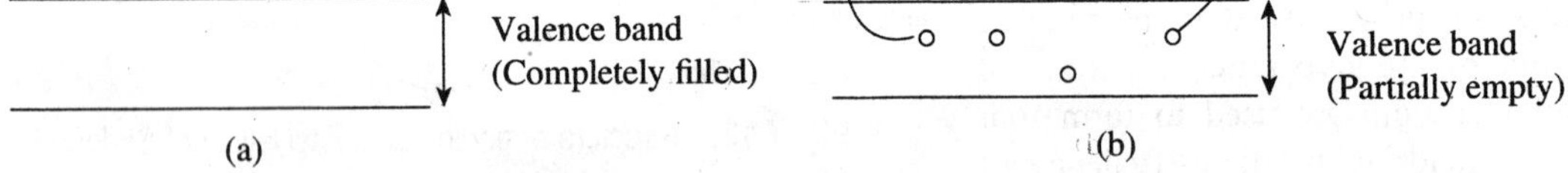

Fig. 7.15 Band diagram of intrinsic semiconductor (a) at 0 K (b) at room temperature

Energy Band Structure of Extrinsic n-type Semiconductor

Figure 7.16 (a) shows that the fifth electron of the donor impurity atom (e.g. phosphorus) remains loosely bound to it because at 0 K, energy is not available to detach it from the parent atom. The energy required to detach this fifth electron from donor atom is called *ionization energy* which is very small (approximately 0.05 eV) as compared to the energy gap E_g (i.e. the energy needed to break the covalent bond). At room temperature, the thermal energy is sufficient to detach the electron which then goes to conduction band and becomes available for conduction. The corresponding ionized positively charged donor ions remain immobile in the donor energy and do not contribute to conduction as shown in Fig. 7.16 (b).

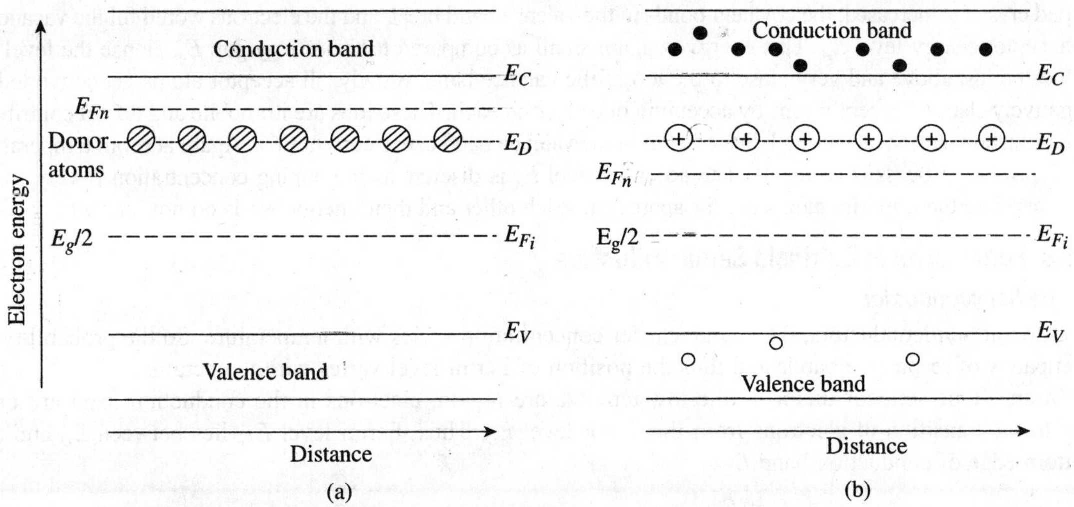

Fig. 7.16 Energy band diagram of an n-type semiconductor (a) at 0 K (b) at 300 K

As the ionization energy is very small compared to E_g, the donor energy level E_D is located just below the bottom of the conduction band and close to it. This donor level E_D is discrete because of low concentration of donor atoms. They are located far apart from each other and hence their energy levels do not overlap or interact to cause splitting and formation of corresponding bands.

At room temperature all the donor atoms get ionized. If the temperature is increased continuously, then at sufficiently high temperatures electrons are contributed by the intrinsic process also, that is, the breaking of covalent bonds to create electron–hole pairs.

Energy Band Structure of Extrinsic p-type Semiconductor

Each acceptor impurity atom (e.g. boron) has one vacancy (hole). At 0 K impurity atoms remain at acceptor energy level with no free charge carriers in both valence and conduction band. When the temperature of such a

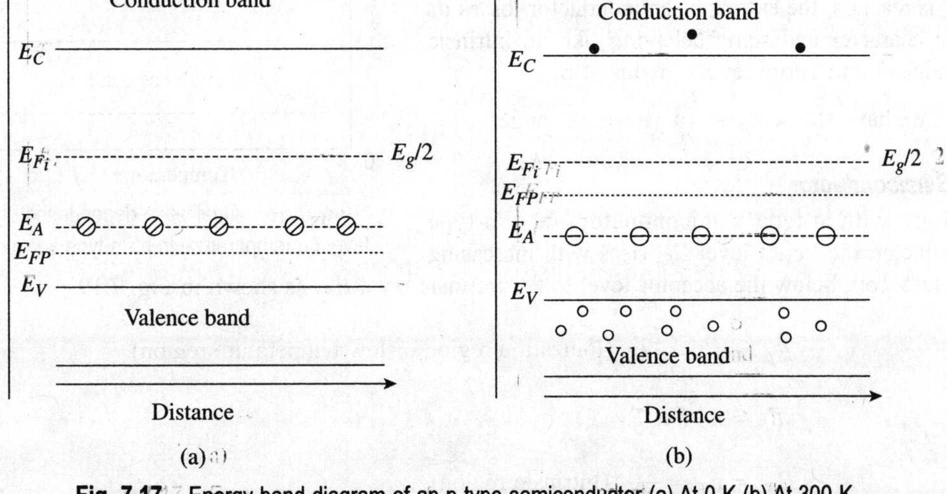

Fig. 7.17 Energy band diagram of an p-type semiconductor (a) At 0 K (b) At 300 K

doped crystal is increased, the covalent bonds in the valence band break and the electrons would fill the vacancies at acceptor energy level E_A. This energy is again small as compared to the energy gap E_g. Hence the level E_A is located just above and very close to the top of the valence band. Slowly, all acceptor atoms get converted to negatively charged acceptor ions by accepting one electron each. These ions are immobile and do not contribute to conduction. But the resulting holes created in the valence band cause current flow. Again at room temperature the ionization process is complete. The acceptor level E_A is discrete as the doping concentration is very small and therefore the impurity atoms are far apart from each other and their energy levels do not interact.

7.5.8 Fermi Level in Extrinsic Semiconductors

n-type Semiconductor

In extrinsic semiconductors, the charge carrier concentration varies with temperature. So the probability of occupancy of respective bands and thus the position of Fermi level varies with temperature.

In an n-type semiconductor, in the low temperature region, electrons in the conduction band are only due to the transition of electrons from the donor level E_D. Thus, Fermi level E_{F_n} lies between E_D and the bottom edge of conduction band E_C.

$$E_{F_n} = \frac{E_C + E_D}{2} \text{ (Low temperature region or ionization region)} \tag{7.12}$$

As temperature increases, the Fermi level shifts downwards as the donor levels get gradually depleted. At the temperature of depletion, that is, when ionization is complete (where all donor atoms get depleted of loosely attached fifth valence electron and become positive ions), the Fermi level coincides with donor energy level E_D.

$$E_{Fn} = E_D \text{ (At } T = T_d) \tag{7.13}$$

As the temperature further increases, Fermi level shifts downwards approximately linearly as shown in Fig. 7.18. At a higher temperature (called intrinsic temperature T_i) approximately above 50° C, the intrinsic process starts, which significantly contributes to the hole concentration in the valence band. The Fermi level shifts downwards and coincides with the intrinsic value $E_{F_i} = \frac{E_g}{2}$. Thus when T_i is reached, the extrinsic semiconductor losses its extrinsic character and starts behaving like an intrinsic semiconductor and Fermi level remains at E_{Fi}.

Thus, we have $E_{F_n} = E_{F_i} = \frac{E_g}{2}$ (Intrinsic range)

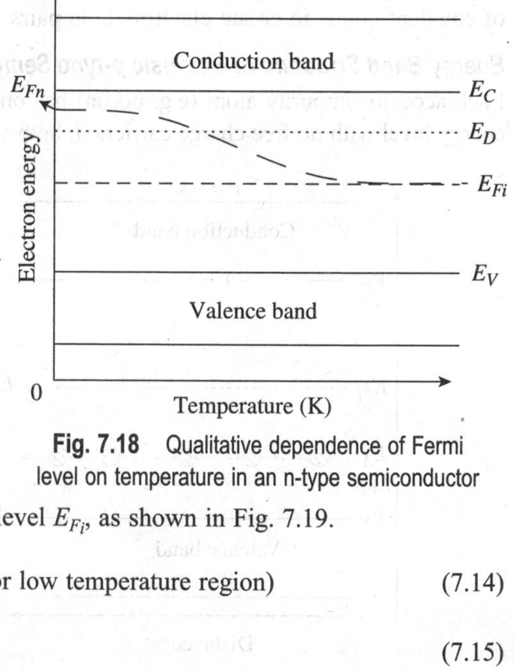

Fig. 7.18 Qualitative dependence of Fermi level on temperature in an n-type semiconductor

p-type Semiconductor

In analogy with n-type semiconductor, in a p-type semiconductor, the Fermi level E_{F_p} rises with increasing temperature from below the acceptor level to the intrinsic level E_{Fi}, as shown in Fig. 7.19.

$$E_{F_p} = \frac{E_A + E_V}{2} \text{ (Ionization region or low temperature region)} \tag{7.14}$$

$$E_{F_p} = E_A \text{ (At } T = T_s) \tag{7.15}$$

$$E_{F_p} = E_{F_i} = \frac{E_g}{2} \text{ (Intrinsic region)}$$

Effect of Variation of Impurity Concentration on Fermi Level in Extrinsic Semiconductors

The addition of donor atoms to a pure (intrinsic) semiconductor leads to the formation of discrete donor levels below the bottom edge of the conduction band as shown in Fig. 7.20(a). At low impurity concentration, the impurity atoms are far apart and hence they do not interact with each other. When the impurity concentration is increased, the atoms come closer and hence there is an increased electrical interaction amongst them. Thus, the donor levels also undergo splitting and form an energy band just below the conduction band as shown in Fig. 7.20 (b).

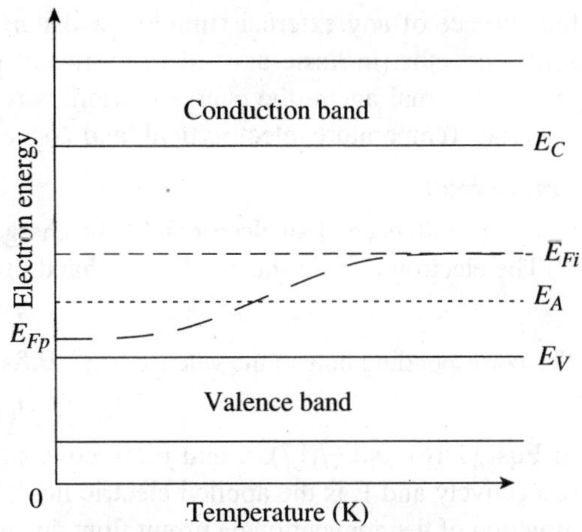

Fig. 7.19 Qualitative dependence of Fermi level on temperature in a p-type semiconductor

For larger donor concentration, the impurity band becomes broader and finally it overlaps with the conduction band, as shown in Fig. 7.20 (c). This is accompanied by a decrease in the width of the forbidden gap and by the upward displacement of the Fermi energy level. With increasing impurity concentration, the Fermi level moves closer and closer to the conduction band and finally moves into the conduction band and becomes an allowed energy level.

In a similar way, in a p-type semiconductor, the acceptor levels just above the top edge of the valence band splits to form the acceptor band. With increasing concentration of acceptor impurity atoms, this band becomes wider. At one stage this band overlaps the valence band and consequently the Fermi level moves into the valence band.

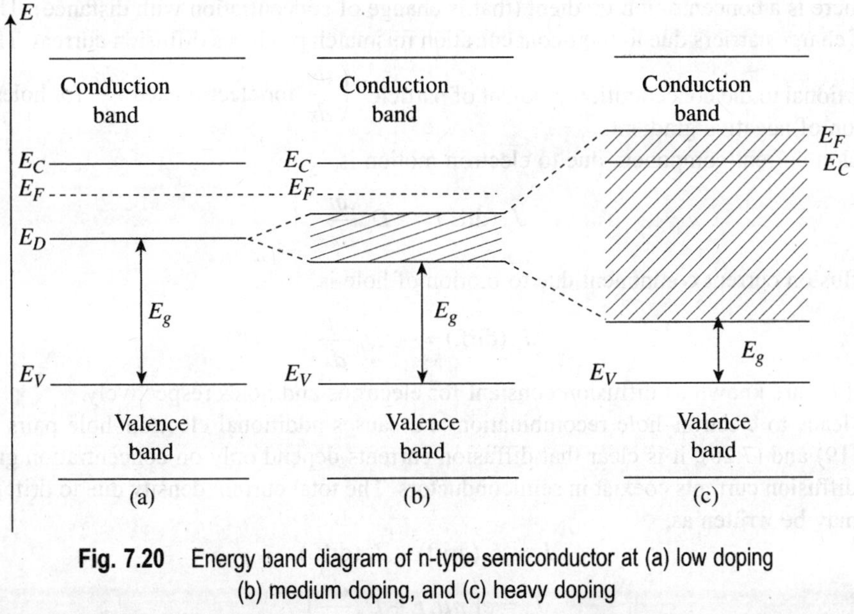

Fig. 7.20 Energy band diagram of n-type semiconductor at (a) low doping (b) medium doping, and (c) heavy doping

7.5.9 Electrical Conduction in Extrinsic Semiconductors

The current flow in semiconductors can be explained through two mechanisms—drift and diffusion. Drift current results from the movement of electrons and holes under the effect of electric field, whereas diffusion current is caused by the gradients in the carrier concentrations. In thermal equilibrium condition and in the absence of any external stimulus or disturbing agent, the average velocity of electrons and holes in a semiconductor (intrinsic as well as extrinsic) is zero and no current flows. If this condition is disturbed by any external agent, the charge carriers may acquire a directional motion leading to current flow in the material. Temperature, electric field, and concentration gradients are examples of such disturbing agents.

Drift Current

Under the influence of an electric field, the charge carriers in the semiconductor materials produce drift current.

The electron drift in the conduction band produces electron current density denoted by a component J_e.

$$J_e\,(\text{drift}) = ne\mu_e E \tag{7.16}$$

The corresponding hole in the valence band drifts and produces hole current density denoted by a component J_h.

$$J_h\,(\text{drift}) = pe\mu_h E \tag{7.17}$$

In Eqs. (7.16) and (7.17), n and p are concentrations and μ_e and μ_h are mobilities of electrons and holes respectively and E is the applied electric field. Though electrons and holes flow in opposite directions, the direction of the conventional current flow due to both of them is same. Hence, their contributions are added to yield the total drift current density.

$$J\,(\text{drift}) = J_e\,(\text{drift}) + J_h\,(\text{drift}) \tag{7.18}$$

$$J\,(\text{drift}) = \text{e}\,(n\mu_e + p\mu_h)\,E$$

Equation (7.18) is equally applicable to both intrinsic and extrinsic semiconductors. Equation (7.18) implies that drift current depends on the electric field and carrier concentration.

Diffusion Current

In a solid, if there are more number of charge carriers on the side of an arbitrary surface than on the other, then we say that there is a concentration gradient (that is change of concentration with distance). The directional movement of charge carriers due to their concentration mismatch produces diffusion current. Thus, diffusion flow is proportional to the concentration gradient of particles $\left(\frac{dn}{dx}\text{ for electron and }\frac{dp}{dx}\text{ for holes}\right)$ and occurs in the direction of negative gradient.

The diffusion current component due to electron motion is

$$J_e\,(\text{diff.}) = e\,D_e\frac{dn}{dx} \tag{7.19}$$

Similarly, diffusion current component due to motion of hole is

$$J_e\,(\text{diff.}) = -e\,D_h\frac{dp}{dx} \tag{7.20}$$

where D_e and D_h are known as diffusion constant for electrons and holes respectively.

Diffusion leads to electron–hole recombination and causes additional electron–hole pairs to disappear. From Eqs (7.19) and (7.20), it is clear that diffusion currents depend only on concentration gradient.

Drift and diffusion currents coexist in semiconductors. The total current density due to drift and diffusion of electrons may be written as,

$$J_e = J_e\,(\text{drift}) + J_e\,(\text{diff.})$$

$$J_e = e\left(n\mu_e E + D_e\frac{dn}{dx}\right) \tag{7.21}$$

Similarly, the total current density due to drift and diffusion of holes is given by,

$$J_h = e\left(p\mu_h E - D_h\frac{dp}{dx}\right) \tag{7.22}$$

7.6 p-n Junction Diode

A p-n junction may be formed from a single-crystal intrinsic semiconductor by doping a part of it with acceptor impurities and the remaining part with donors. Hence, a sudden steep change in impurity exists at the junction plane. Thus, a p-n junction diode is produced when a p-type semiconductor is metallurgically joined with an n-type semiconductor.

The p-type region of the resultant structure will have a large number of 'holes' and very few electrons, whereas n-type region will have large number of 'electrons' and very few holes. Thus, due to abrupt change in the carrier concentration at the junction, the charge carriers will start diffusing into the opposite regions. The holes give rise to hole diffusion current density component J_{hp} and the electrons give rise to electron diffusion current density J_{en}. In the subscripts, the first letters h and e denote the carrier and the second letters p and n denote the region of their origin. Thus the net diffusion current density will be,

$$J\,(\text{diff.}) = J_{hp} + J_{en} \tag{7.23}$$

Figure 7.21 (a) shows the picture just when the junction is formed. Figure 7.21 (b) shows the electrical symbol for p-n junction diode.

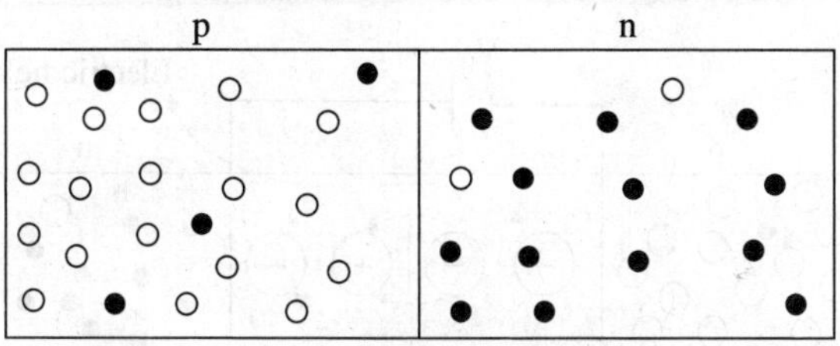

Fig. 7.21 (a) Schematic of a p-n junction just at the time of its formation

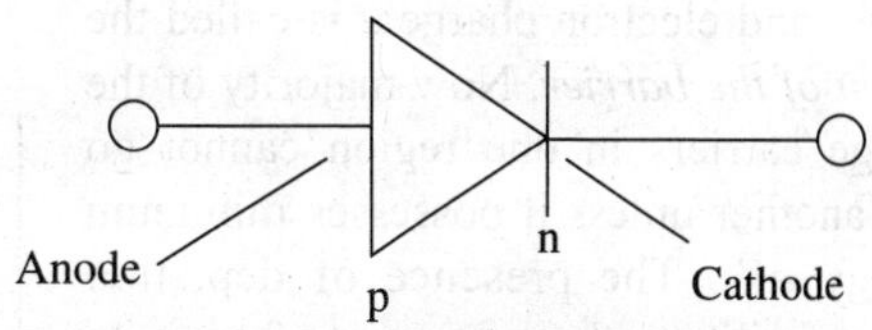

Fig. 7.21 (b) Circuit symbol of a p-n junction diode

The open circles on the left side of the junction in Fig. 7.21 (a) represent holes and the solid or filled circles on the right side denote electrons. Specifically, current will flow readily in one direction (forward biased) but not in the other (reverse biased), creating the basic diode. Hence the resulting structure is known as p-n junction diode.

Depletion Region

As soon as the p-n junction is formed, the diffusion of charge carriers starts. Some free electrons from n-region diffuse across the junction and combine with holes. The result is that the corresponding acceptor impurity atoms nearer the junction on p-side become negative ions. Similarly the diffusion of holes from p-side across the junction causes recombination with electrons on n-side. Therefore, over a very small region near the junction, electrons and holes disappear due to their recombination. The small width near the junction now contains only positive and negative ions on n and p-side respectively. This region is known as *depletion region*, as shown in Fig. 7.22.

The negative ions on p-side are represented by circles with negative sign inside and positive ions on n-side are represented by circles with positive sign inside. These ions are immobile and do not contribute to conduction. The distribution of fixed ions near the junction is also called *space charge*.

Potential Barrier

The space charge in the depletion region produces an electric field spontaneously across the junction and is confined to the depletion region, which is about 1 μm wide. The direction of this built-in electric field is such that it pushes back the holes into the p-region and electrons into the n-region as shown in Fig. 7.23. The creation of electric field reduces the further diffusion of majority charge carriers. Thus, it acts as a barrier to the diffusion of majority charge carriers. The voltage, say V_0, appearing across the depletion region is called the *built-in potential barrier*. The potential V_0 is also referred to as *barrier potential* or *contact potential*.

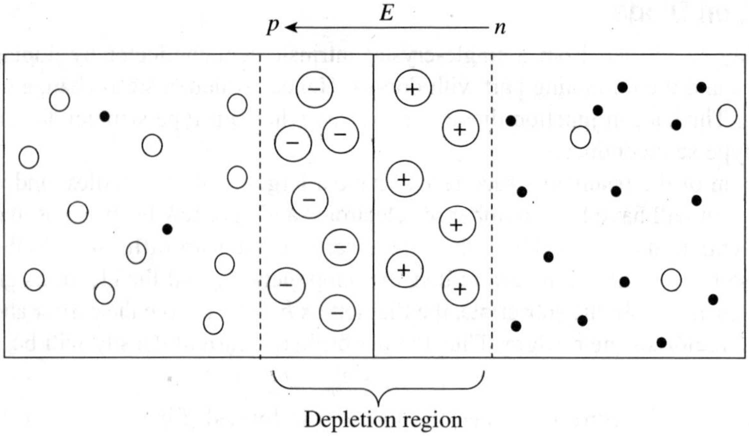

Fig. 7.22 Depletion region in a p-n junction diode

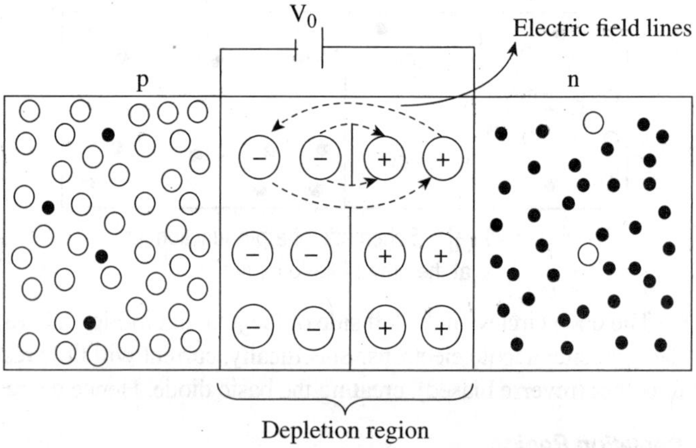

Fig. 7.23 The built-in electric field and potential barrier across depletion region

The product eV_0 of the built-in potential barrier and electron charge e is called the *height of the barrier*. Now majority of the charge carriers in one region cannot go into another unless it possesses minimum energy eV_0. The presence of depletion region facilitates the movement of minority charge carriers across the junction. The current due to minority charge carriers is due to the field which exists over the depletion region and hence is called drift current. This current flows from n-side to p-side and reduces to zero when the depletion region vanishes. Thus, under equilibrium there is no net movement of charge carriers across the junction and hence resultant current through diode is zero. Thus, the function of potential barrier across the junction is to oppose the movement of majority charge carriers and assist the movement of minority charge carriers.

The value of the built-in potential barrier is characteristic of the material and is approximately equal to $E_g/2$. The width of the depletion region depends on the potential barrier and the carrier concentration.

7.6.1 p-n Junction at Equilibrium

The energy bands approach is a useful way to visualize a p-n junction diode at equilibrium. Let us consider two separate p- and n- type doped crystals of the same material as shown in Fig. 7.24 (a). Now they are brought together to form a junction as shown in Fig. 7.24 (b). This process is not physically possible since the p-n junction is usually formed in a single crystal material, but this approach is useful for understanding the concepts.

As soon as the contact is made, the electrons transfer from n-type material into empty levels in the conduction band of the p-type material and cause electron current to flow. Similarly holes transfer from the valence band of the p-type into that of the n-type material. The process stops only when the Fermi level is equalized and is continuous through both materials. Since the Fermi level in each material is fixed relative to its respective band structure, alignment of the Fermi levels is achieved only if all energy levels in one region move relative to those in other. Thus, diffusion of holes from p-region leave it negatively charged,

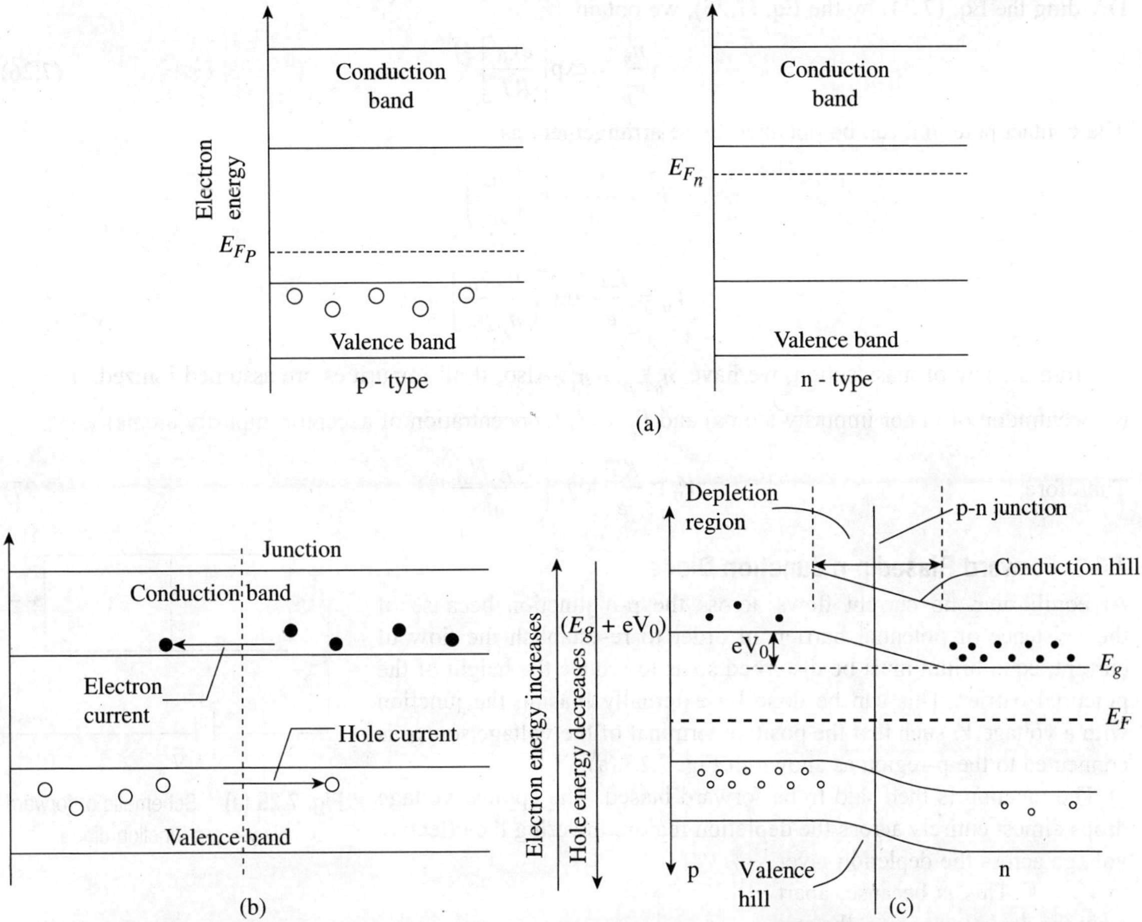

Fig. 7.24 Band structure of p-n junction diode (a) before contact; (b) immediately after contact; (c) in equilibrium

raising all energy levels, and similarly electrons migrating in the opposite direction cause all levels in the n-type material to be lowered, until the Fermi level is continuous across the junction and equilibrium is established as shown in Fig. 7.24 (c).

The equalization of Fermi level symbolizes that the average energy of charge carriers on both the sides is same. At this stage no net current flows across the junction and equilibrium is achieved.

7.6.2 Built-in Potential Barrier or Contact Potential V_0

To derive the expression for the contact potential we assume that all electrons in the valance band on the n-side have zero energy as shown in Fig. 7.24(c). Hence, electrons in conduction band on n-and p-type semiconductor will have energies E_g and $(E_g + eV_0)$ respectively. Using the notation for energy levels as shown in Fig. 7.24 (c), the number of electrons in the conduction band of the *n*-region is (from Eq. (7.3))

$$n_n = N_c \exp[-(E_g - E_F)/kT] \tag{7.24}$$

and the number of electrons in the p-region is,

$$n_p = N_c \exp[-\{(E_g + eV_0) - E_F\}/KT] \tag{7.25}$$

Dividing the Eq. (7.24) by the Eq. (7.25), we obtain

$$\frac{n_n}{n_p} = \exp\left[\frac{eV_0}{KT}\right] \tag{7.26}$$

The contact potential can be obtained by re-arrangement as

$$V_0 = \frac{KT}{e}\log_e\left(\frac{n_n}{n_p}\right)$$

$$V_0 = \frac{KT}{e}\log_e\left(\frac{n_n p_p}{n_p p_p}\right)$$

From the law of mass action, we have $n_p p_p = n_i^2$. Also, if all impurities are assumed ionized, $n_n \approx N_D$ (concentration of donor impurity atoms) and $P_p \approx N_A$ (concentration of acceptor impurity atoms).

Therefore,
$$V_0 \cong \frac{KT}{e}\log_e\left(\frac{N_D\, N_A}{n_i^2}\right) \tag{7.27}$$

7.6.3 Forward Biased p-n Junction Diode

At equilibrium no current flows across the p-n junction because of the existence of potential barrier. In order to re-establish the flow of current, equilibrium must be disturbed so as to reduce the height of the potential barrier. This can be done by externally biasing the junction with a voltage, V, such that the positive terminal of the voltage source is connected to the p-region as shown in Fig. 7.25 (a).

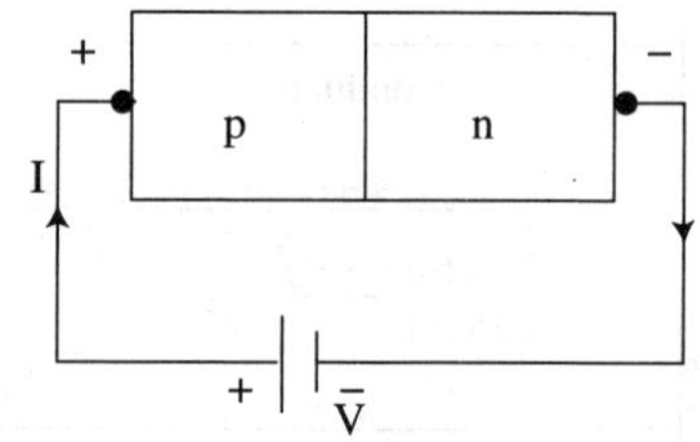

Fig. 7.25 (a) Schematic of forward biased p-n junction diode

The junction is then said to be forward biased. The applied voltage drops almost entirely across the depletion region, lowering the effective voltage across the depletion layer to $V_0 - V$. This is because, apart from the depletion region, the rest has very high conductivity.

Thus, the potential barrier is reduced and once again the current flows across the junction. The following discussion describes the situation in terms of band structure of p-n junction.

When forward bias is applied to a p-n junction, the equilibrium is disturbed. The majority carriers are now able to surmount the decreased potential barriers and diffuse across the junctions. Thus, large currents can flow. A steady-state situation exists as shown in Fig. 7.25 (b). Note that on application of external forward

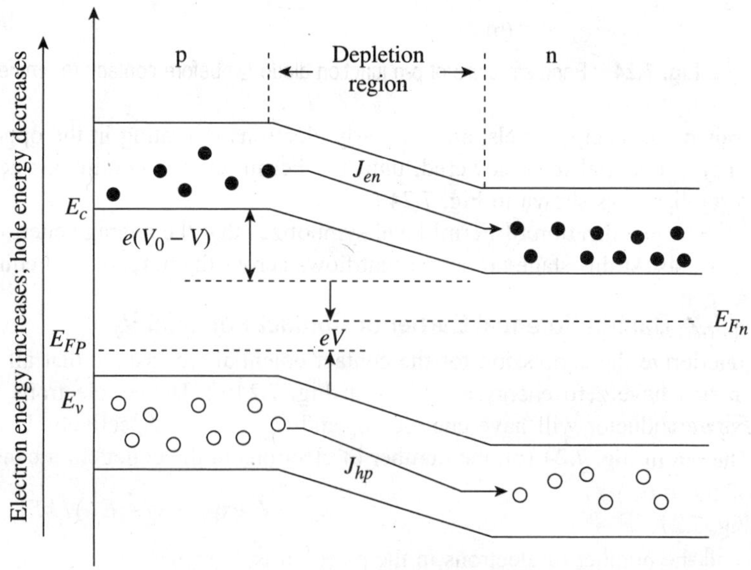

Fig. 7.25 (b) Energy band diagram of p-n junction in forward bias

bias, when the equilibrium gets disturbed, the Fermi level is no longer continuous across the junction. The reduction of barrier height by the applied bias causes holes to be injected from p to n region and electrons to be injected from n to p-region. Current flows in the external circuit in the conventional direction.

7.6.4 Reverse Biased p-n Junction Diode

When the positive terminal of voltage supply (external biasing) is connected to the n-side of the junction, the p-n junction is said to be reverse biased. Fig. 7.26 (a) shows the reverse biased p-n junction diode.

In this case the bias voltage V supports the built-in voltage V_0 and the effective junction voltage becomes greater than the equilibrium value, V_0, by an amount equal to the applied voltage, V. The height of potential barrier preventing diffusion of majority carriers is thus increased in reverse bias. This situation is shown in the band structure diagram of Fig. 7.26 (b). The potential barrier is now of such a height so as to stop almost completely the diffusion of majority charge carriers across the junction; hence, J_{hp} and J_{en} are almost zero. The voltage across the depletion region prevents majority carriers from migrating across the junction but facilitates the flow of minority charge carriers. But as minority charge carriers are less in number, small currents J_{hn} and J_{ep} flow when they cross the junction. Thus large current of the order of mA flows when the junction is forward biased, but a small current of the order of μA flows when it is reverse biased.

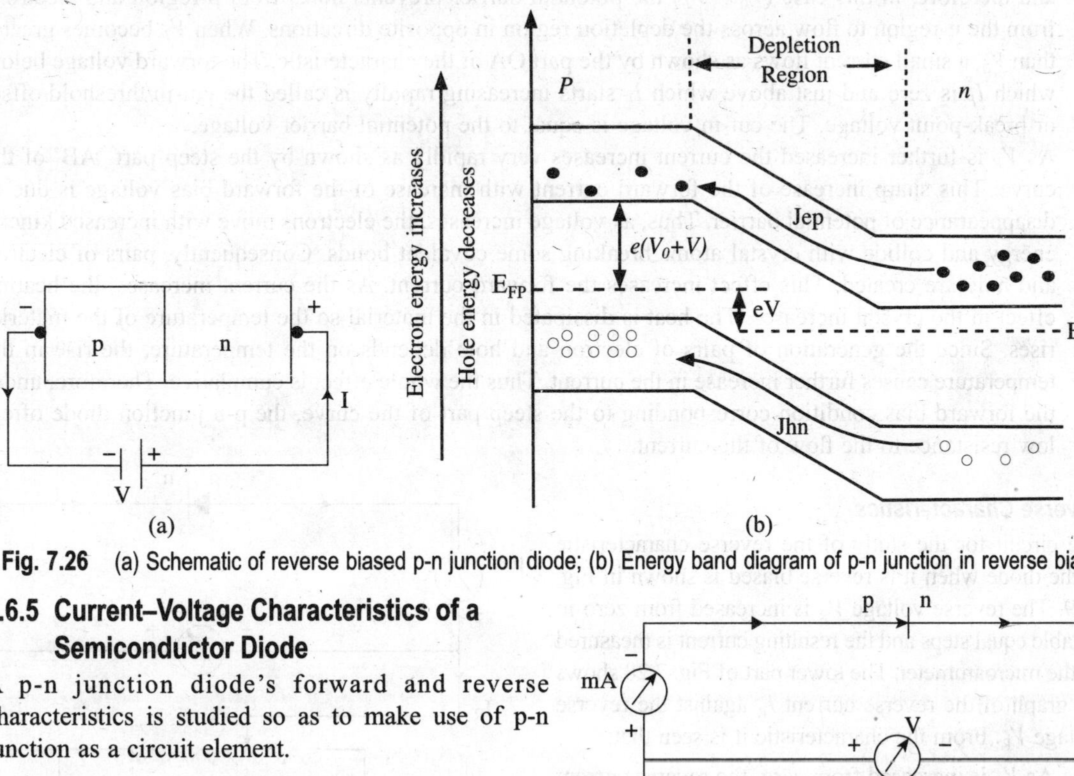

Fig. 7.26 (a) Schematic of reverse biased p-n junction diode; (b) Energy band diagram of p-n junction in reverse bias

7.6.5 Current–Voltage Characteristics of a Semiconductor Diode

A p-n junction diode's forward and reverse characteristics is studied so so as to make use of p-n junction as a circuit element.

Forward Characteristics

The circuit for the study of the forward characteristic of the diode when it is forward biased is shown in Fig. 7.27.

The forward voltage is increased from zero in suitable equal steps and the resulting current is

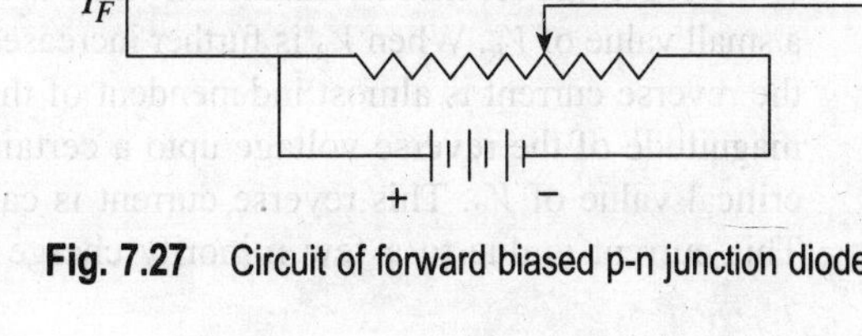

Fig. 7.27 Circuit of forward biased p-n junction diode

recorded by the milliammeter in the circuit. Figure 7.28 shows the graph of the forward current against the applied forward voltage. From the forward characteristic we can infer the following:

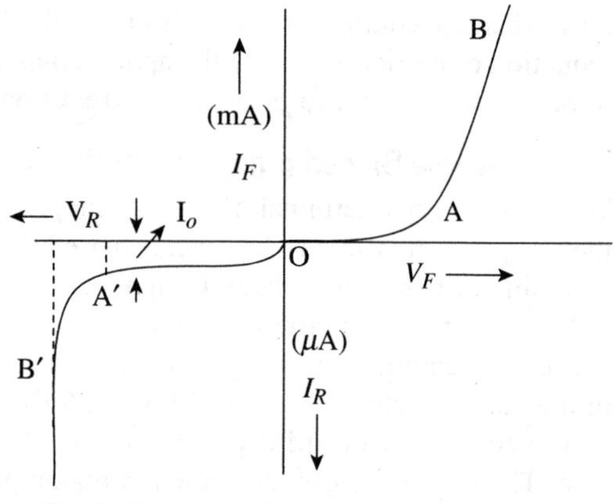

Fig. 7.28 Forward and reverse characteristics of a p-n junction diode

(1) The forward characteristic is not a straight line; hence, the ratio V/I is not constant. Therefore the diode does not obey Ohm's law and a semiconductor diode is a non-linear conductor of electricity. Thus, a p-n junction diode is called a non-linear or non-ohmic device.

(2) For small values of the forward voltage, V_F is less than the internal potential barrier V_0, which is approximately 0.3 V for germanium diodes and 0.7 V for silicon diodes, the forward current, I_F is approximately zero. This is because V_0 opposes V_F and therefore, in this case ($V_0 > V_F$) the potential barrier prevents holes from p-region and electrons from the n-region to flow across the depletion region in opposite directions. When V_F becomes greater than V_0, a small current flows as shown by the part OA of the characteristic. The forward voltage below which I_F is zero and just above which I_F starts increasing rapidly is called the cut-in/threshold/offset or break-point voltage. The cut-in voltage is equal to the potential barrier voltage.

(3) As V_F is further increased the current increases very rapidly as shown by the steep part 'AB' of the curve. This sharp increase of the forward current with increase of the forward bias voltage is due to disappearance of potential barrier. Thus, as voltage increases, the electrons move with increased kinetic energy and collide with crystal atoms breaking some covalent bonds. Consequently, pairs of electron and hole are created. This effect increases the forward current. As the current increases, the heating effect in the crystal increases. The heat is dissipated in the material so the temperature of the material rises. Since the generation of pairs of electron and hole depends on the temperature, the rise in the temperature causes further increase in the current. Thus the whole effect is cumulative. Therefore, under the forward bias condition corresponding to the steep part of the curve, the p-n junction diode offers low resistance to the flow of the current.

Reverse Characteristics

The circuit for the study of the reverse characteristic of the diode when it is reverse biased is shown in Fig. 7.29. The reverse voltage V_R is increased from zero in suitable equal steps and the resulting current is measured by the microammeter. The lower part of Fig. 7.28 shows the graph of the reverse current I_R against the reverse voltage V_R. From the characteristic it is seen that:

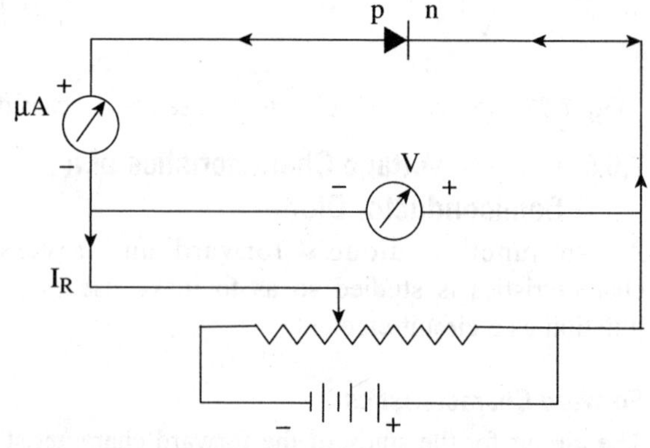

Fig. 7.29 Circuit of reverse biased p-n junction diode

(1) As V_R is increased from zero, the reverse current I_R increases and reaches its maximum value I_0 at a small value of V_R. When V_R is further increased the reverse current is almost independent of the magnitude of the reverse voltage upto a certain critical value of V_R. This reverse current is called the *reverse saturation current*, or *leakage current*. This current is due to a few minority charge carriers (electrons in the P-region and holes in the N

region). Whenever a covalent bond is broken minority charge carriers is generated which gives rise to minority current. As the breaking of covalent bond process is temperature dependent, the minority current depends on the junction temperature. This current increases approximately 7 per cent / °C for both Si and Ge. It approximately doubles for every 10°C rise in temperature.

(2) When V_R is increased beyond the critical value, corresponding to the point A′, I_R increases rapidly. The critical value of V_R is called the *turnover voltage*. After the point, B′, which is just beyond the curvature of the characteristic, is reached, a very small increase in V_R produces a large increase in I_R. The value of V_R corresponding to B′ is called the *breakdown voltage*, and beyond this voltage, the junction is said to be in the *breakdown region*.

Thus regarding the behaviour of a p-n junction diode we observe the following points:

(i) When the p-n junction is forward biased it allows a large forward current to flow through it, that is, it offers a very small resistance.

(ii) When the junction is reverse biased and if the reverse voltage is less than the turnover voltage, it allows a very small reverse current to flow through it, that is, it offers a very high resistance. Hence, a p-n junction diode is a unidirectional device. This characteristic of diode is used for rectifying an alternating voltage (converting an AC voltage into DC voltage).

7.6.6 Application of p-n Junction Diode as Rectifier

The process in which a simple harmonic AC voltage is converted into a unidirectional voltage is known as *rectification* and the circuit used for this conversion is called a *rectifier*. The property of a junction diode that offers a very small resistance and hence allows large current to flow through it when it is forward biased, and does not allow the current to flow through it when reverse biased is used for rectification.

Single Phase Half-wave Rectifier

A simple circuit for half-wave rectification is shown in Fig. 7.30. The alternating voltage to be rectified is connected to the primary coil P_1, P_2 of the power transformer T. The p-region or anode of the diode D is connected to one end S_1 of the secondary of the transformer. The n-region or cathode of the diode is connected to the load resistance R_L and the other end of the resistance is connected to the other end S_2 of the secondary. The power transformer provides the following advantages: (i) It can step up or step down the peak value of the A.C. input voltage. (ii) It isolates the rectifier circuit and the chassis from the power line. Thus chassis shocks are avoided.

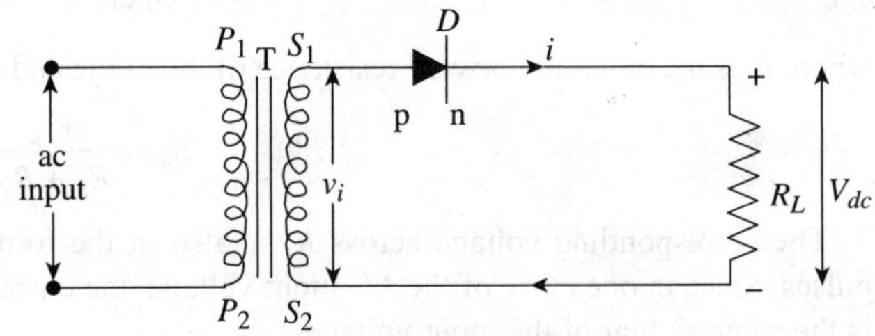

Fig. 7.30 Circuit diagram of half-wave rectifier

Working of the rectifier A sinusoidal AC voltage of frequency $\omega/2\pi$ is applied to the primary of the transformer. The AC voltage developed across the secondary of the transformer, which is applied to the diode in series with R_L is given by

$$V_i = V_m \sin \omega t \tag{7.28}$$

where V_m is the maximum transformer-secondary voltage. The variation of V_i with ωt is shown in Fig. 7.31 (a).

During the half cycles of V_i when the p-region of D is positive with respect to the n-region, the diode is forward biased. Hence a current flows from p to n through the diode in the direction $pnR_LS_2S_1p$. During the half cycles when the p-region is negative with respect to the n-region, the diode is reverse biased and hence no current flows in the circuit. Consequently, the current through R_L is unidirectional and it flows

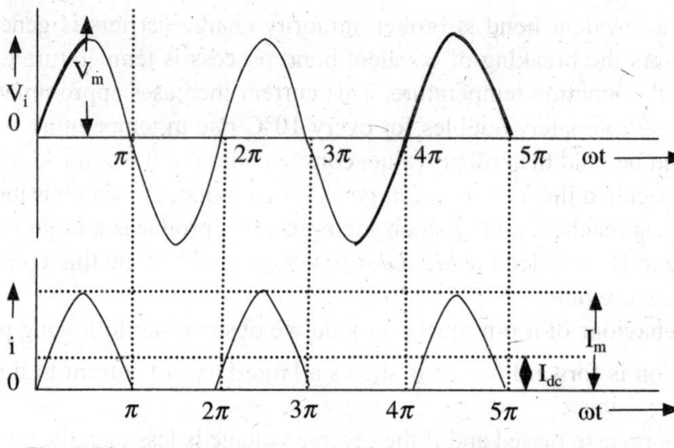

Fig. 7.31 (a) Variation of input signal; (b) Variation of output current

in the form of half-sine waves separated by a period of 180° as shown in Fig. 7.31 (b). The unidirectional current pulses are represented by

$$i = \frac{V_m \sin \omega t}{R_L + R_f} = I_m \sin \omega t \tag{7.29}$$

when $0 \le \omega t \le \pi$

and $i = 0$, when $\pi \le \omega t \le 2\pi$ (7.30)

where R_f is the dynamic forward resistance of the diode and I_m is the maximum value of the current given by

$$I_m = \frac{V_m}{R_L + R_f} \tag{7.31}$$

The corresponding voltage across R_L is also in the form of half sine waves in phase with the current pulses. Since in one cycle of the AC input voltage one current pulse is obtained, the frequency of the pulses is the same as that of the input voltage.

Single Phase Full-wave Rectifier

The p-regions p_1 and p_2 of the diodes are connected to the ends S_1 and S_2 of the secondary of the power transformer. The centre tap 'O' of the secondary is connected to the junction of the N-region, n_1 and n_2 of the diodes through the load resistance R_L.

Working of the rectifier It is assumed that the two diodes are identical and the dynamic forward resistance of each is R_f.

Under the action of a sinusoidal AC voltage of frequency $\omega / 2\pi$ applied to the primary of the transformer, the AC voltages across S_1O and S_2O are given by

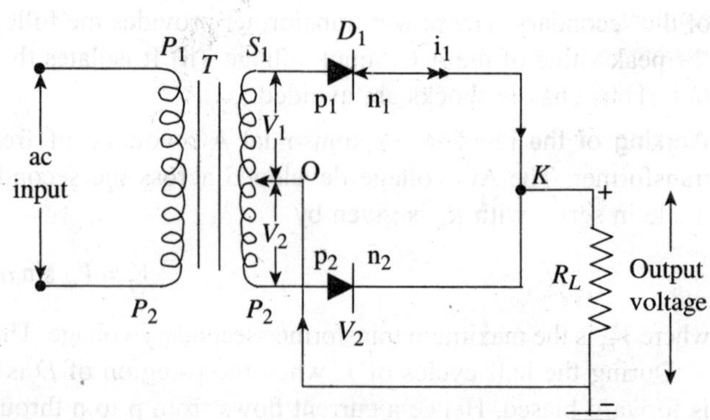

Fig. 7.32 Circuit diagram of full-wave rectifier

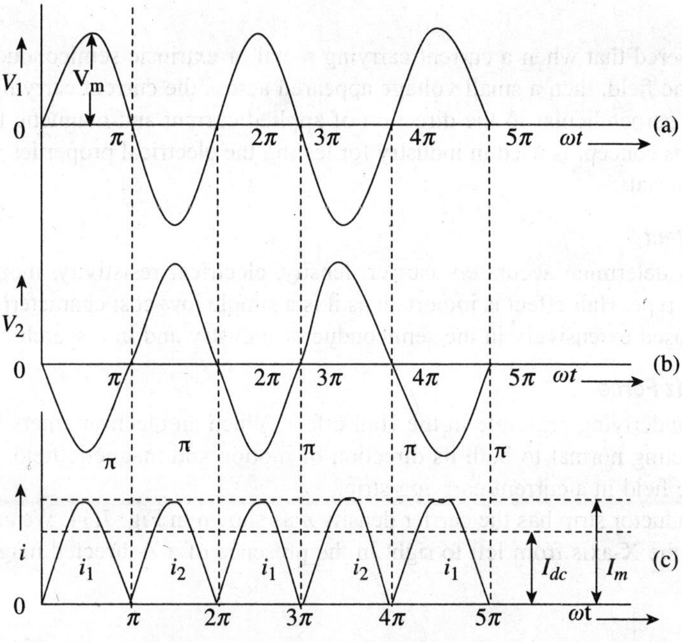

Fig. 7.33 Variations of input and output signals in full wave rectifier

$$V_1 = V_m \sin \omega t \tag{7.32}$$

$$V_2 = V_m \sin(\omega t - \pi) \tag{7.33}$$

where V_m is the maximum value of the voltage from either end of the secondary to the centre tap. These voltages are simultaneously applied across the diodes through the load R_L. Their variations with ωt are shown in Figs 7.33 (a) and (b).

During the half cycle of the AC input voltage when the diode D_1 is forward biased a current i_1 flows in the circuit in the direction $p_1n_1\,R_Los_1p_1$. During this time the diode D_2 is reverse biased. Therefore, no current flows through it. During the next half cycle the diode D_2 is forward biased. Consequently a current i_2 flows in the circuit in the direction $p_2n_2\,R_Los_2p_2$. During this time D_1 is reverse biased and hence it does not conduct. In subsequent half-cycles of the AC input voltages the same process is repeated. Consequently the current through R_L is unidirectional and it flows in the form of half sine waves without separation as shown in Fig. 7.33 (c). The current pulses are represented by

$$i_1 = \frac{V_m}{R_L + R_f} \sin \omega t = I_m \sin \omega t \text{, when } 0 \le \omega t \le \pi \tag{7.34}$$

and
$$i_2 = 0 = \frac{V_m}{R_L + R_f} \sin \omega t = I_m \sin \omega t \text{, when } \pi \le \omega t \le 2\pi \tag{7.35}$$

The corresponding voltage across R_L is also in the form of half sine waves in phase with the current pulses. Since in one cycle of the AC input voltage, two current pulses are obtained, the frequency of the pulses is two times the frequency of the input voltage.

7.7 Hall Effect

Edwin H. Hall discovered that when a current carrying metal or extrinsic semiconductor strip is placed in the transverse magnetic field, then a small voltage appeared across the current carrying strip. The direction of induced voltage is perpendicular to the direction of applied current and magnetic field. A measurement technique based on this concept is used in industry for testing the electrical properties and quality of almost all semiconductor materials.

Importance of Hall Effect

Hall effect is used to determine accurately carrier density, electrical resistivity, mobility of carriers in a semiconductor and its type. Hall effect is important as it is a simple low-cost characterization technique with fast turnaround time used extensively in the semiconductor industry and in research laboratories.

Hall Effect and Lorentz Force

Lorentz force is the underlying principle in the Hall effect. When an electron enters in a magnetic field it experiences a force acting normal to both its direction of motion and magnetic field. This force gives rise to the internal electric field in a current-carrying strip.

An n-type semiconductor strip has the carrier density n as shown in Fig. 7.34. We assume that a constant current I flows along the X-axis from left to right in the presence of a Z-directed magnetic field.

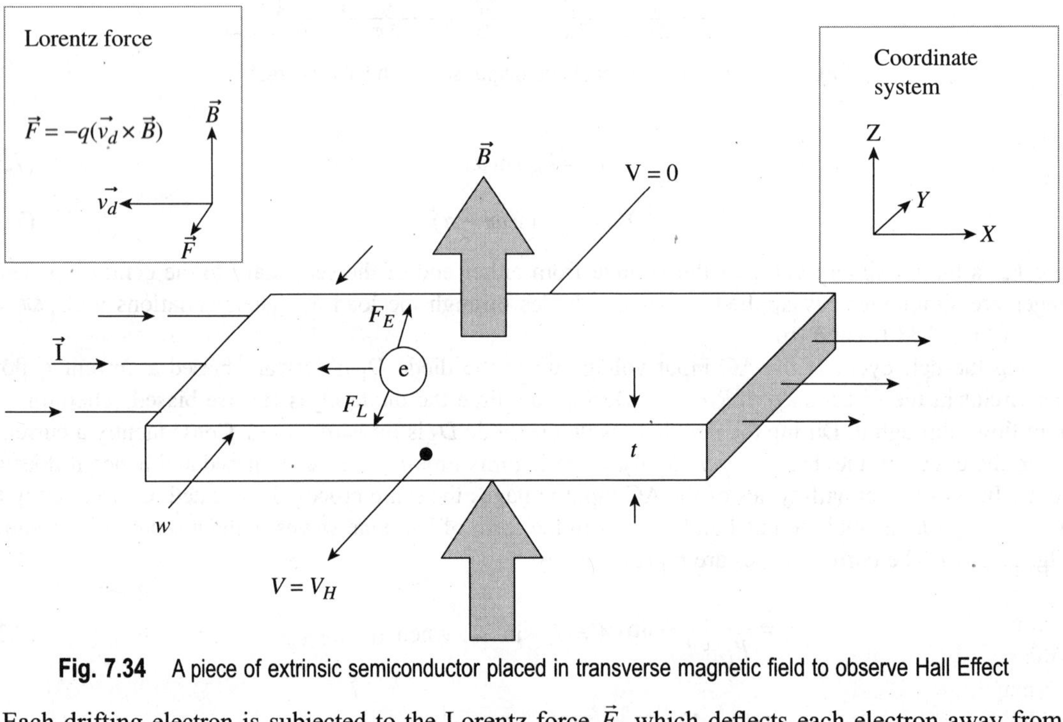

Fig. 7.34 A piece of extrinsic semiconductor placed in transverse magnetic field to observe Hall Effect

Each drifting electron is subjected to the Lorentz force $\vec{F}_L$ which deflects each electron away from its path towards the negative Y-axis (front edge of the n-type strip).

Thus, electrons pile up on the front edge of the strip which makes it negatively charged. Thus, the potential difference is created between the front and the back face of the strip which produces an electric field $\vec{E}$ within the strip, pointing in the direction of negative Y-axis, that is, from back edge to front edge of the strip. This induced electric field also called Hall field E_H exerts force $\vec{F}_E$ on each electron, pushing it in the opposite direction of the magnetic force. An equilibrium quickly develops in which the electric force on each electron builds up until it just cancels the force due to magnetic field. Thus, the current is restored in the circuit as soon as the force due to magnetic field and the force due to electric field balance each other. The drifting electrons again start moving along the X-axis with velocity $\vec{v}_d$, with no further collection of electrons on the front edge of the strip and thus no further increase in the electric field $\vec{E}$. A Hall potential difference V_H is associated with the electric field across the strip of width 'w'. The magnitude of this potential difference is

$$V_H = E_H \times w \tag{7.36}$$

A voltmeter connected across the width can measure the potential difference between the two edges of the strip. The voltmeter can tell us which edge is at higher potential. For the situation shown in Fig. 7.40, we would find that the back edge is at higher potential.

The Hall voltage V_H is negative for *n*-type semiconductors and positive for *p*-type semiconductors. (Note that for a *p*-type semiconductor, the direction of magnetic force on holes is towards the same front edge because the direction of magnetic force does not change as long as the direction of current and magnetic field remains same).

Hall Coefficient R_H

Hall-coefficient R_H is defined as Hall field (E_H) per unit current density (J) per unit magnetic induction (B). It is given by

$$R_H = \frac{E_H}{J \times B} \tag{7.37}$$

The Hall coefficient can be used to estimate the concentration and mobility of charge carriers of the given extrinsic semiconductor.

Derivation of R_H

Under equilibrium condition inside the n-type strip, the electric and magnetic forces are balanced; therefore we have,

$$F_E = F_L$$

$$eE_H = ev_d B \tag{7.38}$$

The drift velocity v_d is given by,

$$v_d = \frac{J}{ne} \tag{7.39}$$

In Eq. (7.38), substituting for v_d from Eq. (7.39), and rearranging the equation, we obtain

$$\frac{E_H}{J \times B} = \frac{1}{ne}$$

Thus, from Eq. (7.37) we get $$R_H = \frac{1}{ne} \tag{7.40}$$

For an *n*-type semiconductor, R_H is negative since charge carriers are electrons which are negatively charged. Thus the concentration of charge carriers can be estimated if Hall coefficient is known.

The magnitude of Hall coefficient is less for metals than semiconductors. In semiconductors, the Hall coefficient decreases sharply with increase in temperature, as the carrier density increases with the temperature.

Hall coefficient can be expressed in terms of measurable parameters as follows. In Eq. (7.37) substituting for $J = I/A$ where A is the cross-sectional area of the strip ($A = t \times d$) and using Eq. (7.36) for E_H, we get

$$R_H = \frac{E_H}{J \times B} = \frac{V_H \times t \times w}{w \times I \times B} = \frac{V_H \times t}{I \times B} \tag{7.41}$$

Thus, in Hall effect experiment by measuring induced Hall voltage V_H for known values of magnetic field strength and current through the sample, we can calculate Hall coefficient. The induced Hall voltage in an extrinsic semiconductor by using Eqs (7.40) and (7.41) can be written as

$$V_H = \frac{BI}{net} \tag{7.42}$$

Mobility of charge carriers can be obtained by using Eqs (7.16) and (7.40) as

$$J = ne\mu_e E$$

$$\mu_e = \frac{J}{neE} = R_H \times \frac{J}{E}$$

By using microscopic form of ohm's law $J = \sigma_n \times E$, where σ_n is the conductivity of the n-type sample, mobility of electrons is expressed as

$$\mu_e = \sigma_n \times R_H \tag{7.43}$$

Thus, by knowing the conductivity of the n-type semiconductor and Hall coefficient, mobility of the charge carriers can be estimated. For p-type semiconductor Eqs (7.40), (7.42), and (7.43) can be written as

$$R_H = \frac{1}{pe}$$

$$V_H = \frac{BI}{pet}$$

$$\mu_h = \sigma_p \times R_H$$

where 'p' is the concentration of holes and σ_p is the conductivity of the p-type semiconductor.

Checkpoint

1. In conductor Fermi energy level is an
   (a) allowed energy level
   (b) not allowed
   (c) none of these
2. The n-type semiconductor is
   (a) negatively charged
   (b) positively charged
   (c) neutral
   (d) none of these
3. The extrinsic process is different from intrinsic process because
   (a) there is no simultaneous generation of charge carriers
   (b) charge unbalance takes place
   (c) simultaneous generation of charge carrier takes place
4. In p-type semiconductor the acceptor energy level lies
   (a) near the conduction band
   (b) near the valence band
   (c) midway between conduction and valence bands
   (d) none of the these
5. The cause of drift current is
   (a) concentration gradient
   (b) more number of charge carriers
   (c) electric field
   (d) none of these

6. The number of charge carrier remains constant in
   (a) intrinsic region
   (b) depletion region
   (c) ionization region
   (d) none of the above
7. If the value of $f(E_c)$ is 0.2 then the probability that the electron can be found in conduction band is
   (a) 80%
   (b) 20%
   (c) 50%
   (d) 0%
8. In depletion region of p-n junction diode
   (a) majority charge carriers exist
   (b) minority charge carriers exist
   (c) positive and negative ions exist
   (d) none of these
9. The electrical behaviour of material is determined by
   (a) size of the material
   (b) forbidden energy gap
   (c) electronic configuration
   (d) none of the above
10. At 0 K in case of n-type semiconductor the donor level is
    (a) completely empty with electrons
    (b) completely filled with electrons
    (c) partly filled with electrons
    (d) none of the above
11. The mobility of an electron in a metal is expressed as
    (a) σ/ne
    (b) $eE/n\sigma m$
    (c) $e\sigma E/n$
    (d) $Ee/\sigma n$
12. The electrical conductivity of semiconductors is of the order of
    (a) 10^{-15} S/m
    (b) 10^{6} S/m
    (c) 10^{-4} S/m
    (d) 10^{4} S /m
13. Impurity atoms in semiconductors
    (a) reduce the energy gap
    (b) increase the kinetic energy of valence electrons
    (c) Inject more charge carriers
    (d) Neutralize recombination of electrons and holes
14. Which of the following is a semi-conducting material?
    (a) Aluminium
    (b) Rubber
    (c) Carborundum
    (d) None of these
15. The mobility of a charge carrier is defined as
    (a) drift velocity per unit applied electric field
    (b) drift velocity multiplied by applied electric field
    (c) drift velocity in absence of applied electric field
    (d) drift velocity opposite to the applied electric field
16. The depletion region of p-n junction under no external bias consists of
    (a) electrons
    (b) holes
    (c) neutral atoms
    (d) immobile ions
17. Silicon doped with phosphorus is
    (a) intrinsic semiconductor
    (b) extrinsic semiconductor
    (c) p-type semiconductor
    (d) n-type semiconductor
18. In germanium, at 0 K, the forbidden gap in eV is equal to
    (a) 1.11
    (b) 0.72
    (c) 1.24
    (d) 1.56
19. In metal, the conductivity is determined by
    (a) density of the metal
    (b) the electronic configuration and proximity of sub-shells in atoms
    (c) electronic concentration and mobility of electrons which are free
    (d) absence of potential in the metal

20. As temperature is lowered, the resistivity of all metals
    (a) tends to decrease and approaches zero
    (b) remains constant or tends to a constant value
    (c) decreases at first and then increases
    (d) increases and then decreases
21. The average drift velocity of an electron under the action of applied electric field is of the order of
    (a) 10^7 m/s
    (b) 10^{-3} m/s
    (c) 10^{-7} m/s
    (d) 10^{-14} m/s
22. The height of the potential barrier at a p-n junction depends upon
    (a) physical dimensions of diode
    (b) density of the semiconductor
    (c) mobility and drift velocity of charge carriers
    (d) densities of impurity atoms and temperature of crystal
23. The thickness of depletion layer in a p-n junction is of the order of
    (a) 10^{-6} m
    (b) 10^{-10} m
    (c) 10^{-4} m
    (d) 10^{-2} m
24. In silicon, at 0 K, the forbidden gap in eV is equal to
    (a) 0.785
    (b) 1.12
    (c) 1.24
    (d) 1.56
25. In forward bias the contact or barrier potential
    (a) increases
    (b) decreases
    (c) does not change
26. In a forward biased p-n junction diode the electrons move from
    (a) valence band of p-type to conduction band of p-type
    (b) valence band of n-type to conduction band of n-type
    (c) valence band of p-type to conduction band of n-type
    (d) conduction band of n-type to conduction band of p-type
27. In a forward bias P-N junction diode the holes move from
    (a) valence band of p-type to valence band of n-type
    (b) valence band of p-type to conduction band of n-type
    (c) valence band of n-type to valence band of p-type
    (d) conduction band of p-type to conduction band of n-type
28. When the energy of holes increases it moves to
    (a) conduction band
    (b) higher energy levels in valence band
    (c) Fermi-energy level
    (d) none of these
29. When the energy of conduction electrons increases, it
    (a) moves into valence band
    (b) moves to higher energy level in conduction band
    (c) escapes from the material
    (d) none of the above
30. In p-n junction diode the reverse saturation current is due to the movement of
    (a) majority charge carriers
    (b) minority charge carriers
    (c) majority and minority charge carriers
    (d) acceptor ions
31. Reverse saturation current depends on
    (a) temperature
    (b) concentration of minority charge carrier
    (c) applied voltage
    (d) temperature and concentration of minority charge carrier.

32. Diffusion current is due to
    (a) applied electric field
    (b) concentration gradient
    (c) both (a) and (b)
    (d) none of above
33. In semiconductors the holes
    (a) move from conduction band to valance band
    (b) move from valence band to conduction band
    (c) does not move
    (d) move within the valance band
34. Under the action of electric fields $\vec{F}$ holes move
    (a) in the direction of $\vec{E}$
    (b) opposite to $\vec{E}$
    (c) does not move

Solved Numericals

1. Calculate the fraction of electrons in the conduction band of diamond (E_g = 5.6 eV) at room temperature (27 C). Hence explain its insulating behaviour.

Solution: The fraction of electrons in conduction band is given by

$$\frac{n}{N} = e^{-\left(\frac{E_g}{2kT}\right)}$$

Given: E_g = 53.6 eV, k = 8.6 × 10^{-5} eV/K, and T = 27 °C = 300 K

Substituting these values, the value of the fraction of electrons in the conduction band comes out to be

$$\frac{n}{N} = 1.69 \times 10^{-47}$$

Since the fraction of electrons in the conduction band of diamond is very less at room temperature, it acts as an insulator.

2. In a solid, consider the energy level lying 0.01 eV below Fermi level. What is the probability of this level not being occupied by an electron?

Solution: The Fermi–Dirac distribution function is given by

$$f(E) = \frac{1}{1 + \exp[(E - E_F)/kT]}$$

This equation gives the probability of energy level E being occupied by electrons. Hence the probability that any energy level E is ***not*** occupied by electrons would be given by

$$1 - f(E) = 1 - \frac{1}{1 + \exp[(E - E_F)/kT]}$$

$$1 - f(E) = \frac{e^{(E-E_F)/kT}}{1 + e^{(E-E_F)/kT}}$$

Dividing by $e^{(E-E_F)/kT}$, we get

$$1 - f(E) = \frac{1}{1 + e^{(E_F-E)/kT}}$$

Substituting the known values, we solve:

$$1 - f(E) = \frac{1}{1 + e^{0.01/0.026}} = \frac{1}{1 + 1.469} = 0.405$$

Thus the probability of the energy level 0.01 eV below Fermi level not being occupied by electrons is 0.405 or 40.5%.

3. The Fermi level for potassium (metal) is 2.1 eV. Calculate the velocity of the electrons at the Fermi level.

Solution: We have $E_F = \frac{1}{2} m v_F^2$

$$v_F^2 = \frac{2E_F}{m} = \frac{2 \times 2.1 \times 1.6 \times 10^{-19}}{9.1 \times 10^{-31}}$$

$$= 0.738 \times 10^{12}\, \text{m}^2/\text{s}^2$$

$$\therefore v_F = 8.59 \times 10^5\, \text{m/s}$$

4. Estimate the fraction of electron in the conduction band at 300° K of
   (i) germanium (E_g = 0.72 eV)
   (ii) silicon (E_g = 1.1 eV) and
   (iii) diamond (E_g = 5.6 eV)

What is the significance of the results?

Solution: At room temperature KT = 0.026 eV.

(i) Germanium:

$$\frac{N_c}{N} = e^{-E_g/2kT} = e^{-\left(\frac{0.72\text{eV}}{2 \times 0.026}\right)} = e^{-13.85}$$

$$= 9.66 \times 10^{-7}$$

(ii) Silicon:

$$\frac{N_c}{N} = e^{-E_g/2kT} = e^{-\left(\frac{1.1\text{eV}}{2\times 0.026}\right)} = e^{-21.15}$$

$$= 6.5 \times 10^{-10}$$

(iii) Diamond:

$$\frac{N_c}{N} = e^{-E_g/2kT} = e^{-\left(\frac{5.6\text{eV}}{2\times 0.026}\right)} = e^{-107.7}$$

$$= 1.7 \times 10^{-47}$$

The above results show that the larger the bandgap the smaller the electron in the conduction band, at a given temperature.

5. The forbidden gap in pure silicon is 1.1 eV. Compare the number of conduction electrons at temperatures 37°C and 27°C. (Given: Boltzmann constant K = 8.6 × 10^{-5}eV/K)

Solution: (i) At 37°C, T = (37 + 273) K = 310 K

$$\frac{N_c}{N} = N_1 = e^{-E_g/2kT} = e^{-\left(\frac{1.1}{2\times K\times 310}\right)}$$

(ii) At 27°C, T = (27 + 273) K = 300 K

$$\frac{N_c}{N} = N_2 = e^{-E_g/2kT} = e^{-\left(\frac{1.1}{2\times K\times 300}\right)}$$

Taking ratio we get

$$\frac{N_1}{N_2} = e^{\frac{1.1}{2\times K}\left(\frac{1}{300}-\frac{1}{310}\right)}$$

$$= e^{\frac{1.1}{2\times 8.6\times 10^{-5}}\left(\frac{1}{300}-\frac{1}{310}\right)}$$

$$= e^{0.068} = 1.7$$

6. Find the resistivity of intrinsic germanium at 300 K. Given that the intrinsic density of carriers is 2.5 × 10^{19}/m^3. (μ_e = 0.39 m^{2}V^{-1}s^{-1} and μ_h = 0.19 m^{2}V^{-1}s^{-1})

Solution: $\sigma_i = en_i(\mu_e + \mu_h)$

$$\sigma_i = 1.602\times 10^{-19} \times 2.5\times 10^{19} \times (0.39 + 0.19)$$

$$\sigma_i = 2.32 \text{ S/m}$$

$$\rho_i = \frac{1}{\sigma_i} = 0.43\ \Omega\text{ m}$$

7. In a doped semiconductor there are 4.52 × 10^{24} holes and 1.25 × 10^{14} electrons per cubic metre. What will be the carrier density in un-doped specimen? Electron and hole mobility are 0.38 m^2/ Vs and 1.18 m^2/ Vs respectively. Calculate the conductivity of intrinsic and the doped semiconductors.

Solution: Intrinsic carrier density, $n_i = \sqrt{p_p n_p}$

Given: p_p = 4.52 × 10^{24} holes/m^3, n_p = 1.25 × 10^{14} electron/m^3μ_e = 0.38 m^2/V.s, μ_h = 0.18 m^2/Vs

$$n_i = \sqrt{(4.52\times 10^{24}/m^3)(1.25\times 10^{14}/m^3)}$$

$$n_i = \sqrt{5.65\times 1038} = 2.38\times 10^{19}/\text{m}^3$$

Conductivity of the doped semiconductor is,

$$\sigma_p = ep_p\mu_h$$

$$= \left(1.602\times 10^{-19}C\right)\left(4.52\times\frac{10^{24}}{m^3}\right)\left(0.18\frac{m^2}{Vs}\right)$$

$$= 1.30\times 10^5\frac{S}{m} = 130\times 10^3 \text{ S/m}$$

8. A sample of intrinsic germanium at room temperature has a carrier concentration of 2.4 × 10^{19}m^{-3}. It is doped with antimony at a rate of one antimony atom per million atoms of germanium. If the concentration of the germanium atoms is 4 × 10^{28}/m^3, then determine the hole concentration.

Solution:

Given: n_i = 2.4 × 10^{19} m^{-3}, N = 4 × 10^{28}/m^3 and $N_p : N$ = 1: 10^6

$$N_D = \frac{4\times 10^{28}\text{atoms/m}^3}{10^6\text{atoms/m}^3} = 4\times 10^{22}\text{donors/m}^3$$

$$n_n = N_D = 4\times 10^{22}$$

$$p_n = \frac{n_i^2}{n_n} = \frac{(2.4\times 10^{19}\text{carriers/m}^{3})^2}{(4\times 10^{22}\text{electrons/m}^3)}$$

$$p_n = 1.4\times 10^{16}\text{ holes/m}^3$$

Thus the minority concentration, that is, holes in an n-type semiconductor can be found out by the above procedure.

9. An n-type germanium sample has a donor density of 10^{21}/m^3. It is arranged in a Hall experiment having magnetic field of 0.5 T and the current density is

500 A/m². Find the Hall voltage if the sample is 3 mm wide.

Solution:

$N_D = 10^{21}/m^3$, $B = 0.5T\ m^2$, $J = 500\ A/m^2$, $w = 3$ mm $= 3 \times 10^{-3}$ m.

Hall voltage, $V_H = \dfrac{BI}{net} = \dfrac{BJA}{net} = \dfrac{BJwt}{net} = \dfrac{BJw}{ne}$

As $n = N_D$

$$V_H = \frac{BJw}{N_D e} = \frac{(0.5T)(500\text{A/m}^2)\ (3 \times 10^{-3}\text{m})}{\left(10^{21}\text{m}^3\right)(1.602 \times 10^{-19} C)}$$

$$= 4.7 \times 10^{-3}\ \text{V}$$

$$V_H = 4.7\,\text{mV}$$

Summary

- The band theory of solids classify the solids into three categories taking into account the arrangement of electrons within the bands and the gap between the filled and empty states, that is, width of forbidden energy gap denoted by Eg. Thus, it successfully explains the differing electrical properties of conductors, semiconductors, and insulators.
- The electrons in the outermost orbits of atoms are known as valence electrons. The band formed by the splitting of orbitals or energy levels containing valence electrons is known as valence band. The valence band is the highest filled energy band in the solid.
- The empty band above the valence band is known as conduction band. The conduction band is the lowest unfilled energy band.
- According to Band Theory, conductors are the materials in which there is no energy gap between valence band and conduction band. The part of the valance band itself is the conduction band or both overlap on each other.
- In insulators, a large energy gap exists between valence band and conduction band whereas semiconductors have a small energy gap between the valence band and the conduction band.
- A pure semiconductor is called an intrinsic semiconductor as at room temperature covalent bonds make and break, which is called an intrinsic process. There are equal number of holes and electrons in an intrinsic semiconductor as for every bond broken an electron–hole pair is created. As both electrons and holes contribute to conductivity, the semiconductor is called bipolar.
- To enhance the conductivity of intrinsic semiconductor, it is doped with impurities such as boron, gallium, phosphorous, and arsenic. Depending on the type of impurity added to an intrinsic or pure semiconductor, extrinsic semiconductors can be of two kinds: n-type semiconductor and p-type semiconductor.
- The current in n-type semiconductor crystals is mostly due to electrons whereas in p-type it is due to holes.
- The Fermi distribution function f(E) is a probability function which determines the occupancy of energy states by electrons given by

$$f(E) = \frac{1}{1 + \exp[(E - E_F)/kT]}$$

- In semiconductor Fermi energy EF is defined as the highest energy which the electrons can have at zero temperature. It is also defined as the average energy of charge carriers at room temperature.
- In an intrinsic semiconductor, the Fermi level lies midway between the conduction band and the valence band.
- As temperature increases the Fermi level in an n-type semiconductor shifts downwards as the donor levels get gradually depleted. At the temperature of depletion, that is, when ionization is complete Fermi level coincides with donor energy level E_D. As the temperature further increases, Fermi level shifts downwards approximately linearly and at a higher temperature (called intrinsic temperature T_i) the Fermi level shifts downwards and ultimately coincides with the intrinsic level. For p-type semiconductor the Fermi level drifts upwards gradually and finally coincides with the intrinsic level for higher temperatures.
- In an n-type semiconductor as the impurity concentration increases, the donor levels undergo splitting and forms an energy band just below the conduction band. For larger donor concentration, the impurity band becomes broader and finally

it overlaps with the conduction band. With increasing impurity concentration, the Fermi level moves closer and closer to the conduction band and finally moves into the conduction band and becomes an allowed energy level.

- In a similar way, in a p-type semiconductor, the acceptor levels just above the top edge of the valence band splits to form the acceptor band. With increasing concentration of acceptor impurity atoms, this band becomes wider. At one stage this band overlaps the valence band and consequently the Fermi level moves into the valence band and becomes an allowed energy level.
- The current flow in semiconductors is of two types, drift and diffusion. Drift current results from the movement of electrons and holes under the effect of electric field, whereas diffusion current is caused by the gradients in the carrier concentrations.
- A p-n junction diode is produced when a p-type semiconductor is metallurgically joined with an n-type semiconductor.
- As soon as the p-n junction is formed a small width near the junction which contains only positive and negative ions on n and p-side respectively is created due to diffusion of majority charge carriers. This region is known as depletion region.
- The potential developed across the depletion region is called as the built-in potential which is the characteristic of the diode material and depends on the doping concentration. It is approximately equal to $E_g/2$.
- The p-n junction diode is forward biased to restore the current due to majority charge carriers in the circuit. This can be done by externally biasing the junction with a voltage,V, such that the positive terminal of the voltage source is connected to the p-region and negative terminal to n-region of the diode. The diode conducts linearly only after a minimum forward voltage called cut in voltage is applied. Thus, diode is a non linear device.
- When the positive terminal of voltage supply (external biasing) is connected to the n-side of the junction, the p-n junction is said to be reverse biased and a very small current due to minority charge carriers flow in the circuit. This current is called reverse saturation current.
- Thus large current of the order of mA flows when the p-n junction diode is forward biased, but a small current of the order of μA flows when it is reverse biased. Thus, diode is a unidirectional device. The property of a junction diode is used for rectification of an ac signal.
- When a current carrying metal or extrinsic semiconductor strip is placed in the transverse magnetic field, then a small voltage appeares across the current carrying strip. The direction of induced voltage is perpendicular to the direction of applied current and magnetic field. This effect is called as Hall's effect.
- Hall Effect is used to determine accurately the type of extrinsic semiconductor, carrier density, electrical resistivity and mobility of charge carriers.

Self-study Questions

1. Explain the formation of energy bands in solids on the basis of band theory.
2. According to band theory, a completely filled band or empty band is not associated with electric conduction. Only partially filled band is responsible for electrical conduction. Explain why.
3. Explain the classification of solids on the basis of energy band diagrams.
4. Write down the expression for Fermi distribution function $f(E)$. Show graphically and analytically that f (E) as function of E always passes through a point $E_F = 0.5$ at different temperatures.
5. Draw a neat sketch of a band diagram of intrinsic semiconductor at room temperature and show that Fermi level in an intrinsic semiconductor lies in the middle of the energy gap. What is the effect of temperature on Fermi level in an intrinsic semiconductor?
6. What is the effect of temperature on Fermi level in an extrinsic semiconductor?
7. Draw a neat energy band diagram for n-type semiconductor at 0 K and T K. Explain the effect of high doping concentration on the Fermi level.

8. Draw the energy band diagram for (a) intrinsic semiconductor (b) n-type semiconductor (c) p-type semiconductor at 0 K and at room temperature.
9. Draw a typical band diagram of an intrinsic semiconductor. In this diagram show an electron which is (a) free (b) bound (c) very close to the nucleus.
10. Differentiate between drift and diffusion currents.
11. Explain the formation of potential barrier across the junction region of semiconductor diode.
12. Draw the energy band diagram for the p-n junction diode in equilibrium, and hence obtain an expression for the height of potential energy barrier.
13. Draw the energy band diagram for the p-n junction diode in forward biased and reverse biased mode.
14. Explain Hall effect and obtain an expression for Hall coefficient for an extrinsic semiconductor.
15. Derive the formula for density of charge carriers in a p-type semiconductor.
16. State two applications of Hall effect.

Exercises

1. Calculate the fraction of electrons in conduction band of Ge (Eg = 0.72 eV) at 300 K. Hence, find the temperature at which silicon (Eg = 1.14 eV) will have the same fraction of electrons.
(Ans. T = 466.77 K)
2. The resistivity of intrinsic silicon at 27°C is 2000 Ω m. Calculate the intrinsic carrier density. Assume μ_e = 0.17 m²/V and μ_h = 0.035 m²/V.
(Ans. $n_i = 1.52 \times 10^{16}/\text{m}^3$)
3. A sample of intrinsic silicon at room temperature has carrier concentration of $2.5 \times 10^{16}/\text{m}^3$. A donor impurity is added to the extent of 1 donor atom per 10^8 atoms of silicon. If the concentration of silicon atom is 5×10^{28} atoms/m³ determine the resistivity of the material. (Given: μ_e = 0.135 m²/Vs and μ_h = 0.048 m²/Vs). (Ans. ρ = 0.092 Ω m)
4. The resistivity of intrinsic silicon is 2.5×10^3 Ω m at 300 K. Calculate its resistivity at 100°C. Assume E_g = 1 eV and $K = 1.38 \times 10^{-23}$ J/K.
(Ans. σ = 65.38 S/m)
5. Silicon wafer is doped with 10^{22}phosphorus atom/m³. Calculate (a) the majority carrier concentration, (b) the minority carrier concentration and (c) the electrical resistivity of the doped silicon at room temperature. Assume complete ionization of the dopant atoms, $n(Si)$ $= 2.5 \times 10^{16}/\text{m}^3$ μ_e = 0.135 m²/Vs, and μ_h = 0.048 m²/Vs.
[Ans. (a) $N_D = 10^{22}/\text{m}^3$ (b) $p_n = 6.25 \times 10^{10}/\text{m}^3$ (c) $\rho = 4.6 \times 10^{-3}$ Ω m]
6. What fraction of the conductivity of intrinsic silicon at room temperature is due to (a) electron and (b) holes? The electron and hole mobilities are 0.135 m²/Vs and 0.048 m²/Vs respectively.
[Ans. (a) σ_e = 0.737 (b) σ_p = 0.262]
7. A current density of 10^4 A/m² flows through an n-type Ge crystal which has resistivity 0.05 Ω m. Calculate the time taken by an electron in materials to drift 40 μm distance. The mobility of electron is 0.38 m²/Vs. (Ans. $t = 2.1 \times 10^{-7}$ s)
8. Silicon has a conductivity of only 5×10^{-4} Ω m in its pure form. An engineer wants it to have conductivity of 300 Ω-1 m-1 when it contains aluminum as an impurity (p-type). Calculate the impurity concentration. Assume $\mu_h = 0.05\ \text{m}^2\text{V}^{-1}\text{s}^{-1}$.
(Ans. $N_A = 3.75 \times 10^{22}/\text{m}^3$)
9. Find the resistivity of intrinsic silicon. When it is doped with a pentavalent impurity of one impurity atom for each 50-million silicon atoms, calculate the resistivity of the doped silicon. $N = 5.5 \times 10^{28}/\text{m}^3$, $n_i = 1.5 \times 10^{16}/\text{m}^3$, μ_e = 0.135 m²/Vs and μ_h = 0.048 m²/Vs.
[Ans. (a) $\rho_i = 2.28 \times 10^3$ Ω-m (b) ρ_n = 0.042 Ω m]
10. An electric field of 150 V/m is applied to a sample of n-type semiconductor whose Hall coefficient is –0.015 m³/C. Determine the current density in the sample assuming μ_e = 0.036 m²/Vs.
(Ans. J = 360 A/m²)
11. In Hall coefficient experiment a current of 0.15 A is sent through a metal strip having thickness 0.2 mm and wide 4 mm. The Hall voltage is found to be 0.25 mV when a magnetic field of 3000 Gauss is used. What is (a) carrier concentration and (b) drift velocity of the carriers?
[Ans. (a) $n = 5.62 \times 10^{24}/\text{m}^3$ (b) 0.21 m/s]
12. An n-type Ge sample has a donor density of $10^{22}/\text{m}^3$. It is arranged in Hall effect experiment having magnetic field 0.6 Tesla and current density

400 A/m^2. Find the Hall voltage if the sample is 5 mm wide. (Ans. $V_H = 0.736$ mV)

13. The resistivity of doped silicon material is 9×10^{-3} ohm-m. The Hall coefficient is 4.1×10^{-4} m^{3}c. Assuming single carrier conduction, find the mobility and concentration of charge carriers.
[Ans. (i) $\mu_h = 0.0455$ m^2/V·s (ii) $p = 1.5 \times 10^{22}$/m^3]

14. A flat copper ribbon 0.3 mm thick carries a steady current 40 mA and is located in a uniform 1.5 T magnetic field directed perpendicular to the plane of the ribbon. If a Hall voltage of 8.6 mV is measured across the ribbon, what is the charge concentration of the free electrons?
(Ans. $n = 1.45 \times 10^{23}$/m^3)

15. The current flowing in a certain reverse biased junction at room temperature is 6×10^{-6} A, what will be the current flowing through the junction if forward bias of 0.3 V is applied.
(Ans. $I = 3.05 \times 10^{-4}$ A)

16. The resistivity of two sides of an abrupt Ge diode is 1.5 Ω cm on p-side and 1 Ω cm on n-side. Calculate the height of potential barrier. (Given: ni = 2.25×1025/m^3, $\mu_h = 0.18$ m^2/Vs, $\mu_e = 0.38$ m^2/Vs.
(Ans. $V_0 = 0.48$ V)

17. Determine the barrier potential V_0 for germanium junction at room temperature when both the n- and p-regions are doped to the extent of one atom per 10^6 germanium atoms. Given atomic density of germanium $N = 4.4 \times 10^{28}$ atoms/m^3, intrinsic density $n_i = 2.5 \times 10^{19}$ carriers/m^3 at room temperature T 300 K.
(Ans. $V_0 = 0.388$ V)

18. Hall probe gives an induced voltage reading of 150 mV for a current of 6 mA when it is placed in a magnetic field of 0.5 T. What is the magnetic field in a region where the reading is 250 mV for 8 mA of current? (Ans. $B = 0.625$ T)

Answers

Checkpoint 1

1(a), 2(c), 3(a), 4(b), 5(c), 6(b), 7(b), 8(c), 9(b), 10(b), 11(a), 12(c), 13(a), 14(c), 15(a), 16(d), 17(d), 18(b), 19(c), 20(a), 21(b), 22(d), 23(a), 24(b), 25(b), 26(d), 27(a), 28(b), 29(b) 30(b), 31(d), 32(b), 33(d), 34(a)

8 Advanced Topics

Learning Objectives

After going through this chapter, you will be able to:

- Learn the types and characteristics of nanomaterials
- Understand the applications and disadvantages of nanomaterials
- Learn the fabrication process of integrated circuits (ICs)
- Understand electron–solid interaction and different types of electron microscopy
- Understand the construction and working details of transmission and scanning electron microscopes

8.0 Introduction

In this chapter, a few applications of concepts explained in previous chapters are discussed. Recent developments in material science engineering have resulted in the novel nano materials with widespread applications. The stark difference in properties at nano level can be explained by using quantum concepts. In the first section of this chapter, nanomaterials, their types with emphasis on carbon-based materials, and the applications of these novel materials are discussed. Semiconductor components are built on the same chip by using Integrated Circuit technology. This technology has tremendous advantages and hence steps of fabrication of simple components are discussed in the second section of this chapter.

In 1981 Scanning Electron Microscope (SEM) was invented at the IBM Laboratories in Switzerland, which has essentially provided the basic technology for work at the nanoscale. By using SEM it is possible to visualize individual atoms and molecules, as the surface of material is scanned. The concepts of wave nature of accelerated electrons and electron lens are manifested in electron microscopy. The construction and working of SEM and TEM are explained in Section 8.3 of the chapter.

8.1 Introduction to Nanoengineering

Nanoengineering is an evolving field. It offers ample opportunities for the development of new materials, devices, and products due to atomic scale engineering. Thus, nanoengineering has led to novel nanotechnology applications which range from improved materials for everyday usage, such as paints, sunscreen lotions, cosmetics, food products, disinfectants to the industry and purification materials such as construction material, materials used for water filtration, desalination of water, wastewater treatment, membranes used in fuel cells, and many more.

Engineered nanomaterials are devised at the nanometre scale, that is, at the molecular level; the small size contributes to unique properties, which are generally not seen when the same material is in its bulk form. Thus, nano phase engineering has led to the invention of new structural and functional materials, both organic and inorganic, which allows us to manipulate properties of material viz. physical, chemical, mechanical, electric, magnetic, optical, and electronic. This section gives an outline of different types of engineered nanomaterials and their significant applications.

8.1.1 Nanomaterials

Nanoengineering has resulted in many nano materials; these include nano particles, nano rods, nono tubes, and nano fibres. Nanomaterials are substances in which at least one dimension is less than 100 nanometres where one nanometre corresponds to three atoms in diameter. Thin films, layers such as graphene, or surface coatings are nano materials that have only one dimension in the nanoscale and the other two dimensions are extended. In nanowires and nanotubes two dimensions are in nanoscale and are extended in only one dimension. Nano particles, for example, precipitates, colloids, and quantum dots (tiny particles of semiconductor materials) have all three dimensions at nanoscale. Nanomaterials are important due to their unique optical, magnetic, electrical, and other properties.

8.1.2 Factors Responsible for Properties of Nanomaterials

It was considered that physical properties of material do not depend on the size. But it was observed that a material's properties change at the nano-scale size due to its nanometre size. The sudden change in the properties can be attributed to: (i) the large number of surface atoms as compared to bulk atoms; (ii) high surface energy; (iii) quantum confinement; (iv) reduction in imperfections that do not exist when the material is in bulk form. Out of the four factors responsible for changed properties at nano-scale, increased relative surface area and quantum effects are more significant and hence are discussed below.

Large Fraction of Surface Atoms

When the size of the bulk material is larger than one micrometre (or micron), there are less number of surface atoms than the number of atoms in the bulk of the material. As the size of material decreases, a greater proportion of atoms are found at the surface compared to those inside. At nanoscale there is a significant increase in the percentage of surface atoms compared to the bulk atoms as shown in Table 8.1. The atoms at or near the surface are more reactive as they are weakly bonded. If N_S is the number of atoms on the surface and N_b is the number of atoms in the bulk, then their ratio is roughly estimated to be $3/r$, where 'r' is the size of the particle. Thus with reduction in size of the particle, the number of atoms on the surface is greater than the total number of atoms. Therefore, surface atoms start dominating the properties of the material at the nano scale.

Table 8.1 Bulk atoms vs surface atoms

Full-shell clusters	Total number of atoms	Surface atoms (%)
1 Shell	13	92
2 Shells	55	76
3 Shells	147	63
4 Shells	309	52
5 Shells	561	45
7 Shells	1415	35

Quantum Confinement

If the size of the particle is comparable to the de Broglie wavelength of the moving electron, then quantum confinement takes place. An electron if confined to move in dimensions which are large compared to its wavelength, behaves as a free particle. For a free particle, the energy bands are continuous and the material's band gap remains unchanged. However, as the dimension of the confinement decreases and reaches a nanoscale limit, the energy bands become discrete as shown in Fig. 8.1. As the energy levels in the band are spaced out, the band gap changes and becomes size dependent. Thus, the increase in the energy difference between energy levels leads to the change in the band gap of the material, which consequently leads to changed electrical behaviour as shown in Fig. 8.2.

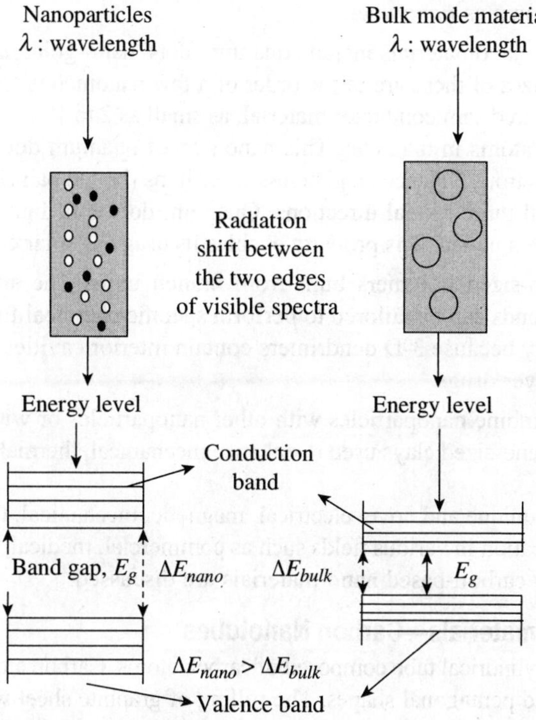

Fig. 8.1 Quantum confinement

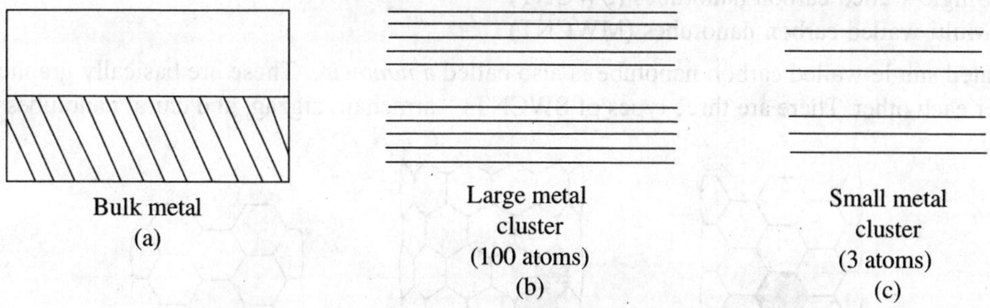

Fig. 8.2 Change in valence energy band levels with size

8.1.3 Types of Nanomaterials

The contemporary nanomaterials can be classified into the following categories:

(i) Carbon-based nanomaterials
(ii) Metal-based nanomaterials
(iii) Dendrimers
(iv) Composites

These can be described as given below.

Carbon-based materials These nanomaterials are primarily composed of carbon in the form of hollow spheres, tubes, or ellipsoids. Spherical and ellipsoidal nanomaterials are known as *fullerenes*. Cylindrically shaped carbon material is known as *nanotubes.*

Metal-based materials These nanomaterials include quantum dots, nano-gold, nano-silver, and metal oxides such as titanium oxide. The sizes of these are of the order of a few nanometers to a few hundred nanometres. A *quantum dot* is a nanostructured semiconductor material, as small as 2 to 10 nanometres, which corresponds to approximately to 10 to 50 atoms in diameter. This nano size of quantum dot leads to confinement of the motion of conduction band electrons, valence band holes, or excitons (bound pairs of conduction band electrons and valence band holes) in all three spatial directions. Quantum dots emit light when excited. The emitted energy is higher when dots are smaller. This property enables its usage as solar cells, LEDs, and diode lasers.

Dendrimers These are nano-sized polymers built from branch units. The surfaces of dendrimers have numerous chain ends. These ends can be tailored to perform specific chemical function. These are useful for catalysis and for drug delivery because 3-D dendrimers contain interior cavities into which other molecules could be placed that can deliver drugs.

Composites Composites combine nanoparticles with other nanoparticles or with large bulk type materials. Some specific examples are nano-sized clays used to enhance mechanical, thermal, barrier, and flame retardant properties.

These nanomaterials have unique and novel electrical, magnetic, mechanical, thermal, or imaging features, which are necessary for application in various fields such as commercial, medical, military, and environmental. In the following section only carbon-based nano-materials are discussed.

8.1.4 Carbon-based Nanomaterials—Carbon Nanotubes

Carbon nanotube is a hollow cylindrical tube composed of carbon atoms. Carbon atoms in the tube are connected to each other in hexagonal and pentagonal shapes. The rolling of graphite sheet with respect to the alignment of C—C bond is used to describe the nature of structure. The tubes are usually 100 μm long and 1–20 nm in diameter. The ends of the tube may not necessarily be closed. Carbon nanotubes are classified into

(i) Single-walled carbon nanotubes (SWCNT)
(ii) Multi-walled carbon nanotubes (MWCNT)

Fabricated single-walled carbon nanotube is also called a *nanowire.* These are basically graphene sheets rolled over each other. There are three types of SWCNTs—armchair, zigzag, and chiral nanotubes as shown in Fig. 8.3.

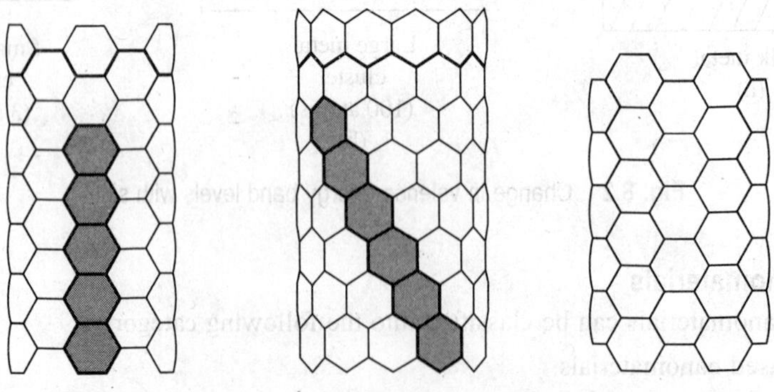

(a) Armchair (b) Zigzag (c) Chiral

Fig. 8.3 Types of SWCNT

Some important properties of carbon nanotubes are as follows which are tabulated in Table 8.2.

- Tensile strength of carbon nanotubes is approximately hundred times greater than steel of the same diameter.
- CNT has Young's modulus nearly five times higher than that of steel.
- CNTs are extremely lightweight with density about one-fourth of steel.
- They have very high thermal conductivity.

Table 8.2 Properties of a few nanotubes

Material	Young's modulus (GPa)	Tensile strength (GPa)	Density (g/cm3)	Bonding	Manufacturing
SWCNT	1054	150	---	sp^2	Difficult
MWCNT	1200	150	2.6	sp^3	Easy
Steel	208	0.4	7.8	---	---

Most of the armchair nanotubes have metallic properties, whereas zigzag and chiral nanotubes have electrical properties like semiconductors. These CNTs conduct electricity better than metals. Whenever an electron travels through a CNT it travels like a wave down a smooth channel with no atoms to offer resistance to electron flow. This quantum movement of an electron within nanotubes is called *ballistic transport.*

Difference between SWCNT and MWCNT Properties of SWCNTs depend on the diameter and the helicity of graphite rings. MWCNTs consist of multiple rolled layers of graphite and has lower mechanical properties than SWCNTs.

8.1.5 Applications of Nanomaterial

Nanomaterials have superior mechanical, electrical, optical, and magnetic properties, which are not found in other materials. It has a wide range of applications in the field of electronics, fuel cells, batteries, agriculture, food industry, and medicines. A few examples are given below.

Optoelectronic devices Nanostructured semiconductors exhibit various non-linear optical properties. A semiconductor quantum dot exhibits luminescence due to quantum confinement. Hence, silicon powders and silicon germanium quantum dots are used in infrared optoelectronic devices. In solar cells, nanostructured semiconductors are used as window layers. Polymer-based composites with a high content of inorganic particles leading to a high dielectric constant are used for photonic band gap structure.

Electronic industry CNT, due to its extraordinary electrical conductivity, has become a favourable material for fabricating transistors. A transistor has three electrodes, namely source, drain, and gate. The voltage applied to the gate electrode controls the electron density of the semiconductor channel between the source and the drain. It is known that the control through the gate electrode can be improved by using a material with a high dielectric constant in the insulation film between the semiconductor channel and the gate electrode, or by fabricating the gate electrode so that it surrounds the semiconductor channel. The insulation film that can be used with ordinary semiconductors such as silicon has limitations due to its unstable surface; but the CNT has a stable surface so that a large variety of insulation film materials can be used with it. In addition, the structure of CNT makes it easy to fabricate the gate electrode so that it surrounds CNT.

Information storage Single nano-sized magnetic particles are mono-domains, that is, they exist in only one domain. Very small particles have distinct atomic structures, which give rise to discrete electronic states. These unique structures of nano particles produce the special property of super-paramagnetism (temporary magnetism). Magnetic nano composites have been used as ferro fluids for mechanical force transfer, high-density information storage, and magnetic refrigeration.

Sensors Nanostructured metal-oxide thin films are used as gas sensors with enhanced sensitivity and selectivity for detection of NO_x, CO, CO_2, CH_4, and aromatic hydrocarbons. Nanostructured metal-oxide (MnO_2) is used in rechargeable batteries for cars.

8.1.6 Disadvantages of Nanomaterials

Instability of the particles Nanomaterials are thermodynamically metastable; hence, retaining the active metal nanoparticles is difficult as they undergo rapid transformations. This leads to poor corrosion resistance, high solubility, and phase changes in nanomaterials. This results in the deterioration of their properties.

Large surface area Nanoparticles have large surface area. Hence, when fine metal particles come in direct contact with oxygen, exothermic combustion can take place causing an explosion.

Impurity Nanoparticles are highly reactive; hence, synthesis of pure nanoparticles becomes highly difficult. While synthesizing nanoparticles it is difficult to avoid formation of oxides, nitrides, etc.

Biologically harmful Nanomaterials are considered to be toxic and carcinogenic due to their high surface area and enhanced surface activity.

Recycling and disposal For disposal of nanomaterials stringent guidelines are not yet available as the issues of their toxicity and harmful effects have not been established.

Checkpoint 1

1. The percentage of surface atoms to bulk atoms with the decrease in the size of the material
   (a) increases
   (b) does not change
   (c) decreases
   (d) none of the above
2. In nanonaterials as the particle size decreases the band gap of the material
   (a) remains unchanged
   (b) increases
   (c) decreases
   (d) none of the above
3. Nanomaterials are broadly classified in to
   (a) 5 types
   (b) 2 types
   (c) 3 types
   (d) 4 types
4. Young's modulus in MWCNTs as compared to SWCNTs is
   (a) less
   (b) more
   (c) almost same
   (d) none of the above
5. The bonding in SWCNT is
   (a) sp^3
   (b) sp
   (c) sp^2
   (d) none of the above

8.2 Semiconductor Device Fabrication

In chapter 7 properties of semiconductor and devices made out of semiconductors, namely p-n junction diode was explained. Semiconductor devices are of two forms—discrete and integrated. The discrete forms include diodes and transistors, whereas integrated circuits, use these discrete components to make one device. Semiconductor device fabrication is a process to produce integrated circuits (IC) which are present in various electrical and electronic devices used in day-to-day life. An integrated circuit (IC) is also called a microelectronic circuit, chip, or microchip. It is a thin substrate of semiconductor material (usually silicon) on which interconnected multiple miniaturized (tiny) electronic components, namely resistors, capacitors, diodes, and transistors are fabricated as a single unit. By massive integration of transistors and other electronic components, complex circuits can be created which are inexpensive and reliable. Introduction of ICs has revolutionized the fabrication technology and finds extensive usage as logic blocks, amplifier, oscillator, counter, computer memory, or microprocessor in consumer electronics, automobiles, medical equipment, and industrial electronics.

8.2.1 IC Classification and Monolithic and Hybrid ICs

Integrated circuits can be classified based on the number of transistors and other electronic components as shown in Table 8.3.

Table 8.3 Classification of ICs

Type of IC	Components
SSI (small scale integration)	10 to 100 electronic components per chip
MSI (medium scale integration)	100 to 1,000 electronic components per chip
LSI (large scale integration)	1,000 to 100,000 (hundred thousand) electronic components per chip
VLSI (very large scale integration)	100,000 to 1,000,000 (1 million) electronic components per chip
ULSI (ultra large scale integration)	Modern IC with more than 1 million electronic components per chip

Further integrated circuits are of two forms, namely monolithic and hybrid. In *monolithic form* transistors, diodes, resistors are fabricated and are interconnected on the same chip. In *hybrid form* circuits, elements are in discrete form and are connected externally with discrete elements, which are formed on the chip. The advantage of hybrid circuits is that components that cannot be included e.g., capacitors of large value, wound components, crystals, inductors, etc. can be used in a monolithic IC.

8.2.2 IC Fabrication Steps

Semiconductor device fabrication is a procedure involving multiple-steps for creating electronic circuits on a wafer made of pure semiconducting material. The steps include wafer preparation, photolithography, and chemical processing viz. etching, thermal, and local oxidation, changing the doping concentration by ion implantation and diffusion, and depositing thin films over the semiconductor's substrate.

In addition to the above-mentioned basic processing steps, a few more fabrication steps are included to produce complex devices and circuits. This step-by-step method of making circuits on a wafer by developing layer-upon-layer on a substrate is called *planar technology*. In the planar process every fabrication step is applied to the entire silicon wafer and hence, it is possible to make and interconnect many devices. This technology not only enables fabrication of complex IC with astronomical precision, but also makes simultaneous fabrication of many IC chips on a single wafer possible. Thus, microminiaturization of electronic circuits not only has reduced the cost with increase in reliability, but also has led to many new applications. This technology is adapted to manufacture flat-panel displays, micro-electromechanical systems (MEMS), and even DNA chips for DNA screening.

The basic important processing steps used in the fabrication of integrated devices like diode or bipolar junction transistor are as discussed below in detail.

Wafer Preparation

The first step in the production of an IC is growth of a large piece of almost perfect crystalline semiconducting material called an ingot. *Ingots* are manufactured by the freezing of a molten liquid (known as the melt) in a mold. Ingots are then sliced using a diamond tipped saw into wafers of approximately 500–1000 μm (0.5 to 1 mm) thickness. The wafers are then cleaned, rinsed, and dried for use in the fabrication of discrete devices and integrated circuits. For manufacturing of integrated circuits, wafers are the starting material and are normally referred to as the *substrate*.

Thermal Oxidation

This is a simple step in IC processing. It consists of thermal oxidation of substrate (silicon wafer) that leads to the growth of a silicon dioxide layer (SiO_2) on the substrate surface. This requires heating of the substrate in an oxidation furnace to 900–1200°C in dry (O_2) or wet (H_2O) steam. As silicon oxidizes quickly and forms highly stable SiO_2 layer that has excellent insulating properties, it is widely used instead of other semiconductors like germanium and gallium arsenide. The relevant chemical reactions are

$$Si + O_2 \Rightarrow SiO_2$$

$$Si + 2H_2O \Rightarrow SiO_2 + 2H_2$$

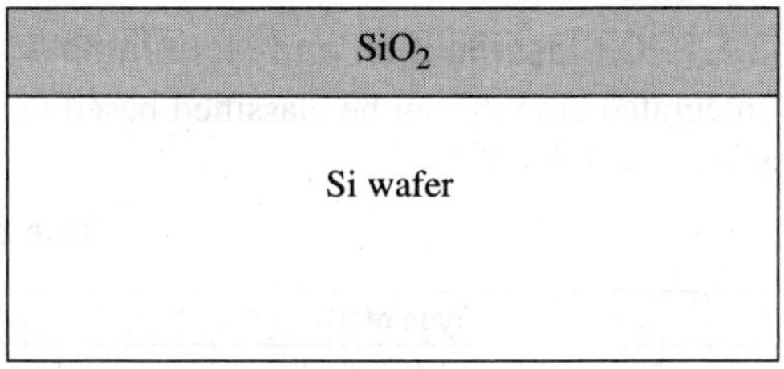

Fig. 8.4 Thermal oxidation

Photolithography

To selectively remove the oxide layer from those areas on the substrate (silicon wafer) where dopant atoms are to be introduced, spatial selection is accomplished using a process called photolithography. In the term photolithography, 'photo' means light (usually ultraviolet light), 'lithos' refers to stone and 'graphein'—to write. Thus, photolithography is the process of transferring geometric shapes designed on a mask to the surface of a silicon wafer by using UV light. The process of photolithography is repeated many times in the manufacture

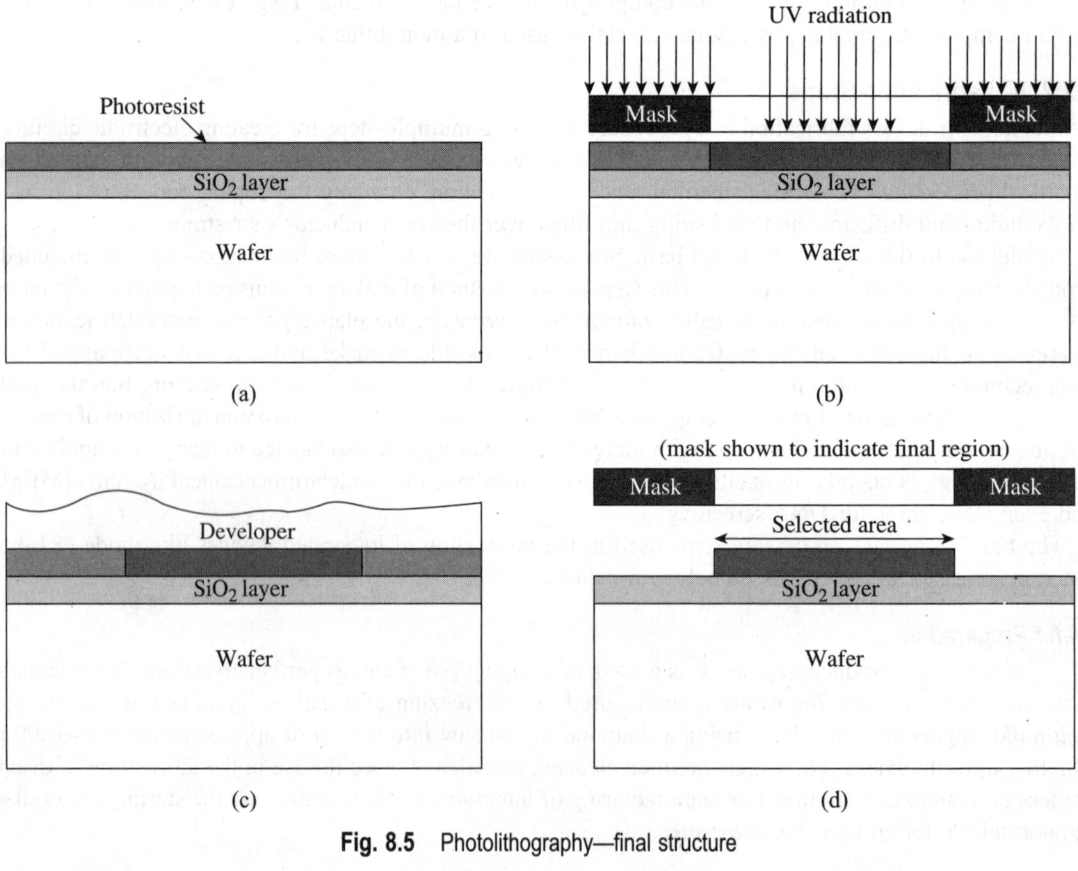

Fig. 8.5 Photolithography—final structure

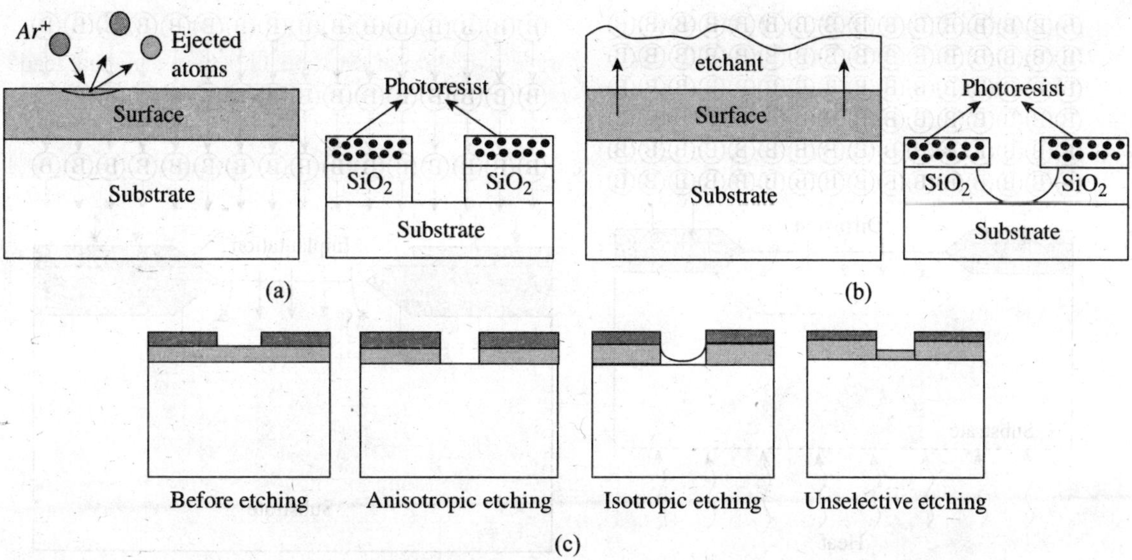

Fig. 8.6 (a) Dry etching (Anisotropic etching) (b) Wet etching (Isotropic etching) (c) Etching–selectivity and anisotropy

of an IC to build up device structures and interconnections. The first step in photolithography is to coat the surface with approximately 1 μm of photoresist (PR) as shown in Fig. 8.5 (a). The PR is then exposed to ultraviolet (UV) radiation through a mask as shown in Fig. 8.5 (b). The mask is a quartz photoplate containing the patterns to be produced generated from the information about device placement and connections. The UV radiation causes a chemical change in the PR where the mask is transparent. The photoresist under the opaque regions of the mask is unaffected. The PR is then developed using a chemical developer as shown in Fig. 8.5 (c). There are two types of PR. When a negative PR is exposed to UV radiation, it gets hardened against the developer and hence remains on the surface. But a positive PR is removed from the regions exposed to the UV. Here a negative PR is assumed. The exposed silicon dioxide is then removed by using hydrofluoric acid in which it dissolves and the obtained structure is shown in Fig. 8.5 (d). This process is similar to that used in developing a photographic film.

Etching

Etching is the process of selective removal of material viz. semiconductor, metal, or silicon dioxide from the chip surface. In dry etching as shown in Fig. 8.6 (a), the wafer to be etched is placed on an electrode and ions of the gas are accelerated toward it. Transfer of tremendous momentum from the gas ions to the atoms at or near the surface of the wafer, causes its ejection. Thus, the atoms at or near the surface of the wafer are etched out. In wet etching as shown in Fig. 8.6 (b), the wafer is immersed in a chemical etchant (hydrofluoric acid) to remove material through chemical reaction. In etching in general, selectivity and anisotropy (non-uniformity) are very important which is illustrated in Fig 8.6 (c). *Selectivity* refers to the ability of an etchant to remove one material on the surface without affecting the other. Isotropic refers to uniformity in etching in all directions. The wet chemical process is not suitable for small dimensions as the material is also removed from the regions where it is to be left which causes serious problems.

Doping (Dopant Diffusion)

A high concentration of dopant atoms can be introduced into the substrate through diffusion. The substrate is heated to a temperature of about 1000° C in the presence of vapour that contains dopant atoms as shown in Fig. 8.7. The dopant atoms then diffuse into the substrate as at high temperature silicon atoms move out

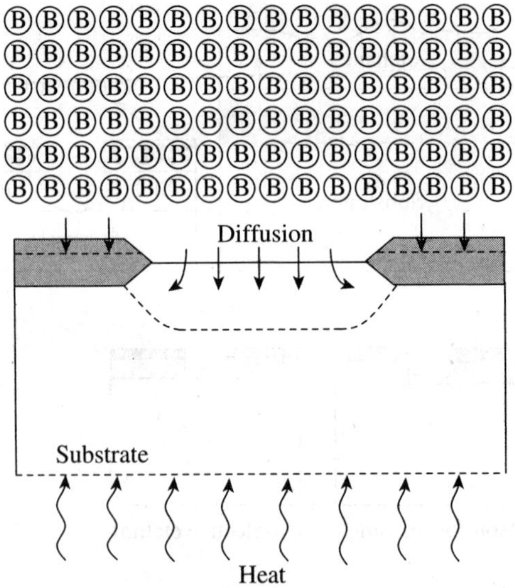

Fig. 8.7 Diffusion of boron atoms

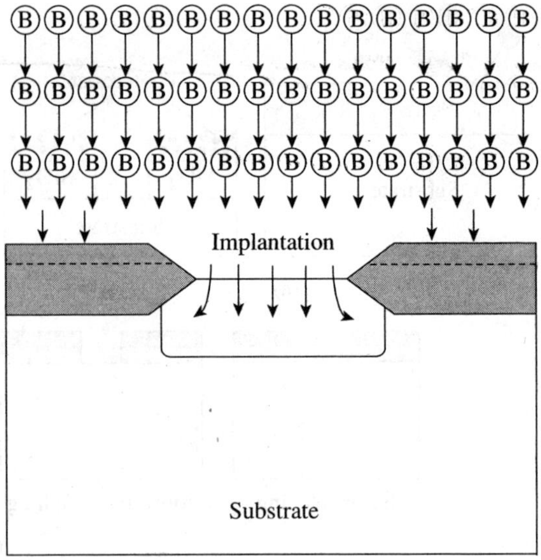

Fig. 8.8 Ion implantation

of their site creating a lot of vacancies. The common dopants used for creating P-type layers are boron and for N-type layers are phosphorus, antimony, and arsenic.

Ion Implantation

This is an advanced technique in which dopant ions are accelerated to several hundred keV and are focused on the substrate surface (Fig. 8.8). This is the preferred technique for introduction of dopant, since it is a low-temperature process; so movement of impurity ions is minimized and hence the amount of lateral diffusion is much lower. The dopant ions are passed through a strong magnetic field before bombarding on the target to remove unwanted impurities. Thus extreme purity of dopant is attained. Another advantage of this process is that by controlling the velocity of the dopant ions, one can easily regulate the depth of its diffusion into the substrate.

Deposition of Thin Film

The process of forming a layer of material on the surface is termed as deposition. Generally, layers of materials such as metal and silicon dioxide may need to be deposited on the surface. There are two types, physical and chemical deposition. In *physical deposition*, a piece of the material to be deposited is bombarded with ions. This leads to the ejection of some atoms of the material to be deposited, which then adheres to the substrate surface. In *chemical deposition* through chemical reaction the layer of the desired material in the form of a precipitate is formed on the substrate surface. A process of the formation of a layer of crystalline-doped semiconductor material on a single-crystal substrate is called *epitaxy*. It is a specialized form of deposition. Epitaxy process is used to deposit N layer on N+ silicon layer, which is otherwise impossible to accomplish by any other diffusion method, (N+ is a heavily doped N-type semiconductor)

Patterning

Patterning is a process where photo resist (PR) deposition, exposure, development, and etching are used several times to create regions of a particular shape. For example, a PR is coated on a newly deposited metal. It is then exposed to UV using a mask. PR is then developed and etched. The method that selectively removes the metal not covered by the PR is referred to as patterning.

8.2.3 Fabrication of Diode

The step-wise fabrication of a planar p-n junction diode is discussed as follows.

(a) An N+ substrate of approximately 150 μm thickness is the starting layer.
(b) On this substrate a layer of N-type silicon of thickness 1–5 μm is grown by epitaxy process.
(c) Silicon dioxide (SiO_2) layer is then deposited by oxidation process.
(d) On SiO_2 surface, a positive photoresist is coated.
(e) Then mask is placed on PR. It is aligned and then exposed to UV light.
(f) The mask is then removed. PR is removed from the exposed area. SiO_2 under the exposed resist is also etched off creating a window for deposition of P-type impurity.
(g) To form a P-region of the diode, boron is diffused. Boron diffuses easily in silicon, but not in SiO_2.
(h) Thereafter a thin aluminum film is deposited over the entire surface.
(i) PR is coated on the metal layer. Mask is used to identify the areas where metal is to be retained. It is then exposed to UV rays. PR is developed. Etchant is then used to remove unwanted metal and PR is also removed by dissolving it in chemical.
(j) To make ohmic contacts, contact metal is deposited on the back surface and is given heat treatment.

Figure 8.9 shows all the fabication steps from (a) to (j).

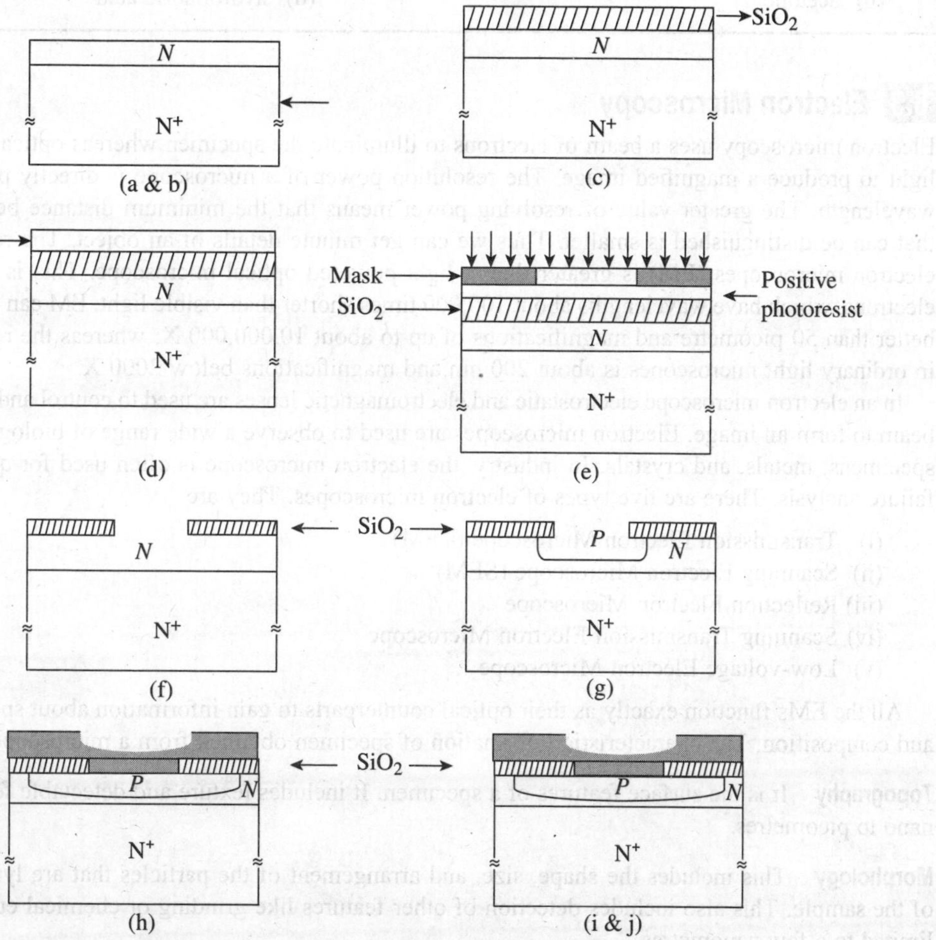

Fig. 8.9 Fabrication steps for p-n junction diode

Checkpoint 2

1. For VLSI technology the number of transistors and other electronic components integrated on a single chip are
   (a) 1 lakh (c) 1 million
   (b) more than a million (d) 1 thousand
2. The method of selectively removal of SiO_2 from a substrate is called
   (a) oxidation (c) photolithography
   (b) etching (d) patterning
3. A photoresist(PR) which is removed from the region exposed to UV light is called
   (a) negative PR (c) positive PR
   (b) neutral PR (d) opaque PR
4. The process for introduction of dopant in IC fabrication is called
   (a) diffusion (c) ion implantation
   (b) etching (d) deposition
5. The chemical used to dissolve SiO_2 and remove it from a substrate is
   (a) sulphuric acid (c) hydrochloric acid
   (b) acetone (d) hydrofluoric acid

8.3 Electron Microscopy

Electron microscopy uses a beam of electrons to illuminate the specimen whereas optical microscopy uses light to produce a magnified image. The resolution power of a microscope is directly proportional to the wavelength. The greater value of resolving power means that the minimum distance between two points that can be distinguished is smaller. Thus we can get minute details of an object. The resolving power of electron microscopes (EM) is greater than a light-powered optical microscope. This is because EM uses electrons, which have wavelengths about 100,000 times shorter than visible light. EM can achieve resolution better than 50 picometre and magnifications of up to about 10,000,000 X; whereas the resolution obtained in ordinary light microscopes is about 200 nm and magnifications below 2000 X.

In an electron microscope electrostatic and electromagnetic lenses are used to control and focus the electron beam to form an image. Electron microscopes are used to observe a wide range of biological and inorganic specimens, metals, and crystals. In industry, the electron microscope is often used for quality control and failure analysis. There are five types of electron microscopes. They are

(i) Transmission Electron Microscope (TEM)
(ii) Scanning Electron Microscope (SEM)
(iii) Reflection Electron Microscope
(iv) Scanning Transmission Electron Microscope
(v) Low-voltage Electron Microscope

All the EMs function exactly as their optical counterparts to gain information about specimen's structure and composition. The characteristic information of specimen obtained from a microscope is as follow:

Topography It is the surface features of a specimen. It includes texture and detectable features within few nano to picometres.

Morphology This includes the shape, size, and arrangement of the particles that are lying on the surface of the sample. This also includes detection of other features like grinding or chemical etching of a sample limited to a few nanometres.

Composition The constitution of the sample in terms of elements and compounds and their relative ratios within areas 1 micrometre in diameter.

Crystallographic information The atomic arrangements in the specimen and detection of atomic-scale defects in the areas of few nanometres in diameter.

8.3.1 Electron–Solid Interactions

Electron microscopy is possible only due to the interactions between the electrons and the specimen. When an energetic electron beam strikes a sample, large numbers of signals are generated as shown in Fig. 8.10. These signals can be divided into two broad categories: (i) electron signals, (ii) photon signals

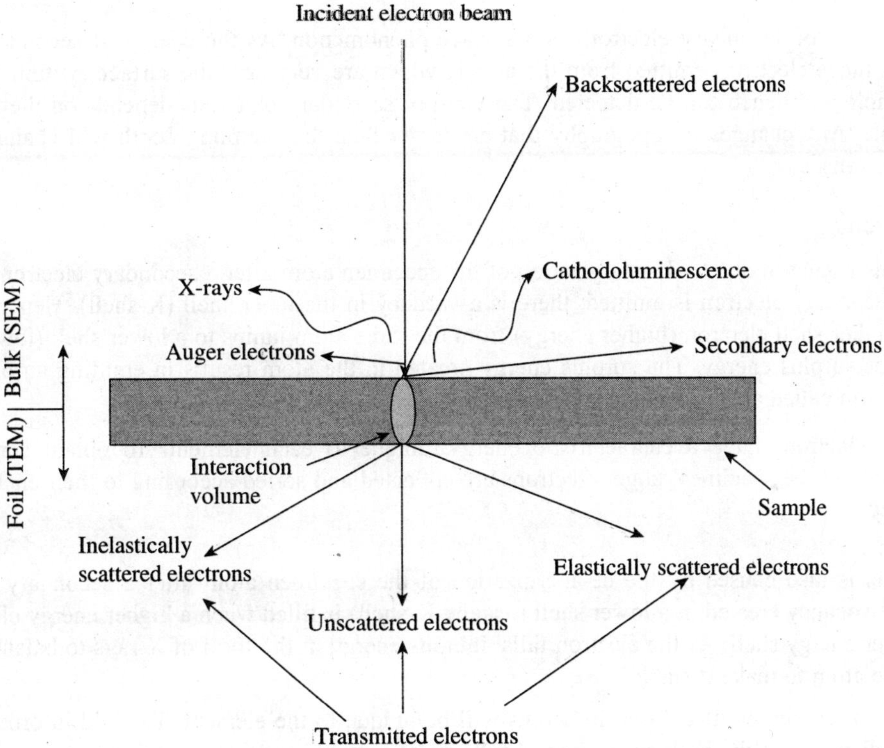

Fig. 8.10 Electron–solid interaction

For thick or bulk specimens, the reactions labelled on the topside of the diagram are used (for SEM) whereas for thin foil specimens the reactions on the bottom side are examined (for TEM). Thus signals produced by SEM includes secondary electrons, backscattered electrons, X-rays, cathodoluminescence, and augur electrons. These signals give various types of information about the specimen. The formation of these bulk specimen interactions and their uses are discussed below.

Backscattered Electrons

Formation These are produced when an incident electron collides with an atom in the specimen, which is nearly normal to the path of the incident electrons. The incident electron is then scattered backward, that is, at 180 degrees.

Use The number of backscattered electrons produced varies directly with the atomic number of the specimen. Thus, elements with higher atomic number will appear brighter than the lower atomic number elements. This information is utilized to identify parts of the specimen having a different average atomic number.

Secondary Electrons

Source These are produced when an incident electron penetrates deep in an atom of the specimen. After penetration the incident electron then imparts some of its energy to an inner shell electron (usually K-shell) of the specimen atom. This results in ionization of the specimen atom and a small energy loss and path change of the incident electron. The electrons coming out from the specimen atoms have very small kinetic energy (5eV) and are called *secondary electrons*. Many secondary electrons are generated from a single incident electron.

Use Production of secondary electrons is a surface phenomenon. As the energy of secondary electron is less, only those electrons emitted from the atoms, which are very near the surface (within 10 nm) can exit the sample and hence can be detected. The yield of secondary electrons depends on the topography of the sample. Any changes in topography that are larger than the sampling depth will change the yield of these electrons.

Auger Electrons

Source This is caused due to de-energization of the specimen atom after a secondary electron is emitted. When the secondary electron is emitted, there is a vacancy in the inner shell (K-shell). Hence, to fill the vacancy a higher shell electron (higher energy) from the same atom jumps to a lower shell (lower energy), which creates surplus energy. The surplus energy created in the atom results in emitting an outer (lower energy) electron called auger electron.

Use Auger electrons have a characteristic energy, unique to each element. To obtain compositional information about the specimen, auger electrons are collected and sorted according to their energies.

X-rays

Source This is also caused by the de-energization of the specimen atom after a secondary electron is emitted. The vacancy created in a lower shell (usually K-shell) is filled when a higher energy electron falls into the lower energy shell. As the electron falls, it emits energy in the form of X-rays to balance the total energy of the atom to make it stable.

Use X-rays or energy emitted from the atoms will be unique to the element. To yield information up to micrometre diameter of the bulk specimens, the X-ray signals are collected and sorted according to their energies.

The signals produced by TEM include elastically scattered electrons, inelastically scattered electrons, and unscattered electrons._The formation of thin specimen interactions and their uses are discussed below.

Unscattered Electrons

Source These are the incident electrons, which are transmitted through the thin specimen, as there is no interaction inside the specimen.

Use The specimen thickness determines the transmission of unscattered electrons. The number of unscattered electrons is inversely proportional to the thickness of the sample. Thus, there will be fewer transmitted unscattered electrons where the area of the specimen is thick and so it will appear dark, whereas the thinner areas will appear lighter as there are more transmitted unscattered electrons.

Elastically Scattered Electrons

Source These are the incident electrons, which do not lose energy after being scattered by atoms of the specimen. These elastically scattered electrons then come out of the remaining portions of the specimen.

Use All incident electrons are scattered in accordance with Bragg's Law, $2d \sin \theta = n\lambda$, where 'd' is the inter planar spacing. As all the incident electrons have the same energy (thus same wavelength) and enter the specimen normal to its surface, they are all scattered at the same angle if interatomic spacing remains same ('d' and inter-atomic spacing 'a' is related through $d = \frac{a}{\sqrt{h^2 + k^2 + l^2}}$, where h, k, and l are Miller indices). The electrons scattered through the same angle are gathered using magnetic lenses to form a pattern of spots. Each spot corresponds to a specific atomic spacing or an interplanar spacing. Thus, this pattern can yield information about the atomic orientation, atomic arrangements, and phases present in the area of the specimen under investigation.

Inelastically Scattered Electrons

Source These are the incident electrons, which lose energy during interaction with specimen atoms. These electrons are then transmitted through the rest of the specimen.

Use In elastically scattered electrons can be used in the electron energy loss spectroscopy. The inelastic loss of energy by the incident electrons when it interacts with the specimen is unique to its element's composition. These energies can give information about the bonding state of each element and hence, can be used to obtain information related to both composition and bonding (i.e., oxidation state) of the specimen region being examined.

8.3.2 Working of TEM and SEM

The basic steps involved in all electron microscopes are summarized below:

(i) A stream of electrons is obtained by using the electron gun.
(ii) The metal apertures are used to confine the stream of electrons and by using magnetic lenses it is converted into a thin, focused, monochromatic beam.
(iii) This beam is then focused onto the sample by using a magnetic lens.
(iv) Interactions of electron beam occur inside the sample, affecting it. These interactions and effects are detected and transformed into an image.

A more specific treatment of the workings of two different types of EMs is described in detail below.

Transmission Electron Microscope

The virtual source at the top represents the electron gun, producing a stream of single wavelength electrons. The electron beam is then focused and converted into a small, thin, coherent beam by the use of condenser lenses as shown in Fig. 8.11. The first condensing lens determines the range of size of the final spot that strikes the sample. The second condensing lens focuses the electron beam on the sample as a pinpointed beam. The condenser aperture restricts the angular spread of the beam by knocking out high-angle electrons (those electrons that are far from the optic axis, i.e., the dotted line drawn at the centre).

Thereafter the electron beam strikes the specimen and part of it is transmitted. This objective lens forms an image from the transmitted portion of the electron beam. The image is passed down the column through the intermediate and projector lenses, which enlarges the image. The image becomes visible when it strikes the phosphor screen. The darker areas of the image represent those areas of the sample from where fewer electrons were transmitted through, whereas the lighter areas are the one which transmitted more electrons.

Applications Applications of TEM include topographical, morphological, compositional, and crystalline information of the specimen and it can also identify flaws, fractures, and damages to micro-sized objects.

Scanning Electron Microscope

The virtual source at the top represents the electron gun, producing a stream of single wavelength electrons. The stream of electron is focused to a small, thin, coherent beam by the use of condenser lenses as shown in Fig. 8.12. The first condensing lens condenses the electron beam. The second condensing lens forms the electrons into a thin, tight, coherent beam. The condenser aperture eliminates some high-angle electrons thus constricting the size of the electron beam. A set of coils then makes the electron beam move in a grid fashion (like a television, where the beam moves back and forth continuously), staying on points for a few microseconds determined by the scan speed.

The objective lens focuses the scanning electron beam onto the required part of the specimen. When the beam strikes the sample for a few microseconds, interactions that occur inside the sample are detected using various instruments. Before the beam moves to the next point these instruments display a pixel on a CRT depending on the number of interactions. The number of reactions occurring inside the sample determines the intensity of each pixel. This process is continuously repeated until the entire grid scan is finished.

Applications Some of the applications of SEM are to detect and analyse surface fractures, examine surface contaminations, reveal spatial variations in chemical compositions, and identify crystalline structures.

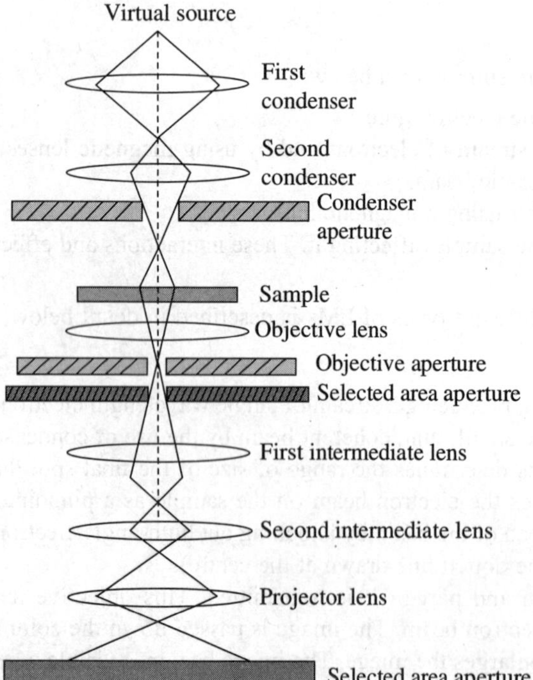

Fig. 8.11 Transmission electron microscope (TEM)

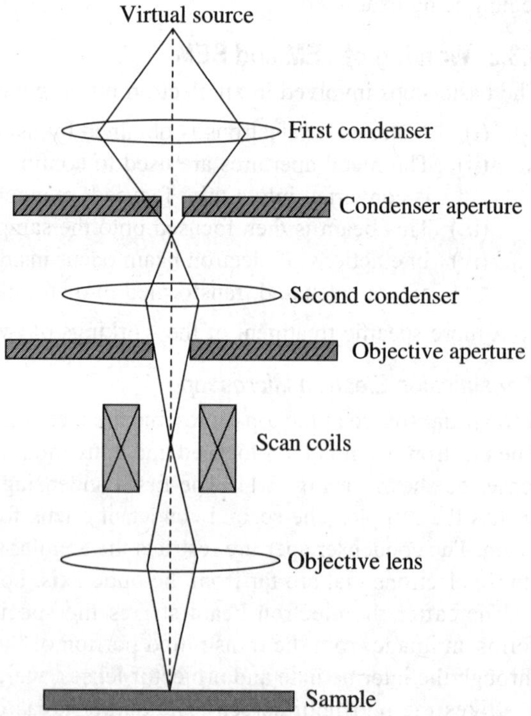

Fig. 8.12 Scanning electron microscope (SEM)

Checkpoint 3

1. The resolving power of an electron microscope is more compared to that of an optical microscope because
   (a) wavelength of light is more than that of electron
   (b) wavelength of electron is more than that of light
   (c) not related to the wavelength
2. To get information about the topography of a sample, the source used in electron microscopy is
   (a) back-scattered electron
   (b) augur electrons
   (c) secondary electrons
   (d) all of the above
3. The source used in electron microscopy to identify parts of the specimen having different average atomic number is
   (a) back-scattered electron
   (b) X-rays
   (c) secondary electrons
   (d) cathodoluminescence
4. To find the composition of a specimen, the signal used is
   (a) back-scattered electron
   (b) X-rays
   (c) augur electrons
   (d) secondary electrons
5. Topography of a sample is
   (a) the information of the sample about its surface features
   (b) shape, size, and arrangement of the particles lying on the surface of the sample
   (c) atomic arrangements in the specimen
   (d) constitutional information of the sample

8.4 Optoelectronics Devices

Optoelectronics is the study and use of electrical devices that emit, detect, transmit, sense and modulate light. Various examples of optoelectronic devices are given below.

8.4.1 Types of Optoelectronics Devices

Few examples of optoelectronic devices are discussed below.

(i) **Laser diodes, super luminescent diodes and light-emitting diodes (LEDs)**: Light is produced by these devices using electrical energy.

Fig. 8.13 (i) Symbol for LASER diode (ii) Symbol for LED

(ii) **Photodetectors (e.g. photodiodes and phototransistors, photovoltaic cell/solar cell)**: These devices converts optical signals into electrical currents.

Fig. 8.14 (i) Symbol for Photodiode (ii) Symbol for Photovoltaic cell

(iii) **Imaging detectors, based on electronic image sensors:** Optoelectronic sensor generates an electrical signal in direct proportion to the amount of light that strikes its active region. Thus, it detect objects with light to trigger function to control, to switch or to regulate. Optoelectronic sensors are used for several medical applications viz. Pulse oximetry, blood glucose and heart rate monitors etc.

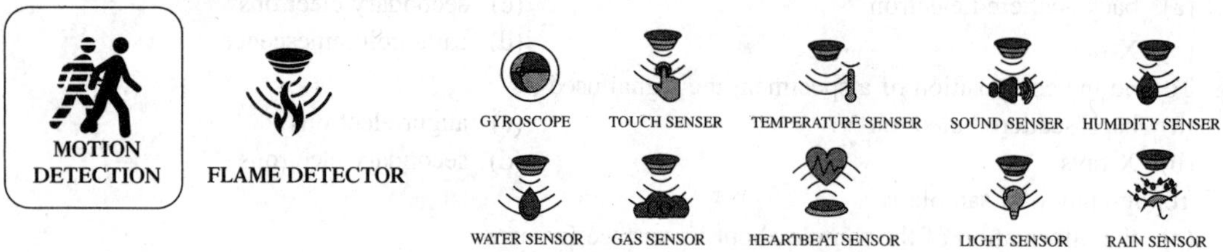

Fig. 8.15 Icons of sensors

(iv) **Electro-optic modulators**: It is used for manipulating the power, phase or polarization of light with an electrical control signal.

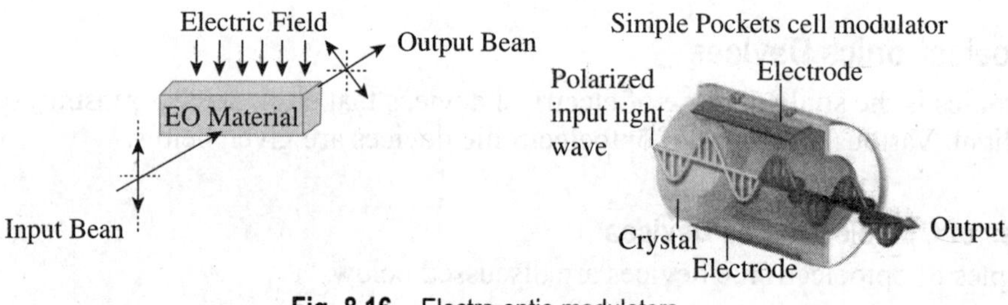

Fig. 8.16 Electro-optic modulators

(v) **Electro-absorption modulators**: This device is a semiconductor-based optical modulator. It can be used for regulating the optical power of a laser beam by changing an electric voltage. Electro-absorption modulators can function with significantly lower voltages as compared to electro-optic modulators which require few hundred volts. These devices are useful for optical fiber communications.

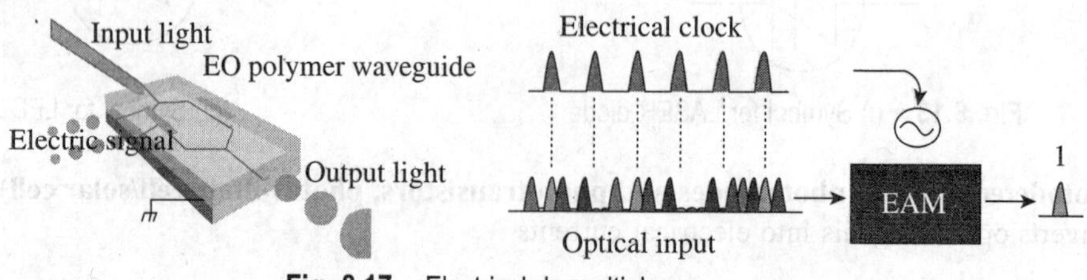

Fig. 8.17 Electrical de-multiplexer

(vi) **Opto-isolators/couplers**: These are employed for ensuring electrical separation/isolation while transmitting analogue or digital signals.

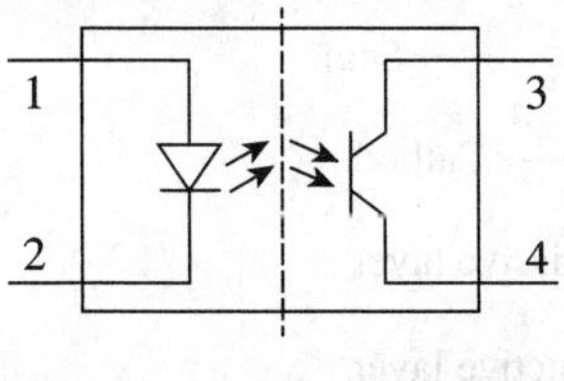

Fig. 8.18 Symbol for Opto-isolator/Coupler showing source of light (LED) on the left, dielectric barrier in the center, and sensor (phototransistor) on the right

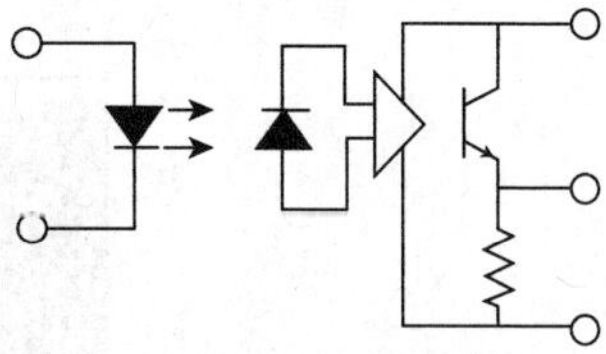

Fig. 8.19 A fast photodiode opto-isolator with an output-side amplifier circuit.

(vii) **Photonic integrated circuits**: A photonic integrated circuit (PIC) or integrated optical circuit is a microchip containing two or more photonic components which form a functioning circuit. This technology detects, generates, transports, and processes light.

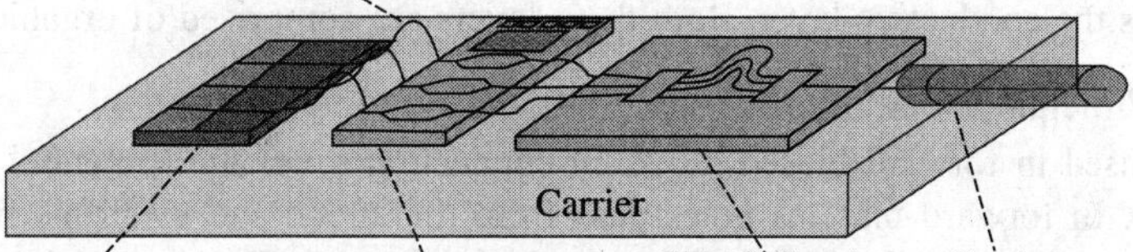

Fig. 8.20 A photonic integrated circuit

(viii) **Photoemissive detectors:** Devices such as phototubes and photomultipliers, also called image intensifiers, makes use of the photoelectric effect, the phenomenon whereby light-sensitive surfaces give off electrons when struck by light.

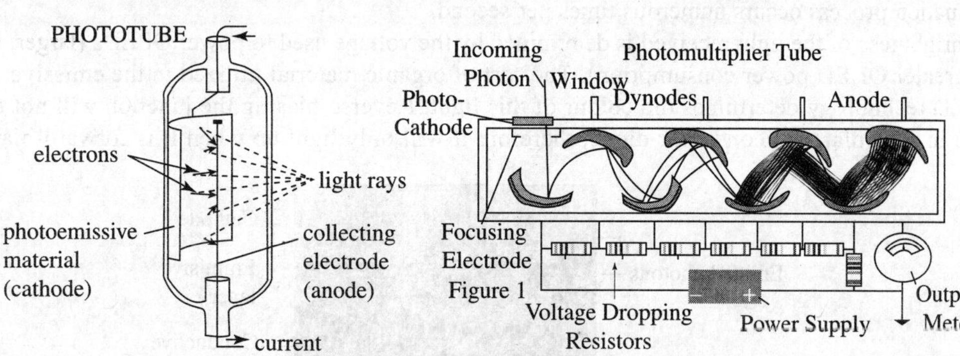

Fig. 8.21 Photoemissive detectors

Optoelectronic devices find application in diverse fields, such as optical fiber communications, laser technology and all kinds of optical metrology.

A few optoelectronic devices are discussed in detail in the following sections.

8.4.2 Organic Light-Emitting Diode-OLED

Structure of OLED

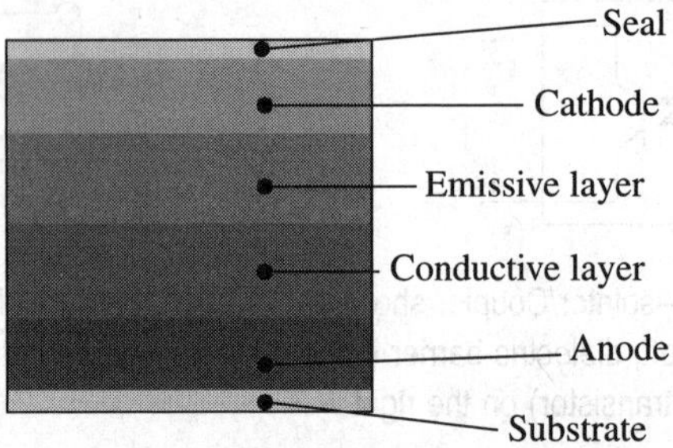

Fig. 8.23 The arrangement of layers in a simple OLED.

As shown in Figure 8.23, OLED consists of basic six layers. The top layer is called seal and the bottom layer is called substrate. A negative terminal, also referred to as the cathode, and a positive terminal, referred to as the anode, are located in between those layers. There are two layers, emissive and conductive, sandwiched between anode and cathode. Adjacent to the cathode is the emissive layer where light is produced. The layer adjacent to anode is the conductive layer. Both these layers are comprised of organic molecules.

OLED – Working (Principle of operation)

OLDs are always used in forward biased mode by connecting a voltage (potential difference) across the anode and cathode. In forward bias, as holes pass from the anode and electrons from the cathode enter the emissive layer, the emissive layer becomes negatively charged. The extra holes, which act as positive charge-carriers, cause the conductive layer to be positively charged. Electrostatic force of attraction causes the electrons and holes in the emissive layer to be attracted to one another. As a result of attraction between two particles with opposing electrical charges, the recombination process takes place. The emission of light, an electromagnetic radiation that can be seen, is a result of the energy released during recombination as shown in Figure 8.24. The OLED provides continuous light for as long as the current flows since the recombination process occurs numerous times per second.

The brightness of the light received is determined by the voltage used to power OLEDs (larger, brighter means greater OLED power consumption). The sort of organic material utilized in the emissive layer of the OLED technology determines the colour of this light. Reverse-biasing the junction will not operate, just like in a regular semiconductor diode, therefore it will only light up when it is forward-biased.

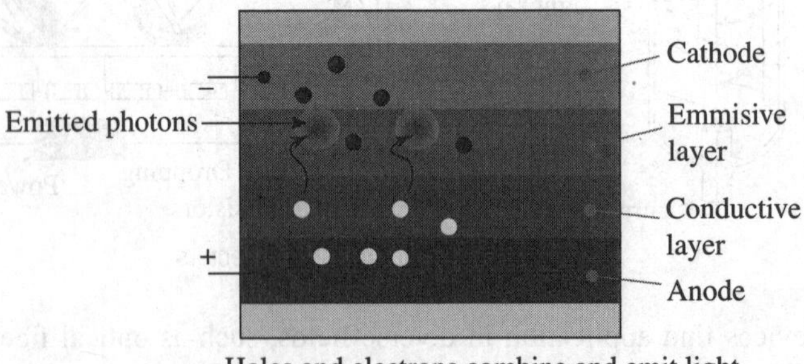

Fig. 8.24 Recombination of charge carriers emit light in OLED

Advantages and disadvantages of OLEDs

There are benefits and drawbacks of using OLED technology in building any display.

Advantages

1. Flexible displays: It is possible by applying the organic material to a flexible substrate.
2. Lightweight: It is very light in weight.
3. Small size: OLEDS are substantially thinner (0.2-0.3mm) than LCDs, which are typically at least 10 times thicker.
4. Lower manufacturing costs: Less manufacturing cost is because of the absence of backlighting. The simple design of the display uses few electronic components and reduced the number of layers.
5. Less Power consumption: OLEDS use less energy than LCDs since they are brighter and do not require a backlight, which results in longer battery life for portable electronics like cellphones and MP3 players. So little energy usage.
6. Faster response time: OLDs have faster response time of ~ 0,01ms as compared to LCD technology which is about 2-12ms. OLEDs respond up to 200 times faster than LCDs. Thus, fast-moving images like in sports on TV or computer games can be displayed flawlessly.
7. True colors: OLEDS produce real colors.
8. Viewing angle: Viewing angle is almost unlimited. Unlike LCDs, where if you glance to one side the colours get darker and disappear.
   OLEDs, which are simpler than LCDs and are currently significantly more expensive since the technology is newer and less widely used, should someday be less expensive to produce than LCDs.

Disadvantages

1. Less Lifetime: OLED displays don't last as long since the organic chemicals in them degrade over time. This meant that early OLED models tended to degrade roughly four times more quickly than traditional LCD or LED displays. The blue OLED has the shortest lifetime of the RGB (Red, Green, and Blue) colours, lasting only 5000–14000 hours. However, red and green OLEDs may operate between 46,000 and 230,000 hours, making blue OLED as a challenge to improve the display's lifespan.
2. Sensitivity to the environmental factors: Organic compounds are particularly vulnerable to environmental elements like water or humidity. OLED organic compounds are extremely water sensitive. For domestic devices like TVs and home computers, that shouldn't be an issue, but it might be more difficult for portable products like cellphones, laptops etc.
3. Limited patents: The firms that own the patents requires the payment of license fees to allow other companies for taking any substantial, further steps, which in turn hinders the development of OLED technology.

Applications of OLEDs

1. OLED displays are used in TV, computer screens, MP3 and mobile phone displays.
2. OLDs can be used to create animated billboards, incredibly thin pages for electronic books and magazines, tablet computers with folding displays that neatly transform into pocket-sized smartphones, and even clothing with constantly changing colours and patterns that is connected to visualizer software that is running from user's iPod.

8.4.3 Photovoltaic Cell (PV Cell/Solar Cell)

A solar cell, also known as a photovoltaic cell, is a type of electronic device that uses the photovoltaic effect to convert solar energy directly into electricity.

Photovoltaic effect: The photovoltaic effect (commonly abbreviated as PV) is the conversion of light (photons) to electricity (voltage).

PV cell: Solar energy is converted into electricity by PV devices. A single photovoltaic (PV) device, referred as a PV cell is constructed by using various semiconductor materials. A PV cell produces average power output of 1 to 2 watts.

Construction of PV cell: Figure 8.25 shows how a Si solar cell is built. A solar cell is a fundamentally a p-n junction semiconductor device. It is created by metallurgical joining of two types of semiconductor materials p type (majority charge carriers are holes) and n type (majority charge carriers are electrons). Most charge carriers (electrons from n-side and holes from p-side) attempt to diffuse at the junction due to concentration gradient. As a result of this diffusion positive ion cores in the n-side and negative ion cores in the p-side are exposed crating a space charge region at the junction. Thus, an electric field is created over the region which contains immobile positive and negative ions. This region at the junction is also called as depletion region as it is devoid of mobile charge carriers.

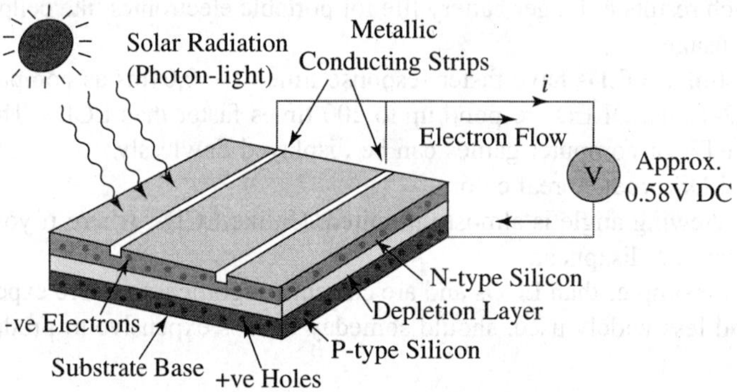

Fig. 8.25 Construction of PV cell

Working of PV cell: When a PV cell is exposed to sunlight, it absorbs photons with energies larger than the semiconductor's band gap. The absorbed energy is used to create an electron-hole (e-h) pair. Due to the electrostatic force of the field across the junction, the free charge carriers, electron-hole (e-h) pair created will be separated. Electrons move to the n-side and holes to p-sides of the p-n junction respectively. Thus, a potential difference between two sides of the cell is established. A photovoltaic or solar cell typically has a positive rear contact and a negative front contact. The p-n junction acts like a battery and if the two sides of the cell are connected by an external circuit then current will start flowing from positive to negative terminal of the solar cell. Figure. 8.16 explains the working principle of a solar cell.

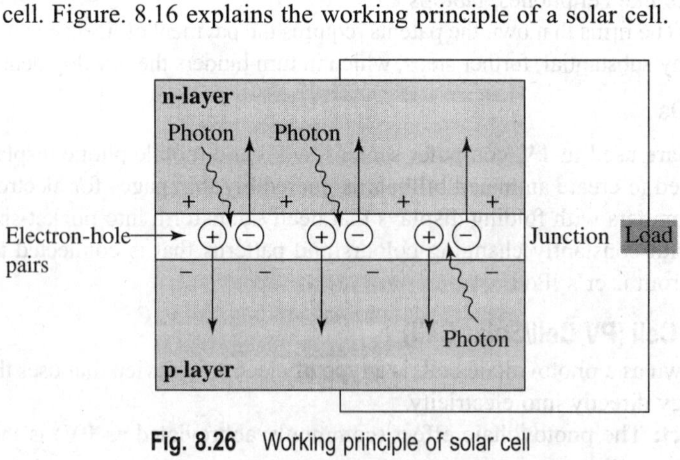

Fig. 8.26 Working principle of solar cell

The band gap of Silicon, which is used to make solar cell is Eg = 1.1 eV at room temperature. Hence, the cut-off wavelength is 1.107 um or 1107 nm or 11,070 Å. Figure 8.27 explains how the photon's wavelength affects movement of electrons to the conduction band.

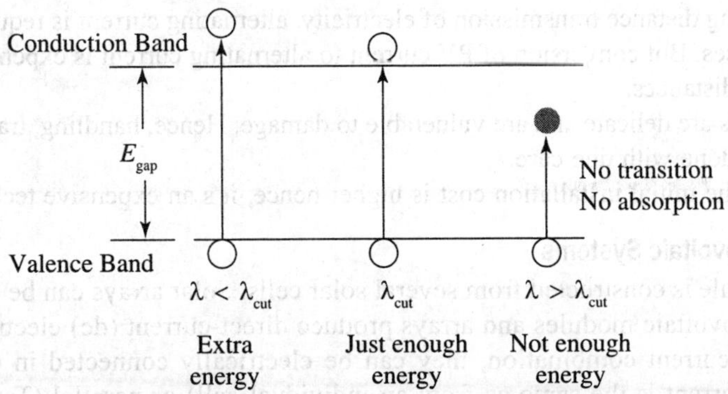

Fig. 8.27 Effect of Photon's wavelength on generation of electrons in conduction band

If wavelengths of the incident photons is less than the cut-off wavelength, then the absorbed photons have enough energy to break covalent bonds. As the covalent bonds break charge carriers are generated. Thus, now generated electrons have sufficient energy to move into the conduction band resulting into generation of current in the external circuit. However, if the wavelength of the absorbed photons is longer than the cut-off wavelength, then enough energy is not available for moving electrons into the conduction band. And hence, there will be no current in the external circuit. As a result, not every photon that enters a photovoltaic cell can be converted into usable electrical power. Energy conversion efficiency for solar cells are typically about 15%. This indicates that 15% of the optical power that was incident is transformed to electricity. The efficiency of some recent photovoltaic devices have reached 40%. Every commercially accessible solar module will have a corresponding energy conversion efficiency; the manufacturer need to include that information with the module.

Advantages

(i) Clean Energy: Energy comes from the sun's rays, which is completely pollution-free. As a result, there is no risk of environmental deterioration, and no contaminants are discharged during the process. Thus, use of solar energy can significantly reduce carbon emissions.

(ii) Durability: The performance of solar cell is consistent and utilize eco-friendly technologies. They have a long lifespan and suffer little efficiency loss.

(iii) Noise Free: They make no noise and don't have any mechanical or moving parts. Therefore, no disruption is created.

(iv) Less maintenance cost: There aren't many expenses associated with operation and maintenance. The efficiency of the Solar panels can be maintained over a period of time by simply cleaning their surface.

(v) Electricity in Remote areas: In rural and distant locations that have not yet been electrified, photovoltaic cells can be quite beneficial. As Solar energy is widely available, so it can be used effectively in remote locations. Thus, it is possible to significantly reduce the power loss that results from the long-distance transmission of electricity.

Disadvantages

(i) Less Efficiency: Reported efficiency of solar cells is still around 40% which is much lower in comparison to the other available renewable sources of energy.

(ii) Performance susceptible to Environment conditions: PV cells need sunlight for its functioning. Hence, its performance is affected by the prevalent weather conditions. In absence of sufficient sunlight light alternative arrangement will be required.

(iii) Short distance transmission: PV produces direct current which is suitable for transmission over short distances. For long distance transmission of electricity, alternating current is required as it involves less transmission losses. But conversion of PV current to alternating current is expensive hence it is mostly used over short distances.

(iv) Delicate: PV cells are delicate and are vulnerable to damage. Hence, handling, transportation and installation has to be done with due care.

(v) Expensive: As the initial installation cost is higher hence, it's an expensive technology.

Applications of Photovoltaic Systems

A solar panel or module is constructed from several solar cells. Solar arrays can be created by connecting many modules. Photovoltaic modules and arrays produce direct-current (dc) electricity. To produce any desired voltage and current combination, they can be electrically connected in either series (Voltage builds up but total current is the same as from an individual cell) or parallel (Total current is the sum of all currents from individual cells with no voltage increase). Some terminologies and specifications are given as below.

- One PV cell = ~0.5 V
- Module = 72, 96, or 128 pre-wired cells in a series in a package
- Array=modules wired in series to increase V or in parallel to increase i.
- For array, modules must optimized in series or parallel for maximum Power (P) = V I as shown in the Figure 8.28.

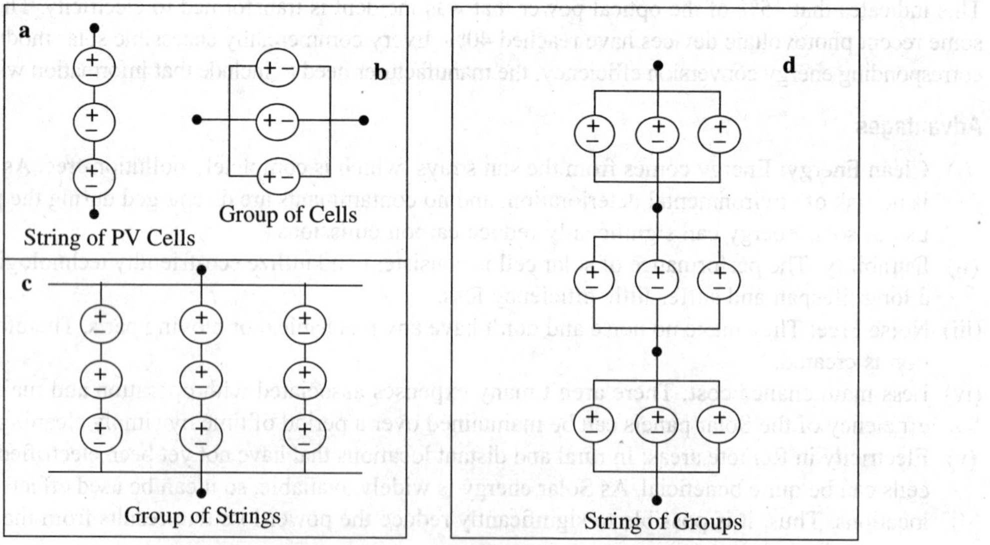

Fig. 8.28 Various types grouping of PV cell

(i) Solar cells can be used to provide power to equipment or for charging batteries.

(ii) PV cells are used in orbiting satellites and other spacecraft to supply power.

(iii) Photovoltaic modules are used for providing power to homes at remote places, boat houses, leisure vehicles, electric automobiles etc.

(iv) The smallest photovoltaic systems can be used to power calculators and wristwatches.

(v) For providing more power larger systems are used to produce electricity. The significant power requirements can be for operating water pump, power equipment used in communication, etc.

Checkpoint 4

1. Solar cells are made up of
   (a) Steel (c) Sulphur
   (b) Silicon (d) Antimony

2. The names of the symbol of optoelectronic devices is

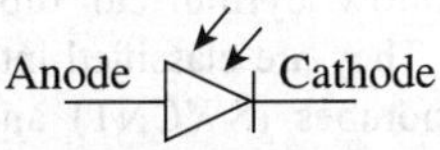

Photodiode symbol

   (a) Laser Diode (c) Solar Cell
   (b) LED (d) Photodiode

3. The names of the symbol of optoelectronic devices is

   (a) Photodiode (c) LED
   (b) Photovoltaic Cell (d) Super luminescent LED

4. The names of the symbol of optoelectronic devices is

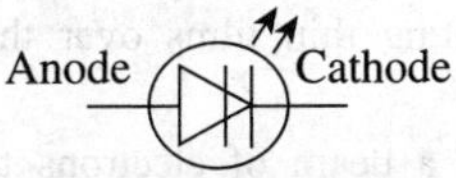

Laser Diode

   (a) Super luminescent LED (c) Laser Diode
   (b) Photovoltaic Cell (d) Photodiode

5. The names of the symbol of optoelectronic devices is

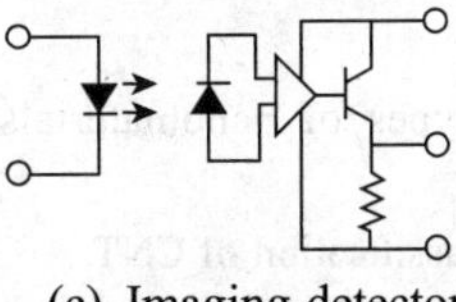

   (a) Opto-modulator (c) Imaging detector
   (b) Phoemissive detector (d) Opto-isolator/coupler

6. OLED is always used in
   (a) Reverse bias (c) Forward Bias
   (b) Biasing not required (d) when exposed to Sunlight

7. When PV cell are connected in parallel
   (a) Voltage adds up but current remains same (c) Voltage as well as current adds up
   (b) Current adds up but Voltage remains same (d) Voltage as well as current remains same

8. When PV cell are connected in Series
   (a) Voltage adds up but current remains same (c) Voltage as well as current adds up
   (b) Current adds up but Voltage remains same (d) Voltage as well as current remains same

Summary

- Nanomaterials are substances in which at least one dimension is less than 100 nanometres, where one nanometre corresponds to three atoms in diameter.
- The factors responsible for changed properties at nano-scale are mainly increased relative surface area and quantum effects.
- There are four classes of nanomaterials—carbon-based nanomaterials, metal-based nanomaterials, dendrimers, and composites.
- Carbon nanotube is a hollow cylindrical tube composed of carbon atoms. They are classified into Single-Walled Carbon Nanotubes (SWCNT) and Multi-Walled Carbon Nanotubes (MWCNT).
- There are three types of SWCNTs—armchair, zigzag, and chiral nanotubes.
- Semiconductor device fabrication is a process to produce integrated circuits (IC). An integrated circuit (IC) is also called a microelectronic circuit, chip, or microchip.
- Semiconductor device fabrication includes wafer preparation, photolithography, and chemical processing viz. etching, thermal, and local oxidation, changing the doping concentration by ion implantation and diffusion, and depositing thin films over the semiconductor's substrate.
- Electron microscopy uses a beam of electrons to illuminate the specimen whereas optical microscopy uses light to produce a magnified image. In an electron microscope electrostatic and electromagnetic lenses are used to control and focus the electron beam to form an image.
- The five types of electron microscopes are Transmission Electron Microscope (TEM), Scanning Electron Microscope (SEM), Reflection Electron Microscope, Scanning Transmission Electron Microscope, Low-voltage Electron Microscope.
- Electron microscopy uses signals (electron and photon signals) generated due to the interactions between the electrons and the specimen.
- For thick or bulk specimens, the reactions used for SEM are secondary electrons, backscattered electrons, X-rays, cathodoluminescence, and augur electrons.
- For thin foil specimens the reactions used for TEM include elastically scattered electrons, inelastically scattered electrons, and unscattered electrons.
- Applications of TEM include topographical, morphological, compositional, and crystalline information of the specimen and it can also identify flaws, fractures, and damages to micro-sized objects.
- Applications of SEM are to detect and analyse surface fractures, examine surface contaminations, reveal spatial variations in chemical compositions, and identify crystalline structures.

Self-study Questions

1. What are the different types of nanomaterials? Explain in brief.
2. What is CNT? Give the classification of CNT.
3. What is SWCNT and MWCNT? Differentiate them.
4. Write a note on the applications of nanomaterials.
5. Highlight some of the disadvantages of nanomaterials.
6. What is an integrated circuit?
7. How do you classify the ICs based on the number of components?
8. What is planar technology of IC fabrication? Why is it preferred?
9. What is photolithography?
10. What is a positive and negative photoresist?
11. What is wet and dry etching? Which type of etching is preferred?
12. What are deposition techniques? Explain in brief.
13. What do you mean by epitaxy?
14. What is ion implantation technique? Discuss in brief.
15. Discuss with neat and well-labelled diagrams, the steps involved in the fabrication of a p-n junction diode.
16. Enlist the different types of electron microscopes.
17. What characteristic information about specimen is obtained in electron microscopy?

18. Name any three formations during bulk and thin specimen interactions with incident electrons.
19. Discuss with a well-labelled diagram the construction and working of TEM.
20. Discuss with a well-labelled diagram the construction and working of SEM.
21. List any four optoelectronic devices. Identify the symbols of the given devices.
22. Describe the working of OLED with a neat and well labelled diagram.
23. State any two advantages and disadvantages of OLED.
24. List any four applications of OLED.
25. Explain the working of Solar Photovoltaic Cell with a neat and well labelled diagram.
26. State any two advantages and disadvantages of Solar Photovoltaic Cell.
27. List any four applications of Solar Photovoltaic Cell.

Answers

Checkpoint 1

1(a), 2(b), 3(d), 4(b), 5(c)

Checkpoint 2

1(c), 2(c), 3(a), 4(b), 5(d)

Checkpoint 3

1(a), 2(c), 3(a), 4(c), 5(a)

Checkpoint 4

1(b) 2(d) 3(b) 4(c) 5(d) 6(c) 7(b) 8(a)

9 Fundamentals of Quantum Computing

Learning Objectives

After going through this chapter, you will be able to:

- Differentiate between quantum bits and classical bits.
- Understand the various types of qubits and their features.
- Learn various operations on qubits
- Distinguish various quantum gates
- Identify limitations of quantum computing

9.0 Introduction

Our world is ruled by three technologies viz. computers, sensors and communication networks. Theoretical advances has brought revolutionary changes in these technologies. The first quantum revolution at the beginning of 20th century gave us computer chips, GPS, lasers, smartphones which modernized and transformed the society. So our everyday life is reigned by quantum technologies. In this century quantum mechanics was used to comprehend the things that already existed. Now in the second quantum revolution we are aggressively using quantum mechanics to change our own design into unnatural states which never existed before thus bringing in most crucial advancement in the existing technologies. Scientists have studied and found that the quantum mechanical effects can be used to create different kinds of computers. Richard Feynman in 1959 suggested the possibility of using quantum effects. He proposed a basic model for a quantum computer that can simulate the evolution of quantum systems as classical computers cannot do it in an efficient way. A computer that makes use of quantum mechanical principles is known as a quantum computer. It has a very basic structure and uses laws of quantum mechanics to perform computing which has a possibility to exponentially outperform classical computers.

9.1 The Basis of Quantum Computing

Quantum computing deals with various operations on the quantum systems. Any physical system, for example an electron or a photon can be expressed as a quantum state. Any quantum system's state is always represented as a vector in the intricate vector space known as a Hilbert space. The attainment of the Hilbert space depends on the nature of the individual system. For instance, in the case of a single particle system in three dimensions, the state can be described by a complex-valued function defined on R3 (three-dimensional space) called a wave function. This wave function has a probabilistic interpretation which is of special significance, as the quantum states can exist in a combination of the basis states. The change of the system state vector with respect to time is considered to be unitary, meaning that it is reversible. The various quantum processes on the system termed as algorithms are nothing but the transformations acting on this vector space.

9.1.1 Bit and the Qubit

The memory of a traditional computer is composed of digital bits, where each bit can either retain a one or a zero. When a computer does a calculation, it manipulates those bits, moving them back and forth between memory and logic gates. In quantum computers a qubit (acronym for "quantum bit') is the fundamental information carrying unit. A unit vector in the 2-dimensional complex vector space represents a qubit's state. These qubits are handled by a quantum computer, which moves them back and forth from memory to quantum logic gates.

A qubit can be symbolized with a wave, an atom, a neutrino or a photon, even the spin of an electron can be used to create a qubit. Thus, any system possessing a minimum two distinct discrete states of an observable quantity which sustains its features with respect to time can be considered for realizing a qubit.

9.1.2 The Quantum Bit-Qubit

Classical computer contains computer chips having millions of transistors and capacitors. Classical computing is based on laws of Boolean algebra. The processing of the data is in an exclusive binary state which are termed as bits having value s either 0 or 1 at any point in time. Thus the logic gates made up of transistors must be in only one state either 0 or1. Now the speed of processing depends on how quickly these devices can be made to switch states. The speed can be increased via smaller and faster circuits. However, as we begin to shrink transistor size to couple of atoms, mystical powers of quantum physics starts to set in which puts the threshold on size reduction. Bits are always in a known state in a traditional computer. For instance, if a three-bit register is utilized, the bits will be in specific states, such as 110. However, the qubits of a quantum computer can be in any combination of the classically permitted states. A qubit is expressed as a vector in a two dimensional quantum system. The vector state of a qubit can be represented as

$$|\phi\rangle = \alpha\,|0\rangle + \beta|\,1\rangle \tag{9.1}$$

Here, α and β are complex numbers such that, $|\alpha|^2 + |\beta|^2 = 1$. Here | and > notations are used as we are dealing with vectors that represent 0 and 1 qubit states. In the ket-notation or the Dirac notation, $|0\rangle = \begin{pmatrix} 0 \\ 1 \end{pmatrix}$ and $|1\rangle = \begin{pmatrix} 1 \\ 0 \end{pmatrix}$. This is the representation of two basis states of a two dimensional vector space. Thus, Eq. (7.1) expresses the fact that the state of the qubit is the two dimensional complex vector $\begin{pmatrix} \alpha \\ \beta \end{pmatrix}$. There is a distinct difference between classical and qubit that he state of a classical bit can be measured without changing it whereas the state of a qubit cannot be measured without changing it. Measuring a qubit's state as represented by Eq. (9.1) will result in the classical value of either zero ($|0\rangle$) with probability $|\alpha|^2$ or one ($|1\rangle$) with probability $|\beta|^2$.

9.1.3 System of Qubits

For higher dimensional quantum system the mathematical structure of a qubit can be generalized to higher dimensions. The state of any quantum system is always a normalized vector in a complex vector space. The normalization is necessary to ensure that the total probability of all the outcomes of a measurement should be equal to one.

A quantum computer involves many number of qubits. So it is essential to understand how from the given states of an individual qubits the combined state of a system of qubits can be constructed. A mathematical process known as the tensor product, or, is used to explain the combined state of a system of qubits. In mathematics, combining two states into a tensor is equivalent to combining their respective vectors into a Kronecker product. For example if we have two single qubit states $|\phi\rangle = \begin{pmatrix}\alpha \\ \beta\end{pmatrix}$ and $|\phi'\rangle = \begin{pmatrix}\alpha' \\ \beta'\end{pmatrix}$. Then, using two independent qubits as a system, the total state is given by

$$|\phi\rangle \otimes |\phi'\rangle = \begin{pmatrix}\alpha \\ \beta\end{pmatrix} \otimes \begin{pmatrix}\alpha' \\ \beta'\end{pmatrix} = \begin{pmatrix}\alpha\ \alpha' \\ \alpha\ \beta' \\ \beta\ \alpha' \\ \beta\ \beta'\end{pmatrix} \tag{9.2}$$

For simplicity the tensor product symbol ⊗ is usually dropped.. Thus, the states are written inside a single ket. For example, $|\phi\rangle\otimes|\phi'\rangle|$ is written as $|\phi\phi'\rangle|$ and $|0\rangle\otimes|0\rangle\otimes|0\rangle$ as $|000\rangle$. The distributive property of the Kronecker product can be used to compute the tensor product more methodically for bigger systems using the Dirac notation.

For example, a system of three qubits with each qubit in the state $|\gamma_j\rangle = \alpha_j\,|0\rangle + \beta_j\,|1\rangle$ for j = 1,2,3 the combined state is given by

$$\llcorner\gamma_1\gamma_2\gamma_3 = \llcorner\gamma_1\rangle \otimes |\gamma_2\rangle \otimes |\gamma_3\rangle \tag{9.3}$$

$$\begin{aligned}\gamma_1\gamma_2\gamma_3 &= \alpha_1\alpha_2\alpha_3\,|000\rangle + \alpha_1\alpha_2\beta_3\,|001\rangle + \alpha_1\beta_2\alpha_3\,|010\rangle \\ &+ \alpha_1\beta_2\beta_3\,|011\rangle + \beta_1\alpha_2\alpha_3\,|100\rangle + \beta_1\alpha_2\beta_3\,|101\rangle \\ &+ \beta_1\beta_2\alpha_3\,|110\rangle + \beta_1\beta_2\beta_3\,|111\rangle\end{aligned} \tag{9.4}$$

Thus, any of the eight (2^3) potential bit-strings connected to the eight basis vectors could result from a measurement of all three qubits. Two bits can be represented by one qubit. Four-bit values can be represented by two qubits and hence, n qubits can typically take values of 2^n.

Therefore, it is evident from the example mentioned above that the state space's dimension increases exponentially as the number of qubits n increases. Thus, there are 2^n basis vectors in such a space.

9.1.4 Types of Qubits

Qubits are classified into several different types depending on how one can change states of qubits and how do they interact. Some qubits occurs naturally and others are engineered. Some of the most common types include:

(a) Superconducting qubits
(b) Photonic qubits
(c) Defect based qubits
(d) Rydberg Atom Qubits
(e) Trapped Ions Qubits

(a) **Superconducting qubits**: Superconductors are the materials which allow current to flow without resistance. A loop of superconducting materials having zero resistance offered to the flow of current at lower temperatures behaves as a qubit. These loops are referred to as "Josephson junctions," which are nonlinear inductance device with zero energy dissipation. This junction allows simultaneous clockwise and anticlockwise flow of current which leads to superposition of states. Thus, a superconducting loop is a qubit. The quantum state of this qubit can be changed by using electromagnetic pulses to control the magnetic flux, the electric charge or the phase difference across a Josephson junction. The other types of superconducting Qubit are transmon Qubit, the fluxonium, charge qubits; flux qubits; and phase qubits. Superconducting qubits are currently the most advanced qubit technology. The major advantage of this qubit is the faster operation time and the disadvantage is the need of very cold temperatures (below 100 mK, or 0.1 degrees above absolute zero) which can be expensive and inconvenient.

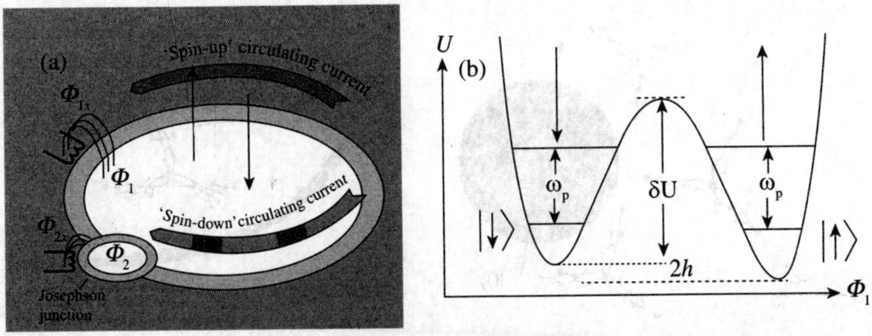

Fig 9.1 Superconducting qubits

(b) **Photonic qubits**: These qubits are utilized in photonic computers, where each photon stands in for a qubit. The qubit's value, which might be 0 or 1, is determined by the photon's path around the storage ring. The loss of propagating photons due to absorption, diffraction or scattering is avoided by a nondestructive photonic qubit detector. The advantage of these qubits is that it can work at room temperature and hence, there is no requirement of maintaining low temperatures. However, the drawback of using this type of system is the unreliability of single-photon generation.

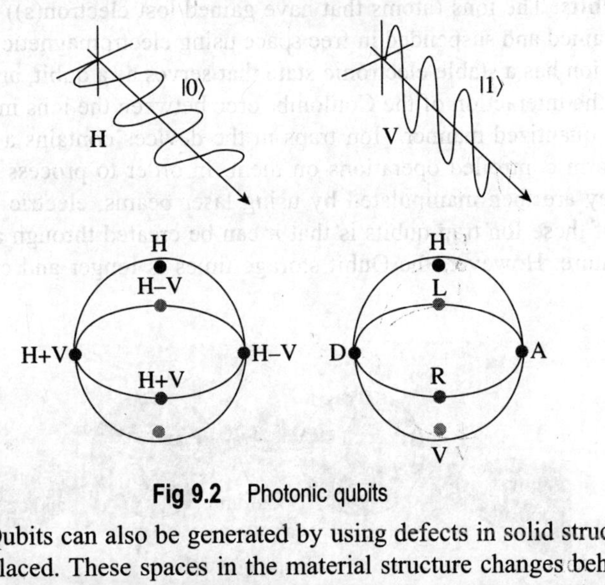

Fig 9.2 Photonic qubits

(c) **Defect based qubit**: Qubits can also be generated by using defects in solid structure where atoms are either missing or misplaced. These spaces in the material structure changes behavior of electrons. In some quantum materials, these spaces trap electrons, thus allowing one to gain access and control of their spins. The advantage of these qubits is that they seldom need ultra-low temperatures and have long consistency times. A qubit can be produced by a number of methods, including dopant phosphorous atoms in semiconductors, colour centres in diamond, nitrogen or phosphorus atoms in molecules C60, unpaired electrons in radiation-induced defects, or free radicals in molecular crystals. In the spin qubit quantum computer controlling the spin of charge carriers (electron and electron hole carriers) in semiconductor devices is the foundation of these quantum computers. Single spin qubits, donor spin qubits, singlet-triplet spin qubits, and exchange-only spin qubits are the four main varieties of spin qubits.

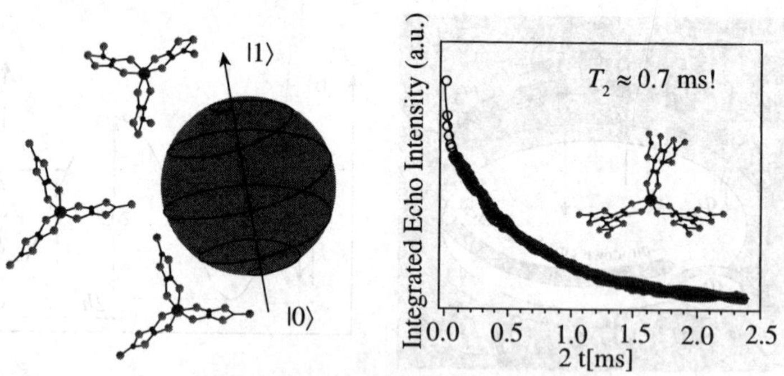

Fig 9.3 Defect-based qubits

(d) **Rydberg atom qubits**: Atoms with only one valence electron is called a Rydberg atom which can be used to create a qubit. The state of these qubits can be changed by moving the valence electron between its original shell and its next excited shell. These qubits are controlled by using optical tweezers. Optical tweezers are used to apply extremely small forces by using a highly focused laser beam. The advantages of this type of system is that they can manage a large number of qubits and have high stability.

(e) **Trapped ion qubits**: The ions (atoms that have gained/lost electron(s)) or charged atomic particles that may be contained and suspended in free space using electromagnetic fields are known as trapped ion qubits. Each ion has a stable electronic state that serves as a qubit, and quantum information can be conveyed by the interaction of the Coulomb force between the ions in a shared trap as they move collectively in a quantized manner. Ion traps in the devices contains a single ion or multiple ions are used to perform controlled operations on them. In order to process information by using these trapped ions, they are then manipulated by using laser beams, electric fields, and magnetic fields. The advantage of these Ion trap qubits is that it can be created through a natural process and exists at room temperature. However, the Qubit storage times is longer and can be few minutes or even hours.

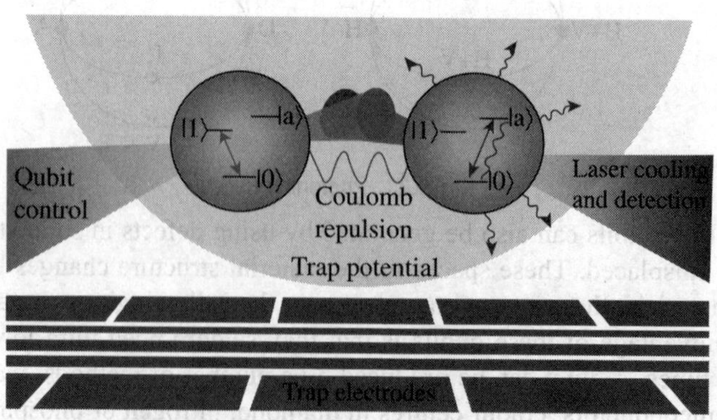

Fig 9.4 Trapped ion qubits

9.2 Superposition and Entanglement of Qubits

Researchers must locate a region in a material where they can access and manipulate these quantum features in order to make a qubit. Once they have access to them, they can then produce superposition, entanglement, and other qualities using light or magnetic fields.

Superposition is the property that any linear combination of two normalized quantum states is likewise a real quantum state. As a result, any quantum state can be described as a linear combination of a small number of base states. For instance, Eq. (9.1) demonstrated that any qubit state may be described as a linear combination of $|0\rangle$ and $|1\rangle$. Similar to this, the 2n bit-string states (states created by the tensor products of $|0\rangle$'s and $|1\rangle$'s) can be expressed as a normalized linear combination of the state of any n qubit system. The computational basis is the orthonormal basis created by the 2^n bit-string states.

It's important to note that Eq. (9.3) defined a system of three qubits whose total state was the tensor product of three distinct single qubit states. But we cannot always write the tensor product of three single qubit states for certain states in which three qubits are present. One illustration of such a state is

$$|\psi\rangle = \frac{1}{\sqrt{2}}(|000\rangle + |111\rangle) \tag{9.5}$$

Such states called entangled states which are states of a system that cannot be described as a tensor product of the states of each of its separate subsystems. This indicates that an entangled state cannot be expressed as a tensor product of n single qubit states for a system of n qubits. Entanglement has crucial implications for both quantum computing and quantum information processing in general, hence it is significant in the context of quantum computing. The exponential upswing in computing power is the result of superposition and entanglement of qubits in quantum computing. For conventional computers, the computing power is doubled by doubling the bits. That means power doubles for a change from 32 bits to 64 bits and so on. However, the computing power for a quantum computer doubles for every addition of a single qubit; which is astonishing. With just n physical qubits, entanglement enables us to construct a whole 2^n dimensional complex vector space for our computations.

9.3 Measurements of Qubits

Measurements are done on the system for transforming the quantum information held in a quantum system into classical information In simple words measurements are done to find whether the qubit is 0 or 1. But as per fundamental principle of quantum mechanics the measurement will always give probabilistic results.

The probability of obtaining $|0\rangle$ after measurement is $|\langle 0|\phi\rangle|^2$ and the probability of obtaining $|1\rangle$ after measurement is $|\langle 1|\phi\rangle|^2$ for a single qubit state as represented by Eq. (9.1). So the measurements results into the square of absolute values of the overlaps. Simplifying this, for a measurement on a n qubit state, $|\phi\rangle$, the probability of getting the bit string $|x_1 \ldots x_n\rangle$ is $|\langle x_1 .. x_n|\phi\rangle|^2$

For a three-qubit system, $|\psi\rangle$ for instance, if we simply want to measure the first qubit while leaving the other two unaffected, the likelihood of finding a $|0\rangle$ in the first qubit is given by

$$\sum_{(x_2 x_3)\in|0,1|^2} |\langle 0x_2x_3|\phi\rangle|^2 \tag{9.6}$$

After this measurement, the state of the system will be obtained by normalizing the state

$$\sum_{(x_2 x_3)\in|0,1|^2} \langle 0x_2x_3|\phi\rangle\, |0x_2x_3\rangle \tag{9.7}$$

The chance of reaching |0⟩ in the first qubit will be 0.5 when applying this exemplar to the state in Eq. (9.5), and once this result is attained, the system's ultimate state will shift to |000⟩. Instead, if |1⟩ in the first qubit must be measured, the measurement will produce the state |111⟩.

9.4 Unitary Transformations on Qubits

Quantum algorithms are always specified as transformations acting on a vector space, as was already mentioned in the part above. A suitable series of unitary operations fundamentally implements a quantum algorithm.

Through a series of unitary transformations or operations, the state of a qubit or a system of qubits can be altered. A matrix U with complicated entries can be used to describe a unitary transformation. If the matrix U is unitary, then

$$UU^\dagger = U^\dagger U = I \tag{9.8}$$

Where I is the identity matrix and $U^\dagger$ is the transposed, complex conjugate of U (also known as its Hermitian conjugate).

A qubit state as represented by Eq. (9.1), $|\phi\rangle = \alpha\,|0\rangle + \beta|\,1\rangle$ changes under the action of the 2 × 2 matrix U as given by following equation.

$$|\phi\rangle \rightarrow U\,|\phi\rangle = \begin{pmatrix} U_{00} & U_{01} \\ U_{10} & U_{11} \end{pmatrix}\begin{pmatrix} \alpha \\ \beta \end{pmatrix} = \begin{pmatrix} U_{00}\alpha + U_{01}\beta \\ U_{10}\alpha + U_{11}\beta \end{pmatrix} \tag{9.9}$$

Now if we want operators to act on different qubits, then we can combine the result of such operation by using Kronecker product. The Kronecker product is an operation that converts two matrices into a larger matrix that contains all the possible products of the entries of the two matrices.

For instance, if U1 and U2 are two different qubit-acting operators, then $U_1 \otimes U_2$ will provide the whole operator for the combined two-qubit system.

The allowable set of physical transformations without measurements, thus, for a system of n qubits, consists of all $2^n \times 2^n$ unitary matrices. Thus, as the number of qubits increases, the size of a general transformation grows exponentially.

9.5 Quantum Gates

The fundamental components of a classical computer are classical gates. Similar to this, the fundamental components of a quantum computer are quantum gates. A basic quantum circuit using a few qubits is known as a quantum gate or quantum logic gate. Mathematically, unitary matrices explain the quantum logic gates. Gates are the name given to these fundamental unitary transformations that serve as the foundation for more intricate n qubit transformations. As a result, quantum gates are unitary transformations, and to satisfy Eq. (9.8), the number of input qubits must always match the number of output qubits. Additionally, it is always feasible to have a gate $U^\dagger$ that cancels the transformation for any gate U. So their action is always logically reversible hence, quantum gates are reversible unlike the classical logic gates. Reversibility means that the inputs of the gate may always be recreated from the outputs of the gate. Because an output of 1 means the input was 0 and vice versa, a traditional NOT gate is reversible. However, for a classical AND gate, the output is 1 only if both of its inputs are 1, but output can be 0 if the inputs were 00, 01, or 10. Hence, it doesn't give enough details to identify the inputs, so AND gate is not reversible.

Table 1 gives an explanation of the most popular gates. The quantum counterpart to the NOT gate is the X gate. Only when the control bit is 1 does the CNOT, also known as "controlled NOT," negate a target bit. It is a two-qubit operation, with the control qubit as the first qubit and the target qubit as the second. The target's qubit state will change from |0> to |1> or vice versa if the control qubit is |1>. The three qubit

Toffoli gate, also known as "controlled-controlled NOT" or CCNOT, can be thought of as the quantum (reversible) AND gate. If both control bits are 1, it just negates a target bit.

The breakdown of multi-qubit quantum algorithms into a series of single- and double-qubit quantum gates creates a universal quantum gate set similar to that seen in classical computers. The quantum computer might therefore run any algorithm if quantum gates function consistently when scaled to numerous qubits.

Table 1 Qubit gates

One Qubit gates	Multi Qubit gates
Hadamard = H = $\frac{1}{\sqrt{2}}\begin{pmatrix} 1 & 1 \\ 1 & -1 \end{pmatrix}$	CNOT = C **X** = $\begin{pmatrix} 1 & 0 & 0 & 0 \\ 0 & 1 & 0 & 0 \\ 0 & 0 & 0 & 1 \\ 0 & 0 & 1 & 0 \end{pmatrix}$
$I = \begin{pmatrix} 1 & 0 \\ 0 & 1 \end{pmatrix}, S = \begin{pmatrix} 1 & 0 \\ 0 & i \end{pmatrix}$	C **Z** = $\begin{pmatrix} 1 & 0 & 0 & 0 \\ 0 & 1 & 0 & 0 \\ 0 & 0 & 1 & 0 \\ 0 & 0 & 0 & -1 \end{pmatrix}$
$T = \begin{pmatrix} 1 & 0 \\ 0 & e^{pi/4} \end{pmatrix}$	Controlled-U = CU = $\begin{pmatrix} 1 & 0 & 0 & 0 \\ 0 & 1 & 0 & 0 \\ 0 & 0 & U_{00} & U_{01} \\ 0 & 0 & U_{10} & U_{11} \end{pmatrix}$
$NOT = x = \begin{pmatrix} 0 & 1 \\ 1 & 0 \end{pmatrix}$	SWAP = $\begin{pmatrix} 1 & 0 & 0 & 0 \\ 0 & 0 & 1 & 0 \\ 0 & 1 & 0 & 0 \\ 0 & 0 & 0 & 1 \end{pmatrix}$
$Y = \begin{pmatrix} 0 & -i \\ i & 0 \end{pmatrix}, z = \begin{pmatrix} 1 & 0 \\ 0 & -1 \end{pmatrix}$	**Toffoli (CCNOT)** = $\begin{pmatrix} 1 & 0 & 0 & 0 & 0 & 0 & 0 & 0 \\ 0 & 1 & 0 & 0 & 0 & 0 & 0 & 0 \\ 0 & 0 & 1 & 0 & 0 & 0 & 0 & 0 \\ 0 & 0 & 0 & 1 & 0 & 0 & 0 & 0 \\ 0 & 0 & 0 & 0 & 1 & 0 & 0 & 0 \\ 0 & 0 & 0 & 0 & 0 & 1 & 0 & 0 \\ 0 & 0 & 0 & 0 & 0 & 0 & 0 & 1 \\ 0 & 0 & 0 & 0 & 0 & 0 & 1 & 0 \end{pmatrix}$
$R(\theta) = P(\theta) = \begin{pmatrix} 1 & 0 \\ 0 & e^{i\theta} \end{pmatrix}$	

9.6 Quantum Decoherence

The loss of superposition and entangled qubits or the decay of quantum behavior of qubit is termed as Quantum decoherence. This results from the spontaneous interaction of a quantum system with its surroundings. Even a slightest change in temperature or a miniscule vibration which is termed as noise is responsible for decay of quantum behavior. The unitary nature (and more especially, the reversibility) of quantum computational steps is violated by this phenomenon. This is viewed as system information lost to the environment. Hence, to protect the states of qubits, the whole system has to be in complete isolation from the environment. The unitary nature of quantum processing steps—more precisely, their reversibility—is contradicted by this effect. In current scenario of quantum computers, however strong the isolation is, still the error creeps in due to decay of qubits. The ratio of operating time to decoherence time often determines error rates. Hence any operation must be completed in time less than decoherence time. Quantum error correction, which corrects errors caused by decoherence, can be used if the error rate is low. The overall computation time will therefore exceed decoherence time. Each gate must be able to do its work 1,000 to 10,000 times faster than the system's decoherence time because each gate has an error rate of about $10^{-3} - 10^{-4}$.

Therefore, it will require more numbers of standard qubits to create a single error free highly reliable single qubit which will strengthen the computation power with implementation of smart quantum algorithms.

9.7 Challenges

There are a significant challenges in creating a quantum computer. Following are the requirements for a practical quantum computer:

(a) Scalability- To increase the quantity of standard qubits
(b) Longer coherence-Quantum gates work faster than the decoherence time
(c) Error Correction- A prominent challenge
(d) Economical for the full system.
(e) Feasibility of operations at room temperatures
(f) Hardware Development
(g) Software Development

Building a system that is totally cut off from everything except the measurement and manipulation mechanism is the difficult task. Moreover, measurements must avoid destroying qubits while being performed on them.

Checkpoint

1. The smallest piece of information in a quantum computer is called
   (a) Bits
   (b) Qubits
   (c) Bytes
   (d) Qubytes
2. What is the quantum state called if the information is between 0 and 1?
   (a) Superposition
   (b) Same position
   (c) Ordinary position
   (d) Different position
3. When two objects are entangled and it is observed that one has its spin up, how long does it take for the other object to change it's spin to the down state?
   (a) 1 microsecond
   (b) 1 nanosecond
   (c) However long it takes light to travel between them
   (d) Instantaneously
4. In quantum mechanics, particle's angular momentum of a is due to which motion of the particle?
   (a) Orbital (b) Swirl
   (c) Spin (d) Rotation

5. What is the size of smallest computer parts?
   (a) A pin head
   (b) An atom
   (c) A water molecule
   (d) A neutrino
6. Which of the following statement described true state of quantum entanglement?
   (a) Entangled qubits cannot affect each other
   (b) Qubits can affect each other, no matter how far apart
   (c) Entanglement cancels out superposition
   (d) A qubit is bound to at least two other qubits in the system
7. The state of a qubit in a quantum register is always
   (a) Probabilistic
   (b) Entangled
   (c) Superimposed
   (d) Negated
8. What will happens if you observe a quantum component, such as a qubit or photon?
   (a) Its state changes to 0
   (b) Its state changes to 1
   (c) Nothing
   (d) its state changes
9. Which of the following refers to the ability for a qubit to be both in 1 and 0 state?
   (a) Cryptography
   (b) Paradox
   (c) Entanglement
   (d) Superposition
10. What is the function of the qubits?
    (a) Store more memory than bits
    (b) Store less memory than bits
    (c) cannot store memory
11. Which of the following phenomena of quantum mechanics are used in Quantum computing?
    (a) superposition
    (b) interference
    (c) entanglement
    (d) All of the above
12. A qubit can be in a_________ quantum state
    (a) 0 (b) 1
    (c) 1 or 0 (d) Can not say
13. In Quantum circuit, a memory consisting of n bits of information has how many possible states?
    (a) 2n (b) n*n
    (c) n (d) 2^n
14. What is quantum computing ?
    (a) Quantum computing is the phenomena of quantum mechanics in computation to solve certain problems.
    (b) Quantum computing is an old technology used in computation.
    (c) Quantum computing used in computation to solve certain physics problems only.
    (d) None of the above
15. When the two members of a Qubit pair exist in a single quantum state, it is known as ________.
    (a) Engagement
    (b) superposition
    (c) entanglement
    (d) None of the above
16. The interaction of qubits with their environment in ways that cause their quantum behavior to decay and ultimately disappear is known as ___________.
    (a) decoherence (b) superposition
    (c) entanglement (d) subposition
17. Qubit stands for _________.
    (a) Quality bits
    (b) Question bit
    (c) Quantum bit
    (d) None of the above
18. A qubit is a __________ quantum-mechanical system.
    (a) one-state (b) two-state
    (c) three-state (d) four-state
19. ________ is a quantum system that can exist in any quantum superposition of two independent quantum states.
    (a) one-state system
    (b) two-state system
    (c) three-state
    (d) None of the above
20. What hurdles do researchers face in developing a quantum computer?
    (a) A large number of qubits into working together efficiently.
    (b) Qubits are finicky.
    (c) Qubits stop functioning at the slightest disturbance.
    (d) All of the above

Summary

- There are two challenges for researchers and engineers across globe to make quantum technology commercially viable.
- The first being the creation of individual physical qubits having greater fidelity. This is achievable through advanced engineering and by optimizing circuit layout and combination of components.
- Second is formation of logical qubits with longer decoherence. There are rigorous research and development initiatives carried out globally to refine the hardware and software requirements.

Self-study Questions

1. Define qubit.
2. What are the different types of qubits?
3. State different features of qubits.
4. Explain superposition, entanglement and decoherence of qubits.
5. Differentiate between classical bit and quantum bit (qubit).
6. What are merits and demerits of superconducting bits?
7. What are unitary transformations?
8. What is a quantum gate?
9. State various quantum gates.
10. State the limitation of quantum computing.

Answers

Checkpoint

1(b) 2(a) 3(d) 4(c) 5(a) 6(b) 7(a) 8(d) 9(d) 10(a) 11(d) 12 (c) 13(d) 14(a) 15(b) 16(a) 17(a) 18(b) 19(b) 20(d)

10 Merits, Demerits and Applications of Quantum Computing

Learning Objectives

After going through this chapter, you will be able to:

- Understand the functional differences between quantum and classical computers.
- Recognize various milestones in the journey of quantum computers and
- Describe the power of quantum computers.
- Discuss the advantages and limitations of quantum computers.
- Explain applications of quantum computers in different areas.

10.0 Introduction

The quantum computing technology is in an evolutionary stage with active involvement of more than 600 renowned companies and almost 30 national laboratories and government agencies world-wide as per the report published by *The Quantum Insider*. The efficacy of any emerging technology is assessed through three factors: time, cost and quality. Parallel computing is possible in quantum computers due to superposition and entanglement of qubits which offers exceptional speed advantage in cracking complex problems. Thus the two requirements of time and cost are fulfilled. However, the quality aspect is lagging on account of a few constraints. This technology is poised to become a disruptive technology as lot of research and development activities are being carried out globally to improve the quality and accuracy of the results to the acceptable standards. In the next decade, this exotic technology will transform the entire industry and turbocharge innovation globally.

10.1 Quantum Computer

With the help of quantum technology one can organize and control various components of a complex system governed by principles of quantum physics. The quantum technology has two requirements. The first one is miniaturization where devises are built at nanoscale. As per Moore's law in a compact integrated circuit the number of transistors doubles in an approximate span of every two years. So for classical computers there is an exponential increase in the processing power. However, further miniaturization of components cannot continue as integrated circuits are limited by the atomic scale and there are issues of electron tunneling. The second imperative was improving the performance. So if we imagine to build the absolute most powerful classical computer then every transistor or gate has to be made from only a single atom which is impossible. So scientists have studied and found that the quantum mechanical effects can be used to create different kinds of computers. Richard Feynman in 1959 suggested the possibility of using quantum effects. He proposed a basic model for a quantum computer that can simulate the evolution of quantum systems as classical computers cannot do it in an efficient way. Hence, he stated the possibility to exponentially outperform classical computers. A quantum computer is a computer that exploits quantum mechanical phenomena.

Quantum computers are sophisticated machines that contains a processor similar to the one used in laptop. They are much smaller and the requirement of energy is also less in comparison with supercomputers. But the hardware system is bigger in size as it houses cooling system to keep the superconducting processor at its ultra-cold operational temperature. The processing power of quantum computer gains is due to extraordinary feature of quantum bits to be in the multiple states at one time. The number of computation steps required in quantum computer is 2^n where n represents number of qubits used. So for a quantum computer using 200 qubits, the number of calculations performed in one step are 2^{200}. So if more qubits are added then the speed of the computing increases exponentially as compared to linear increase for a classical computer. Hence, quantum computers are intended to perform complex calculations on huge data base. Imagine a qubit as a spin of an electron in a magnetic field. Classically, there can be two possibilities either spin aligned i.e. spin up state or antiparallel to the magnetic field i.e. spin down state. To change the spin state, a unit energy of laser pulse is used. Now if half a unit of laser energy is used for the completely isolated system then the spin exists in a superposition state i.e. as if it exists in both the states at given point in time. Thus, it can have multiple spin directions before being measured. For entangled pair of qubits the spin state of one if changed the other assumes the same state. This superposition of qubits is what gives quantum computers their natural parallel processing powers. But for computing we need to manipulate qubits which is done by quantum logic gates.

Similar to the gates used for classical computing, there are analogues quantum gates which forms the basic building blocks of quantum computation. So series of quantum gates are used to manipulate states of qubits and at the end of the computation a measurement is carried out to know the result. Thus, quantum superposition and entanglement of qubits together create enormously enhanced computing power as a qubits can compute every version of the inputs in only one cycle of the algorithm. With a powerful algorithm a quantum computer can do an arbitrary reversible classical computation on all the numbers simultaneously, which a binary system cannot do. Quantum computers utilize a variety of algorithms to conduct measurements and observations. An open-source Python package Qiskit (Quantum Information Science Kit) is a IBM's actual hardware on which one can implement, simulate and run quantum algorithms. The physical form of a real quantum computer consists of mainly three parts. The first part is a conventional computer and infrastructure. This part will run programs creating instructions for the qubits. These instructions will be passed on to the qubits as signals by the second part. The last part will be the storage unit for the qubits which has to be maintained at near zero degrees in a vibration free environment.

10.2 Milestones in the Evolution of Quantum Computers

Quantum computers and the quantum theory have made some progressive advancement in last four decades. Taking a look back on the development it's conspicuous that till 1980, foundation of quantum computer was laid. Thereafter, till 2010, concepts were put in practice. From then onwards more thrust was on commercialization and applications of quantum computing technology. A brief compendium of the evolution of all reported pertaining activities are given as below.

Richard Feynman in 1981 proposed to use quantum computing to model quantum systems. He also proposed theoretical model of quantum computer. David Deutsch in 1985 described first universal quantum computer. First algorithm for quantum computer was developed by Peter Shor in 1994. The most incredible Quantum Bit or qubit was proposed by Schumacher in 1995. In 1996 LovGrover developed an algorithm for search in unsorted database. The very first two qubits quantum computer based on NMR was reported by

IBM, MIT, and Stanford in 1998. A similar 7 qubit quantum computer was reported in 2000. During 2005-06 experiments reported photons; quantum dots; fullerenes and nanotubes as "particle traps". In 2007 a 16 qubit followed by a 84 qubit quantum computer in 2012 was claimed by D-Wave Systems Inc., a Canadian quantum computing company. In 2017 the same company announced 2000 qubits, "2000Q quantum annealer" and a well-known Microsoft company came up with 32 bit "Q Sharp". In 2018, Google announced creation of a 72 qubit quantum chip. In the following year, Google claimed 54 qubits and IBM announced its 53 qubits quantum computer. In 2020, a 76 photonic qubit quantum computer was claimed by Chinese researchers. In 2023 IBM will come up with 1121 qubit quantum computer.

10.3 Benefits of Quantum Computing

Even though the benefits of Quantum computing are speculative but has the immense potential and offers following over its counterpart:

(a) **Faster computations**: Quantum computers have the capability to solve problems at exceptionally fast rate. They can perform the calculations almost 1000 times faster than even the supercomputers.
(b) **Solve complex processes**: Quantum computers have the inherent ability to simultaneously perform multiple complex calculations.
(c) **Complex simulations**: The high-speed computing enables simulations of intricate systems with more accuracy. This simulation power will help scientist to design new molecules which will revolutionize drug industry.
(d) **Optimization**: With the remarkable ability of quantum computers to process big data via parallel computing with unusual speed, it has the potential to solve problems related optimization which can transform fields of Artificial and Machine Learning.

10.4 Power of Quantum Computer

For big integers that are the product of two prime numbers that are about equal in size (for example, products of two 300-digit primes), integer factorization is thought to be computationally impossible on a standard computer. In contrast, a quantum computer would be able to resolve this issue fairly quickly. A quantum computer with slightly over 2n qubits can use Shor's technique to determine factors for a number with n bits (number which is n digits long when represented in the binary numeral system). A quantum computer algorithm called Shor's algorithm can find the prime factors of an integer in polynomial time. The discrete logarithm problem, a related issue, can also be resolved by it. In the sense that there would be a relatively quick algorithm (polynomial time in n) for solving the problem, this capability would allow a quantum computer to "break" many of the cryptographic systems currently in use. Two algorithms that implement a public key cryptosystem are RSA (Rivest-Shamir-Adleman) and ElGamal. However, these algorithms could be cracked much more quickly. These safeguard encrypted emails, secure Web pages, and numerous other types of data. There would be serious repercussions for electronic privacy and security if these were broken. A larger key size and the hope that an opponent lacks the resources to create and utilize a powerful enough quantum computer are the only ways to boost the security of an algorithm like RSA. Building classical computers with more bits than the biggest quantum computer's qubit count seems to be a constant possibility. If so, then it would be possible to make algorithms like RSA secure by guaranteeing that key lengths are greater than the storage capabilities of quantum computers, but at a severe cost in terms of computing time.

Quantum computers may be beneficial for executing simulations of quantum mechanics, which is perhaps not as shocking. This concept has its roots in Richard Feynman's observation that there is currently no known algorithm for modelling quantum systems on a classical computer from 1982, when he proposed researching the use of quantum computers for this purpose. Quantum computers may speed things up just as much as factoring did. Physics, chemistry, materials science, nanotechnology, biology, and medicine could all benefit greatly from this because they are now constrained by the slow pace of quantum mechanical simulations. A quantum computer would only need a few seconds to complete some contemporary simulations that take IBM's Blue Gene supercomputer years to complete.

Factoring, discrete logarithm, and quantum physics simulations are few tasks for which the dramatic advantage of quantum computers has been discovered so far. No matter which of these issues can be demonstrated to have an advantage on a quantum computer, they will all always have the benefit of being good research subjects for quantum mechanical interactions, which is of immeasurable importance to the scientific community.

10.5 Limiting Features of Quantum Computing

It is projected that quantum computers will be revolutionizing many technologies due its exceptional entity of it computing powers. However, the basic problem lies in generating and maintaining states of qubits. The superposition and entangled quantum states are extremely delicate. They are lost easily due to change in their environment. The irony is you cannot copy it. Hence, there are some limitations which have become a roadblock in feasibility of these technology.

- **Decoherence:** That is the decay of qubits states remains the biggest challenge in building quantum computers. This is caused by a minute disturbance in the qubit environment which results in complete destruction of the computation system. So, during the computing stage, the whole system has to be in complete isolation from all external interferences.
- **Error correction:** Even if high accuracy is targeted, a single mistake can completely corrupt the results. During the computation stages complete error correction has not yet been achieved which makes the results unreliable. The only solution for this is to make the qubits and hence the quantum circuit more tolerant.
- **Security:** Security and quantum cryptography is not yet fully perfected.
- **More qubits:** Thousands of standard qubits are required to create one high fidelity logical qubit. IBM is poised to come up with 1000 qubit quantum computers this year.

10.6 Applications of quantum Computers

Quantum computers is an emerging technology which outperform traditional and supercomputers due to exceptional computing speeds. It processes big data remarkably faster and predictions are more precise. As quantum computing technology is not ready for full-scale commercial use cases, most of the applications are still experimental. Quantum computing is invincible and could increase efficiency in various fields. There is a wide spectrum of viable applications of quantum computing. Here are a few applications of quantum computing from the vast possibilities.

- **Cryptography:** In the traditional computers it is possible to break encryption. With the advent of quantum computers and novel quantum computing it's now feasible to have advanced cryptography. This will lead to much stronger protections of our digital assets.
- **Aviation:** The powerful algorithms used in quantum computing will enable complex computer modelling which will be highly used in the aeronautical industry. The scheduling and routing of aircrafts can be optimized reaping commercial benefits of saving time and cost. Large aviation industries are already investing into quantum technology for getting benefits of extraordinary computing and optimizing power to help flourish this industry.

- **Data analytics:** Quantum computers can solve large scale problems with dramatic speed and accuracy. The so called impossible topological analysis and calculation due to the weird data set used boils down to comparatively simple calculations by implementing quantum computing. This feature is used to analyze huge data gathered by satellites about the entire universe to have safer space missions.
- **Forecasting:** The precision in forecasting and predicting increases if simulations are based on large and complex data sets. However, more complexity in data sets constraints the classical computing due to its limited processing capacity. More input of variants improves the results of simulations. But it takes longer to arrive at the results jeopardizing the economy of the region which depends on such forecast. Quantum computing has capability to accurately forecast boosting the economic benefits.
- **Medical research:** In medical research it is of prime importance to know how any drug administered to a human body reacts with every cell and tissue. With a slight change in the drug physiology there are numerous possibilities in which the reactions happen in a human body. It takes more than a decade and huge financial investment to invent new drug and lot many trials to bring it in the market for use. To reduce time and cost involved, pharmaceutical companies are leveraging on powers of quantum computing to create new drugs faster leading to treatment of a range of diseases.
- **Self-driving cars:** Driverless cars are manufactured by Car companies like Tesla and tech companies like Apple and Google. These cars will not only improve the living standard of people but will also reduce pollution, traffic congestion are accidents in addition to lot of other benefits.

Checkpoint

1. In which way the quantum computers could help in simulating?
   (a) proteins for medicine
   (b) carbohydrates for medicine
   (c) fats for medicine
2. Who is the father of quantum computing?
   (a) Isaac Chuang
   (b) Neil Gershenfeld
   (c) Philip Glass
   (d) David Deutsch
3. How can quantum computing improve security over optical data transmission?
   (a) It breaks the encryption keys of current computers
   (b) The keys are built upon blockchain
   (c) Even an observation of the data triggers a change
   (d) Unwanted changes in the data are blocked
4. ________ is a future quantum technology that works to mirror the human brain's function.
   (a) AI
   (b) Machine learning
   (c) Quantum cognition
   (d) Quantum entanglement
5. Which of the following is a significant challenge to the future of quantum computing?
   (a) RSA encryption methods
   (b) Big data analytics
   (c) Quantum error correction
   (d) Quantum computers use more qubits
6. Which of the following is a barrier to quantum computing?
   (a) Data storage
   (b) Cryptographic algorithms
   (c) Cost
   (d) Security
7. Which of the following is the most appropriate assessment of cryptography and quantum computers?
   (a) Traditional computers will be rendered useless
   (b) Cryptography issues cannot be solved with quantum computing
   (c) Advancements must always be made in cryptography
   (d) Quantum computing will develop the final encryption method

8. ________ is the point at which a quantum computer can complete a mathematical calculation that is demonstrably beyond the reach of even the most powerful supercomputer.
   (a) Quantum entanglement
   (b) Quantum supremacy
   (c) Quantum energy
   (d) None of the above
9. What are the applications of quantum computers?
   (a) Cybersecurity & Cryptography.
   (b) Artificial Intelligence & Machine Learning.
   (c) Drug Design & Development.
   (d) All of the above
10. Obstacles for building Quantum computers are ________.
   (a) Universal gate set
   (b) Qubits that can be initialized to arbitrary values
   (c) Quantum gates that are faster than decoherence time
   (d) All of the above
11. _______ was the first company to put a quantum computer on the cloud.
   (a) Google
   (b) Microsoft
   (c) IBM
   (d) Apple
12. What hurdles do researchers face in developing a quantum computer?
   (a) A large number of qubits into working together efficiently.
   (b) Qubits are finicky.
   (c) Qubits stop functioning at the slightest disturbance.
   (d) All of the above
13. A fully functioning quantum computer has the potential to be ________.
   (a) slow
   (b) constant
   (c) transformative
   (d) All of the above
14. What is quantum computing?
   (a) Quantum computing is the phenomena of quantum mechanics in computation to solve certain problems.
   (b) Quantum computing is an old technology used in computation.
   (c) Quantum computing used in computation to solve certain physics problems only.
   (d) None of the above
15. Quantum key distribution QKD, use a series of
   (a) Byte
   (b) Photons
   (c) Both (a) and (b)
   (d) None of the Above

Summary

- There are two challenges for researchers and engineers across globe to make quantum technology commercially viable.
- The first being the creation of individual physical qubits having greater fidelity. This is achievable through advanced engineering and by optimizing circuit layout and combination of components. Second is formation of logical qubits with longer decoherence. There are rigorous research and development initiatives carried out globally to refine the hardware and software requirements.
- Many businesses like IBM working with Mercedes-Benz, ExxonMobil, CERN, and Mitsubishi Chemical are already implementing quantum computing into their respective products and services.
- Quantum computing is poised to become a disruptive technology with further technological progression and has a promising future in real-world applications.
- Quantum computing can b used in many areas such as cryptography, aviation, data analytics, forecasting, medical research and weather forecast.

Self-study Questions

1. What distinguishes a quantum computer from a conventional computer?
2. What challenges must researchers overcome to create a quantum computer?
3. What distinguishes quantum computing from classical computing?
4. Give some examples of quantum computing applications.

Answers

Checkpoint

1(a) 2(d) 3(c) 4(c) 5(c) 6(c) 7(c) 8(b) 9(d) 10(c) 11(c) 12 (d) 13(c) 14(c) 15 (b)

About the Author

Dr Bhavana Butey is Professor in Physics at G H Raisoni College of Engineering, Nagpur. She has completed her Doctoral research from well-known Visvesvaraya National Institute of Technology (VNIT), Nagpur in 2004. She has three decades of teaching experience. She has worked in various capacities during her tenure at GHRCE. Currently she is a member of Academic Council and Board of Studies (BoS) of GHRCE, Nagpur. Her research area is Material Science. She has to her credit more than 20 Papers published & presented in referred International Journals & Conferences. 'Best Teacher' award for the session 2003-2004 was conferred on her by G H Raisoni College of Engineering, Nagpur.

Dr Butey is an approved supervisor in Physics and Mathematics. One candidate has been awarded doctor degree under her supervision and one is ongoing. She has attended more than 30 FDP (Online/Offline Faculty Development Program), STTP (Short Term Training Program), Workshops and MCEP (Management capacity Enhancement Program). She has two copyrights to her credit and is a recipient of Elite Gold & Silver for two NPTEL online certification courses. She was General Chair of IoP International conference on Research Frontiers in Sciences, ICRFS-21. She is subject expert on BoS of three autonomous colleges of Nagpur and Jalgaon. She is reviewer for referred conference proceedings. She is a life member of IPA (Indian Physics Association) and ISTE(Indian Society for Technical Education). Dr Butey has visited Universities Abroad, namely TTU, Tennessee USA, National University of Singapore and various universities in Dubai.